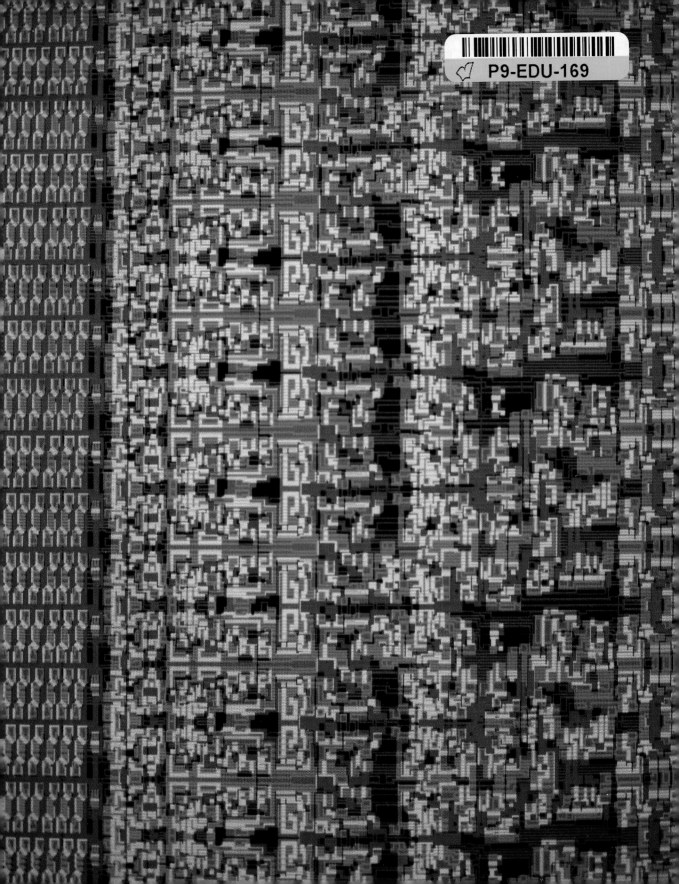

THE DESIGN AND ANALYSIS OF VLSI CIRCUITS

THE DESIGN AND ANALYSIS OF VLSI CIRCUITS

Lance A. Glasser

Massachusetts Institute of Technology

Daniel W. Dobberpuhl

Digital Equipment Corporation

ADDISON-WESLEY PUBLISHING COMPANY

Reading, Massachusetts • Menlo Park, California
Don Mills, Ontario • Wokingham, England • Amsterdam • Sydney
Singapore • Tokyo • Mexico City • Bogatá • Santiago • San Juan

Hugh J. Crawford	*Manufacturing Supervisor*
Marion E. Howe and	
William J. Yskamp	*Production Editors*
Maureen Langer	*Text and Cover Designer*
Herbert Nolan	*Production Manager*
Precision Graphics	*Illustrator*
Marilee Sorotskin and	
Shirley Stone	*Copy Editors*
Martha Stearns	*Managing Editor*
Joseph K. Vetere	*Art Coordinator*

This book is in the **Addison-Wesley VLSI Systems Series**.
Lynn Conway and Charles Seitz   *Consulting Editors*

Library of Congress Cataloging in Publication Data

Glasser, Lance A.
   The design and analysis of VLSI circuits.

   Bibliography: p.
   Includes index.
   1. Integrated circuits—Very large scale integration.
I. Dobberpuhl, Daniel W.   II. Title.   III. Title:
Design and analysis of V.L.S.I. circuits.
TK7874.G573   1985   621.381'73   84–2786
ISBN 0–201–12580–3

To Wendy and Carol

PREFACE

In 1959 Robert Noyce and Jack Kelby changed the technological and economic profile of the world when the independently invented the planar integrated circuit. Since 1961 the number of transistors that can be successfully fabricated on a single chip has doubled almost every year. This has caused a revolutionary and exponential decrease in the cost per unit function for digital electronics. Enticed by the multibillion-dollar market that opened as a result of this technological explosion, the engineering community has developed volumes of specialized knowledge, techniques, and theories to harness the power quiescent in a few milligrams of silicon. In *The Design and Analysis of VLSI Circuits* we will examine one of the fields essential to this technology—integrated circuit design.

The purpose of this text is to bring the reader to the next level of sophistication beyond that offered by introductory VLSI design books. For instance, *The Design and Analysis of VLSI Circuits* is both more advanced and more specialized than the classic work by Mead and Conway, *Introduction to VLSI Systems*. At the Massachusetts Institute of Technology, one of the courses that uses this material has as its prerequisite either an introductory VLSI course in the style of Mead and Conway or a circuit design-oriented course in the style of the book by Hodges and Jackson, *Analysis and Design of Digital Integrated Circuits*. We have made an effort, however, to include the necessary materials to allow this book to be used as a first course in VLSI circuits in cases where the constraints of the curriculum and the student's background warrant it. This text has been successfully used in introductory VLSI courses.

We assume that the reader is familiar with elementary circuits, device physics, and logic design. The text is appropriate for a senior- or graduate-level course on the design of MOS circuits; it will also be useful to the engineer or scientist wishing to become a knowldgeable practitioner in the field of integrated-circuit design. This book was used extensively at Digital Equipment Corporation in the initial training and continuing education of circuit designers, systems designers, technology experts, and CAD professionals.

The book begins with a heuristic treatment of a number of nMOS and CMOS circuit design methodologies that provides a broad base from which the reader can understand what a methodology is, and what is involved in its use or design. Discussions of static and dynamic circuit techniques and the role of regular structures are included. Because we are dealing with circuit techniques, it is possible to understand, qualitatively, how fairly complex circuits are constructed. This is the approach of Chapter 1, in which the context and motivation for the rest of the book are developed. Of the seven remaining chapters, Chapters 2 through 5 emphasize traditional "electrical engineering" issues, while Chapters 6 through 8 are oriented more toward "computer science." While we feel that all of this material is the legitimate domain of the VLSI circuit designer, readers with a computer science background might profit from first reading Chapters 1, 6, 7, and 8, then backtracking to the material in Chapters 2 through 5.

While qualitative discussions are fine for understanding elementary circuit techniques, they are unfortunately insufficient to enable one to actually do quality circuit design. Therefore the emphasis of later chapters shifts to a more quantitative perspective. In Chapter 2 the physics of MOS devices and interconnect is developed. Engineering is largely concerned with models and their appropriate use. In this chapter we develop a number of models of varying levels of complexity and accuracy. Some are useful for "back of the envelope" calculations, while others are useful only when incorporated into sophisticated circuit simulation programs. In fact, starting with Chapter 1, computer-aided design plays an integral part in our approach to VLSI circuits. A computer, however, is no substitute for insight, so we also work to develop a feel for how the various first- and second-order physical phenomena in MOS devices influence the terminal characteristics of capacitors, transistors, and interconnect. The limits of validity of the models are examined. These, and other points in the book, are emphasized in numerous problems and examples. Starting in Chapter 2, we identify numbered equations that are of central importance with a star.

Chapter 3 contains the aspects of fabrication technology that we feel circuit designers must know. A suite of integrated circuit fabrication techniques are presented. We then show how these techniques are used to build MOS circuits. A key section contains a discussion of design rules and their origins. The chapter also includes the development of elementary yield formulas.

Chapter 4 develops the concepts necessary for using the basically analog MOS transistor as a basis in constructing reliable digital systems that meet specifications. Among the concepts discussed are noise, noise margins, macromodeling, and worst-case design. The dynamics of logic gates are carefully investigated. We see examples of the interplay of device physics and circuit design. the inverter, as the archetypical logic element, is the main vehicle of the chapter. We examine two types of nMOS inverters and two types of CMOS inverters.

With these foundations, we present in Chapter 5 a host of specialized circuit techniques including the optimization of large fan-out and fan-in circuits, clock drivers, substrate-bias generation, input protection devices, level conversion, Schmitt triggers, and so forth. Along with our presentation of each circuit technique, we develop the conceptual and analytical tools necessary for studying them. After mastering this chapter, readers will know enough to place new circuit techniques, invented by themselves or someone else, in proper context.

Chapter 6 provides a discussion of clocking methodologies. These methodologies include nonoverlapping clocks, overlapping clocks, and self-timed techniques. The emphasis is on nonedge-triggered systems. Clock generators and issues of skew are addressed. Synchronization circuits are examined quantitatively.

Chapter 7 examines the circuit design of large arrays and regular structures, including decoders, RAM, ROM, and systolic arrays. We see how layout, circuit design, topology, and logic design interact in fascinating ways. The retiming and optimization of systolic arrays are examined. The study of these issues provides a smooth introduction to Chapter 8: "The Microarchitecture of VLSI Systems."

A few years ago microarchitecture would have had no place in a book on circuit design; however, the world has changed. A digital VLSI sicruit is a digital system. In Chapter 8 we concentrate on the points of synergy between microarchitecture and circuit design. The microprocessor is used as the archetypical VLSI logic chip. Techniques for data-path and control-path design are presented. For the data path we emphasize carry propagation as a critical issue. For the control path we emphasize microprogram context switches.

To get the most out of this book, it would be helpful to have access to a computer running SPICE2. To provide realism to the examples and problems and to illustrate points in the text, we have developed hypothetical 2 μm nMOS and CMOS processes that form a common theme. Catagories such as design rules, process specifications, and SPICE models for these processes are listed in the appendixes and are used in many examples. Because these processes are fully specified, it is possible to do significant paper designs as part of a course or self-study program.

The problems at the end of each chapter are categorizd into three levels of difficulty: reasonable, hard, and extremely difficult. Hard problems are denoted by a "$\triangle$" at the end of the problem statement. These problems either involve a lot of work, difficult concepts, or significant innovation. Extremely difficult problems are denoted by a "$\triangle\triangle$" at the end of the problem statement. These problems indicate lines of investigation worth pursuing. If these problems have answers, those answers often depend on information not covered in the book.

Acknowledgements

It is our pleasure and honor to acknowledge helpful and stimulating discussions with Gregory Allen, Cyrus Bamji, Ed Burdick, Jim Cherry, Peter Cook, Bill Herrick, Rich Hollingsworth, C. S. Kim, Tom Knight, Mark Matson, James Mulligan, Jr., Anne Park, Paul Penfield, Jr., John Wroclawski, Andy Wu, and Charles Zukowski. We would expecially like to thank Richard Zippel for an extraordinary amount of help and inspiration. The detailed and highly constructive reviews of Michael K. Maul, AT&T Bell Laboratories; Donald Trotter, Mississippi State University; and Carlo H. Séquin, University of California at Berkeley; have proved invaluable, and we thank them. Don Nelsen has our gratitude for contributing the latchup sections in Chapters 2 and 5. The personal support and enthusiasm of many in the DARPA VLSI community, particularly Paul Losleben, is warmly appreciated. The authors gratefully acknowledge the support and encouragement of the Massachusetts Institute of Technology and Digital Equipment Corporation.

January 1985   L.A.G.
Cambridge, Massachusetts   D.W.D.

CONTENTS

1

CIRCUIT FORMS AND METHODOLOGIES

The domain of the integrated circuit designer stretches from the realms of device physics and nonlinear circuit theory to the regime of digital system microarchitecture. Integrated circuit design plays a central role in VLSI technology and encompasses multiple levels of abstraction: electrostatics, topology, switching theory, and so forth. The VLSI circuit designer is challenged to transcend these disparate disciplines and transform an idea into the detailed specifications for a manufacturable machine.

Each abstraction has a hierarchy of detail. Systems must be decomposed into subsystems, subsystems into modules, and modules into components. With each decomposition the instantiation of the machine becomes more complete. There are many ways to navigate the two-dimensional hierarchy of abstraction and detail, but whatever course is chosen, its ultimate success depends on bottom-up understanding. For this reason we begin our study of VLSI design at the lower levels, with transistors and interconnect.

1.1   A qualitative model of the MOS transistor

In this chapter, our objective is to obtain a good qualitative feel for how a digital MOS circuit works. As part of this task we need to understand how some of the low-level components in a VLSI circuit behave and how

1

they are composed to form interesting functional modules. We discuss a number of different design methodologies, but they should be thought of as parts of a larger, unified whole. In later chapters, these and other circuit techniques will be investigated with more quantitative precision.

Figure 1.1 shows a photomicrograph of an LSI chip [Olsen 81]. The two-dimensional nature of the integrated circuit is clearly evident. Only two primitive elements—transistors and wire—are used to build today's magnificent variety of VLSI circuitry. Most of what we see in Fig. 1.1 is wire, with the transistors taking up only about 10% of the area. The chip contains 13,000 transistors (a small chip by modern standards). It is a few millimeters on a side and only a fraction of a millimeter thick, and all of the interesting structure exists within a few microns of the surface. Several large regular structures are evident. These represent most of the transistors, but only a fraction of the labor. We must take a closer look at the chip to see a single transistor.

A transistor occurs at the intersection of two wires. Figure 1.2 shows two close-up views of some circuitry viewed from the top, but again we have mostly wire; the transistors can be seen only if one knows where to look. We can see why it is often said that in VLSI the transistors

FIGURE 1.1   Photomicrograph of the T11 microprocessor. (Copyright ©Digital Equipment Corporation 1985. All rights reserved. Reprinted with permission.)

Metal

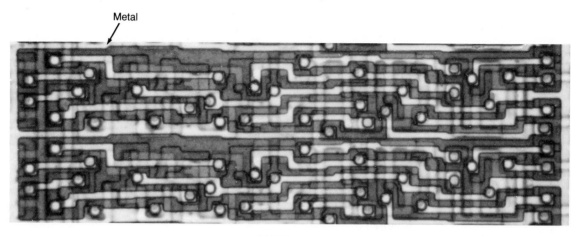

(a) With metal.

Poly | Via (Buried contact) | Active area (gate) | Via (metal contact) | Active area (n+)

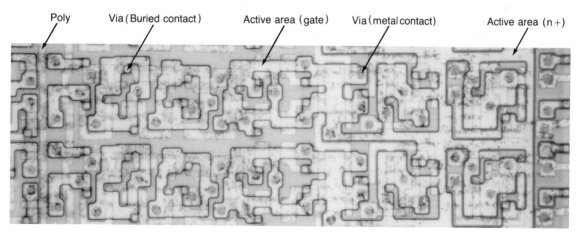

(b) Without metal.

FIGURE 1.2 Close up photomicrographs of a single section of the T11: (a) with metal; (b) with metal removed. (Copyright ©Digital Equipment Corporation 1985. All rights reserved. Reprinted with permission.)

are free; it is the wire that is expensive. There are three layers of interconnect media on this particular chip. The speckled material is aluminum. The two other media are polycrystalline silicon and heavily doped, single-crystal silicon. These two materials have a very harmonious relationship with the transistors, as will be seen shortly.

Figure 1.3 shows the layout of the circuitry shown in Fig. 1.2. The various materials are labeled. The polycrystalline silicon (poly) and metal layers were deposited during the chip fabrication process. The metal acts only as wire while the poly plays the dual role of wire and

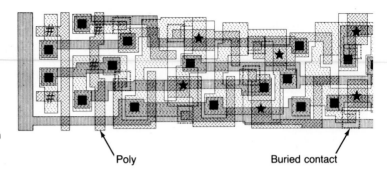

FIGURE 1.3 Layout (mask specifications) of the portion of the T11 illustrated in Fig. 1.2.

Poly

Buried contact

transistor gate material. The active area, which includes the regions of heavily doped single-crystal silicon and transistor gate area, also plays a dual role. It is divided by the poly into two components. Where there is active area and no poly, the silicon is doped n+. The n+ material acts as wire. Transistors are formed at the intersection of the poly and active area regions. Layers of silicon dioxide (glass) separate the three layers of conductor. These oxides are slightly different from each other, as will be discussed in Chapter 3. They include gate oxide, field oxide, CVD oxide, and overglass. In order to interconnect the various conducting media, two types of vias are provided. The first type of via (called a contact cut) connects metal to either poly or n+, and the second type of via (called a buried contact) connects poly to n+. A cross section through a typical interconnect structure is illustrated in Fig. 1.4. When viewed from the top, under a microscope, a patchwork of color is visible. Poly might appear red and n+ green, or vice versa. Most of these colors are "false," caused by the interference of light waves in structures that have a thickness on the order of a wavelength of light.

FIGURE 1.4 Some contacts in nMOS technology.

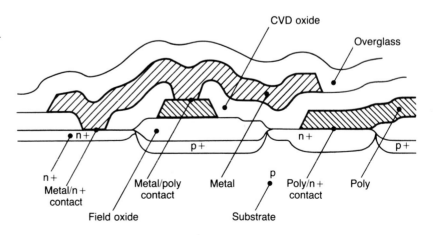

CVD oxide

Overglass

n+

p+

n+

p+

n+
Metal/n+
contact

Metal/poly
contact

Metal

p

Poly/n+
contact

Poly

Field oxide

Substrate

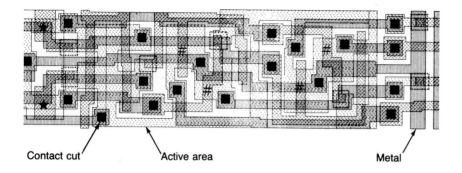

Contact cut   Active area   Metal

The MOS[1] transistor is a four-terminal device. Under normal
operating conditions the gate of the transistor controls the flow of
charge carriers between the source[2] and the drain. The fourth terminal
is the body that in pMOS and nMOS technologies cannot be used
for performing useful logic. This is because in these technologies all
of the body terminals are connected by a common substrate. Figure
1.5 illustrates the cross section of an nMOS transistor. The source and
drain are composed of heavily doped n+ silicon. The gate is separated
from the channel region by a thin layer of silicon dioxide only a few
hundred atoms thick. The gate conductor is made of heavily doped,
polycrystalline silicon. It is important to develop a clear mental picture
of these geometries. Figure 1.6 shows three projections of an nMOS
transistor. Illustrated in Fig. 1.7 are two nMOS devices in a common
substrate. The sources and drains of these devices are isolated from each
other by reverse-biased p–n junction diodes and the associated depletion
regions. (A depletion region is a volume of semiconductor devoid of
charge carriers.)

There are two types of nMOS transistors: normally on and normally
off. In a normally off nMOS transistor there are no mobile charge carriers
in the channel region under the gate when the input voltage is low. This
type of transistor is called an enhancement-mode device because one
must enhance the charge on the gate in order to achieve conduction
between the source and the drain.

When a positive charge Q_G is placed on the gate of an nMOS
transistor an amount of charge almost equal to $-Q_G$ is induced in
the channel. This charge is composed of depleted acceptor ions and

---

[1] MOS stands for Metal Oxide Semiconductor and refers to the physical structure
of the device. The term IGFET (Insulated Gate Field Effect Transistor) is also used.
CMOS, pMOS and nMOS, refer to complementary, p-channel, and n-channel MOS,
respectively. Despite the "M" in MOS, modern MOS devices do not really have metal
gates but use a silicon gate in a similar role.

[2] The source is so called because it is the "source" of mobile charge carriers, which
are electrons for n-channel devices and holes for p-channel devices.

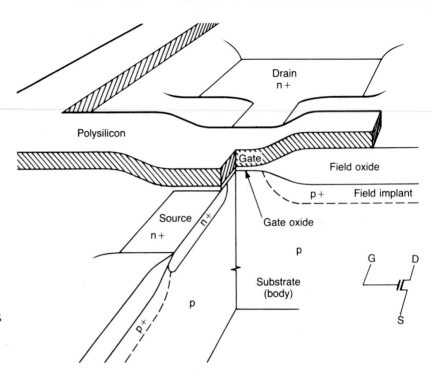

Polysilicon

Drain
n+

Gate

Field oxide

p+   Field implant

Gate oxide

Source
n+

n+

p

Substrate
(body)

p

p+

G   D

S

FIGURE 1.5   Cut-away
view of an n-channel MOS
transistor at the point in
the processing after the
source/drain implants.

FIGURE 1.6   Three
views of an n-channel
MOS transistor: (a) front;
(b) top; (c) side.

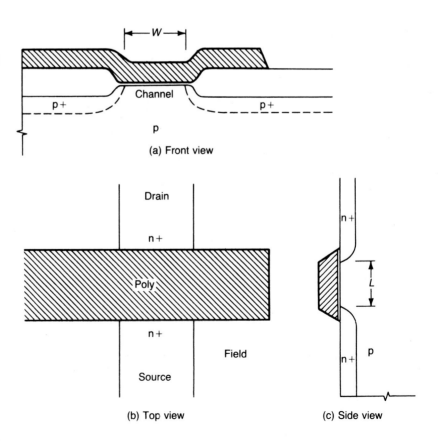

|← W →|

Channel

p+   p+

p

(a) Front view

Drain

n+

Poly

n+

Field

Source

n+

n+

L

n+   p

(b) Top view   (c) Side view

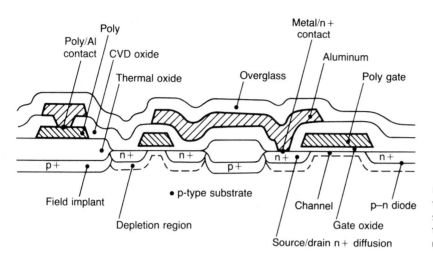

FIGURE 1.7   Two nMOS transistors in a p-type substrate. Isolation between transistors is provided by reverse biased p–n diodes.

free carriers, which flow in from the source and drain. Assume that the transistor has been designed such that when the gate is charged to the positive supply voltage V_{DD} the majority of channel charge is due to mobile charge carriers, that is, electrons. A positive drain-to-source voltage v_{DS} causes a stream of electrons to flow out of the source and be swept into the drain. The time it takes an electron to flow from the source to the drain is the transit time t_τ. It depends on the mobility μ of the charges in the channel, the strength of the electric field $\mathcal{E}$ (which is inducing the carriers to move), and the length of the channel L. One complete changeover or renewal of the channel charge occurs each transit time. The mobility relates the average velocity $\mathcal{V}$ of the carriers to the electric field. We have

$$\mathcal{V} = \mu \mathcal{E}. \tag{1.1}$$

The current i_{DS} is equal to Q_G/t_τ. Up to a point, the higher the drain-to-source voltage v_{DS}, the higher the drain current. The reason is that a higher drain-to-source voltage causes a larger electric field in the channel and hence faster carrier propagation. When the voltage on the gate v_{GS} is low (Q_G small) there is very little or no mobile charge mirrored in the channel. The gate voltage at which mobile charge is first induced in the channel is called the threshold voltage V_T. For gate-to-source voltages below V_T, the resistance between the source and drain is nearly infinite. For gate voltages above V_T, the resistance of a typical MOS transistor is on the order of 20 kΩ. The higher the gate voltage, the lower the effective resistance R because more mobile charges are induced in the channel, increasing its current-carrying efficacy. Thus, for a given drain-to-source voltage, higher gate voltages cause higher

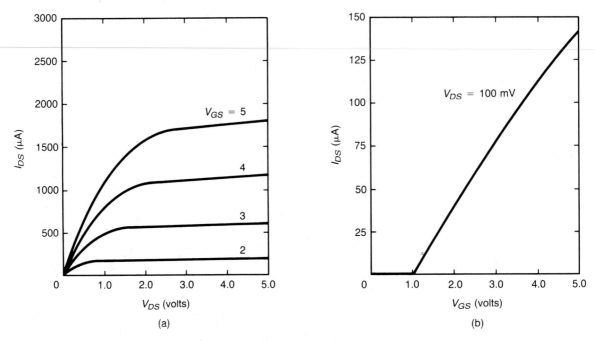

FIGURE 1.8 Current–voltage characteristics of an n-channel MOS device: (a) drain current versus drain-to-source voltage; (b) drain current versus gate-to-source voltage.

drain currents. Roughly then, we have

$$R \propto \frac{1}{v_{GS} - V_T} \qquad (1.2)$$

for gate voltages above threshold. The current–voltage characteristics of a typical nMOS enhancement mode transistor are shown in Fig. 1.8.

Another way to conceptualize the operation of a MOS transistor is in terms of potentials. Figure 1.9 illustrates the voltage potentials along the channel in an enhancement mode nMOS transistor with a positive drain-to-source voltage.[3] We see that electrons are prevented from flowing by the potential barrier at the source when the device is below threshold. This barrier is attributable to the built-in potential of the p–n diode formed between the n-type source and the p-type channel. A positive gate voltage will act to push up the potential under the gate and thus lower the barrier. At threshold the barrier will become low enough to allow current to flow. Current flow is limited by the action of the electrons in the channel, which, by their presence, raise the potential barrier. The higher the gate voltage, the more electrons that can exist

---

[3] We have oriented the diagram so that electrons, which are attracted to the positive drain potential, want to flow uphill.

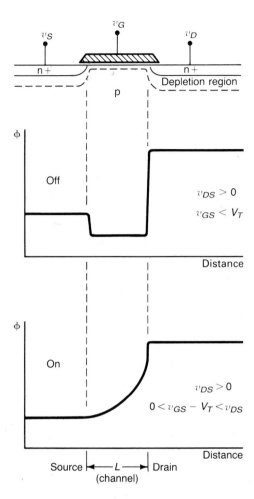

FIGURE 1.9   Voltage potentials under the gate of a MOS transistor. Electrons are attracted uphill to the positive drain voltage.

in the channel at any given time. Once in the channel, the electrons are accelerated by the lateral electric field as they "fall" up the potential hill. When the voltage on the gate is above threshold the carriers in the channel are electrons, despite the fact that the material is p-type. In other words, when the transistor is on, the channel is effectively n-type.[4] This is why transistors with p-type material in the channel are called n-channel transistors.

So far we have discussed only normally off transistors. Since we build integrated circuits out of only wire and transistors, when we need

---

[4] In this text we use the electrical engineering convention that a transistor is on when it is conducting and off when it is nonconducting. Seitz [Seitz 85] has suggested that the terms "cut" and "tied" are more expressive for talking about transistors that are either nonconducting or conducting, respectively. This alternate viewpoint can be helpful for certain circuits.

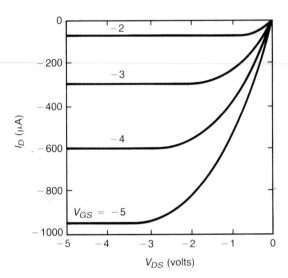

FIGURE 1.10 Current–voltage characteristics of a p-channel MOS transistor.

a resistor we must make it out of a transistor. A normally on transistor works just fine as a somewhat nonlinear resistor. In concept, making a normally on transistor, called a depletion mode transistor, is easy. To turn on an enhancement-mode transistor, we place charges on the gate that we want mirrored by charges in the channel. If we want to turn on the transistor permanently, we can place fixed charges directly in the channel region. These charges act like a battery in series with the gate. We can still turn off the device, but now the charges on the gate must be depleted to do so. Note that even with $v_{GS} = 0$ the region under the gate is effectively n-type.

In addition to depletion- and enhancement-mode n-channel transistors, p-channel devices are possible. In nMOS we have only n-channel devices; in pMOS, only p-channel devices, but in CMOS we have both. The p-channel transistor works just the opposite way from an n-channel device. For an enhancement-mode pMOS device, the transistor is off when v_{GS} is zero (just like in the nMOS case) but it is turned on by applying a negative gate voltage. Figure 1.10 illustrates the current–voltage characteristics of a p-channel enhancement-mode transistor.

The body terminal has only a modest, though important, effect on the electrical characteristics of the MOS transistor. In normal operation, the body of an nMOS device must be nonpositive with respect to the source and drain terminals to keep the isolation diodes reverse-biased. For a pMOS transistor the body must be nonnegative with respect to the source and drain regions. Since the source and drain can be charged up to the positive supply voltage V_{DD}, the body terminal is usually attached to V_{DD}. Thus the body of an nMOS device is usually biased at zero volts or below, while that of a pMOS device is usually biased at V_{DD}.

As the voltage on the body v_B is changed the current–voltage characteristics of the transistor change slightly. We examine an nMOS transistor. Assume that the source terminal is grounded and v_B is made negative. As v_B becomes negative, the potential barrier under the gate rises. For the transistor to turn on, the gate voltage must be large enough to overcome this increased barrier. From a modeling standpoint this means that the threshold voltage of the device is raised by making v_B more negative. Another way of looking at this is that some of the gate charges that previously terminated on mobile electrons in the channel now terminate on the negative charges in the enlarged depletion region in the substrate. More gate charges are needed to achieve the same number of carriers in the channel. We can view the gate and body terminals as both trying to lever the charge in the channel. In a well-designed device, the gate terminal has much more leverage.

The IEEE standard symbols [IEEE 78] for p-channel and n-channel transistors are shown in Fig. 1.11(a). These symbols are somewhat awkward to use. Figure 1.11(b) illustrates the four different types of transistors we will use in this book. They include three nMOS devices (enhancement-, depletion-, and zero threshold-mode devices) and an enhancement-mode pMOS device. Note that the body terminal is suppressed in this notation. The zero threshold device is just what its name implies: a MOS transistor with a threshold as close to zero as can be fabricated. These devices typically have thresholds of a few tenths of a volt. They have also been called "natural" transistors because in some processes they require the minimum amount of processing. The bubble on the pMOS device symbol can be thought of as a logic inversion of the signal on the gate. For a pMOS device, a high voltage on the gate causes the transistor to turn off, rather than on as in the case of nMOS. For a process designed to operate with a 5 volt power supply, enhancement-mode thresholds are typically in the 0.6 to 1.1 volt range for nMOS devices and −0.6 to −1.1 for pMOS devices. Depletion-mode thresholds of nMOS are typically between −2.5 and −4.0 volts, while the zero threshold device has a threshold between 0.2 and 0.6 volts.

Because a very wide transistor is the same as several narrower devices in parallel, the current-carrying ability of a MOS transistor increases as the effective width W of the transistor increases. The current-carrying ability of the transistor decreases as the effective channel length L increases. It is helpful to think of the MOS transistor as a voltage-controlled nonlinear resistor. This resistance is proportional to the length and inversely proportional to the width. We define the shape factor S as

$$S \equiv \frac{W}{L}, \qquad\qquad (1.3)$$

where S is roughly proportional to the conductance of the transistor.

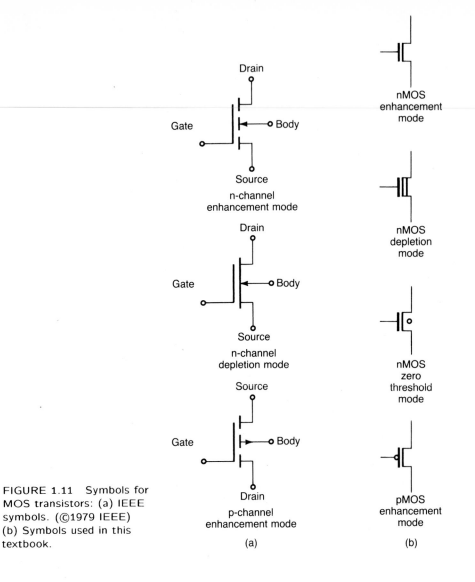

Drain

Gate ——| |——o Body

Source

n-channel
enhancement mode

Drain

Gate ——| |——o Body

Source

n-channel
depletion mode

Source

Gate ——| |——o Body

Drain

p-channel
enhancement mode

(a)

nMOS
enhancement
mode

nMOS
depletion
mode

nMOS
zero
threshold
mode

pMOS
enhancement
mode

(b)

FIGURE 1.11 Symbols for
MOS transistors: (a) IEEE
symbols. (©1979 IEEE)
(b) Symbols used in this
textbook.

FIGURE 1.12   MOS
transistor capacitances.
Ideally we would like C_A to
be the only capacitance.

v_G

C_B C_B

C_A

v_S v_D

C_C

v_B

12

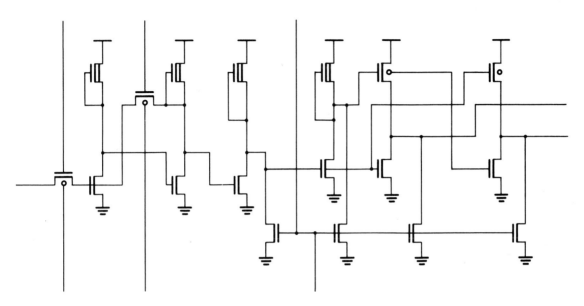

FIGURE 1.13   Schematic of the circuitry seen in Fig. 1.2.

In addition to a threshold and a resistance, a transistor has a variety of capacitances. Figure 1.12 shows a simple model of the MOS transistor. Below threshold, the switches are open. Above threshold, the nonlinear resistances between source and drain come into play. The capacitance C_A is the parallel plate capacitance of silicon dioxide sandwiched between the gate and channel. C_B is the overlap capacitance between the gate and source, and the gate and drain. C_B is important even when the transistor is off. C_C is the capacitance between the channel region and the body. The capacitances C_A, C_B, and C_C are roughly in the proportion 40 : 5 : 1 for a typical MOS device.

Figure 1.13 shows a schematic of the circuit in Fig. 1.2, drawn to conform to the topology of the chip. In VLSI circuit design we try to draw schematics that suggest the chip layout. This is in deference to the enormous importance of wire in determining the performance of a VLSI chip. To actually build a chip, we must generate specifications for the chip geometries. These are called mask specifications or layouts. The layout illustrated in Fig. 1.13 could have been used to build the circuit shown in Fig. 1.2.

In bulk CMOS design we differentiate between the n-channel and p-channel devices by putting one of these devices in a well. For instance, in n-well technology the p-channel devices are placed in the well. Figure 1.14 provides a cross-sectional view of a transistor in a well. Figure 1.15 shows a photomicrograph of a portion of a CMOS chip [Rubinfeld 82].

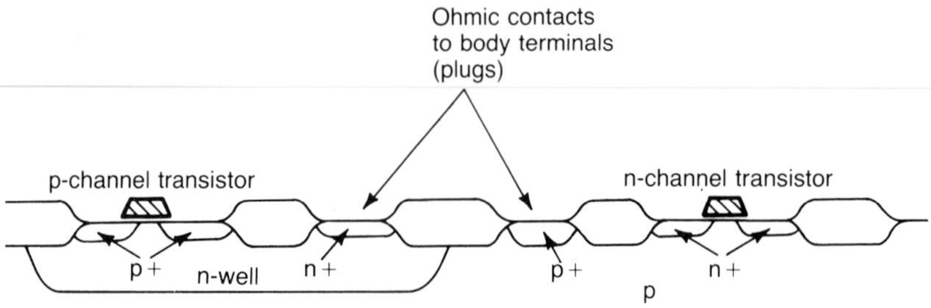

Ohmic contacts
to body terminals
(plugs)

p-channel transistor

n-channel transistor

FIGURE 1.14 Cross section of n-well CMOS transistor.

p+   n-well   n+   p+   n+

p

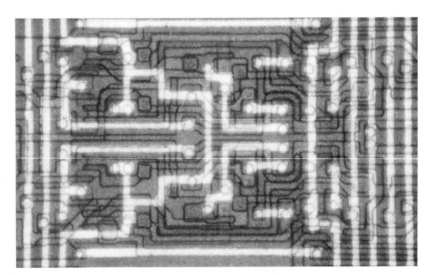

FIGURE 1.15 Photomicrograph of J11 microprocessor done in p-well CMOS technology with two layers of poly. (Copyright ©Digital Equipment Corporation 1985. All rights reserved. Reprinted with permission.)

■■■■■■■■■■■■■ EXAMPLE 1.1

The example processes

Two contrived processes, one for nMOS and one for CMOS, are used throughout the book to illustrate examples and form the basis for problems. While these processes do not represent any one company's process, the parameters are reasonable. The processes are 2 μm processes, meaning that the minimum length of the gate poly is two microns.[5] Design rules, process parameters, worst-case SPICE files, and so forth, are given in the appendixes. SPICE [Vladimirescu 80, 81] is a widely available circuit simulation program from the University of California at Berkeley. The transistor curves illustrated in Figs. 1.8 and 1.10 were generated from SPICE models of the CMOS devices. The device sizes

---

[5] There is no universal agreement on which of the critical dimensions in a process must be 2 μm for a process to be a 2 μm process. Note also that a 2 μm process has a Mead and Conway λ of 1 μm [Mead 80].

TABLE 1.1 SPICE deck for generating Fig. 1.8

```
I-V characteristics for SS model of CMOS devices
********************************************************************
.MODEL NSS NMOS LEVEL=3 RSH=0 TOX=275E-10 LD=0.1E-6 XJ=0.14E-6
+ CJ=1.6E-4 CJSW=1.8E-10 UO=550 VTO=1.022 CGSO=1.3E-10
+ CGDO=1.3E-10 NSUB=4E15 NFS=1E10
+ VMAX=12E4 PB=0.7 MJ=0.5 MJSW=0.3 THETA=0.06 KAPPA=0.4 ETA=0.14
.MODEL PSS PMOS LEVEL=3 RSH=0 TOX=275E-10 LD=0.3E-6 XJ=0.42E-6
+ CJ=7.7E-4 CJSW=5.4E-10 UO=180 VTO=-1.046 CGSO=4E-10
+ CGDO=4E-10 TPG=-1 NSUB=7E15 NFS=1E10
+ VMAX=12E4 PB=0.7 MJ=0.5 MJSW=0.3 ETA=0.06 THETA=0.03 KAPPA=0.4
********************************************************************
M1 1 10 0 0 NSS W=13.2U L=2.25U
VDS 20 0
********************************************************************
* VGS is positive for nMOS and negative for pMOS
********************************************************************
VGS 10 0 5V
* VIDS defines current direction for drain
********************************************************************
VIDS 20 1
.DC VDS 0 5 0.05
.PRINT DC I(VIDS)
.PLOT DC I(VIDS)
.WIDTH IN=75 OUT=75
.END
```

drawn were 15/2, which means that the transistor width specified by
the circuit designer was 15 μm and the length 2 μm. Table 1.1 lists the
SPICE deck used to generate one of the curves.   ■

1.2 Static logic forms

The transistors and interconnect available on an integrated circuit can
be used to build many different types of devices including operational
amplifiers, radios, jewelry, memories, and digital switching networks.
Because our primary interest is in digital switching networks, the
restoring logic gate is the most important of the digital circuit elements
we will study. Logic gates are usually nonreciprocal in that they pass
signals from the input to the output but not vice versa. They are also

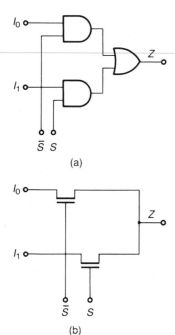

(a)

(b)

FIGURE 1.16 A two-to-one selector circuit implemented with (a) gates and (b) nMOS pass transistors.

FIGURE 1.17 An nMOS all enhancement-mode inverter.

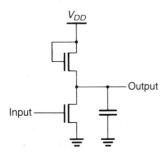

active devices that attenuate the corrupting effects of noise in order to preserve and refresh (restore) the integrity of the logic signals.

A second useful form of switching network is called pass transistor, relay, or steering logic [Shannon 37]. In pass transistor logic, the pass device is used as a switch to conditionally connect two nodes together. Pass transistor logic is often appropriate when the logic function is conveniently conceptualized as signals or tokens being conditionally steered through a network. Figure 1.16 illustrates two versions of a selector circuit, one implemented with conventional restoring logic gates and the other with pass transistors. With the pass transistor implementation it is not obvious from the unlabeled schematic which terminals are inputs and which are outputs. The signals can actually pass in either direction depending on the external circuitry. It is so common to find pass transistor networks that pass signals in the direction opposite from that intended by the designer that such paths have been named "sneak paths."

All enhancement-mode nMOS

The basis for all MOS logic gates is the inverter. The nMOS implementation of this device requires two transistors—one to pull the output high when the input is low and one to pull the output toward ground when the input is high. The "pullup" transistor in static nMOS circuits is typically in the configuration of a two-terminal device. Its resistance depends only indirectly on the input signal. This transistor is called the "load." The gate of the "pulldown" transistor is typically connected to the input terminal. The static nMOS logic gate is a type of ratioed logic, so called because one of the logic levels depends on the controlled ratio of two resistances.

The simplest inverter consists of two enhancement-mode devices. The enhancement-load inverter shown in Fig. 1.17 works as follows. When the input voltage is below V_T, the pulldown transistor is off and the output rises to $V_{DD} - V_T$. Note that it does not rise all the way to V_{DD}, the power supply rail. This is because once the source of the pullup transistor has risen to $V_{DD} - V_T$, the gate-to-source voltage equals V_T and the transistor shuts off. When both transistors are off the output stays at $V_{DD} - V_T$ because of the capacitance on the output node.[6]

When the input is high, both transistors are on. The output voltage is determined by voltage division across the two conducting devices. If we have arranged the ratios of the shape factors of the pullup and pulldown

---

[6] Note that there is nothing to prevent the output from rising to V_{DD} or higher if some other source could provide the current. If it rises higher than V_{DD}, the source becomes the drain and the drain becomes the source. This transistor configuration is called a "diode" connection.

transistors correctly, the output voltage will fall below V_T when the input rises near $V_{DD} - V_T$. This will lead to a self-consistent situation where high inputs cause low outputs and vice versa. We have built an inverter.

Using the inverter as a prototype we can begin to design more complicated logic gates. Figure 1.18 illustrates a three-input NOR gate. This gate consists of a single pullup transistor and three parallel pulldown transistors. The output is pulled down if any one of the pulldown devices are on. It has been shown that the ability to implement NOR gates is all that is required to build any logic function. This does not imply that this is a particularly convenient or efficient implementation method. Figure 1.19 illustrates a full adder implemented with NOR gates. A full adder satisfies the equations

$$S = A\overline{B}\,\overline{C} + \overline{A}B\overline{C} + \overline{A}\,\overline{B}C + ABC \qquad (1.4)$$

and

$$C_o = AB + BC + AC, \qquad (1.5)$$

where A, B, and C are inputs. S is the sum output and C_o is the carry output. A NAND gate is similar to the NOR gate except that the pulldown devices are in series. Thus only if all of the pulldown transistors are on is the output pulled low. Note that the series resistance of the pulldown path increases as transistors are added. In order for the output to still be able to pull low (less than V_T), the ratio of the pulldown to pullup shape factors must be increased. This usually means that the

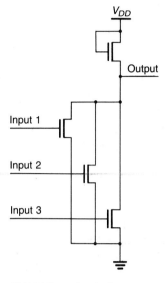

FIGURE 1.18   A three-input NOR gate.

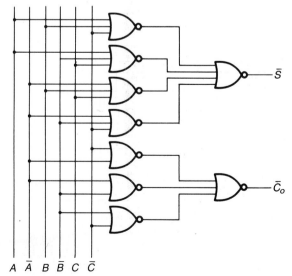

FIGURE 1.19   NOR gate implementation of a full adder.

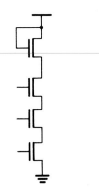

FIGURE 1.20   A three-input NAND gate.

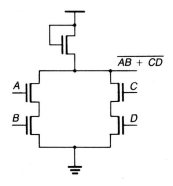

FIGURE 1.21   A complex gate.

shape factor of the pullup must be decreased or the pulldowns must be widened. Figure 1.20 illustrates a three-input NAND gate.

More complicated logic gates can be built. Figure 1.21 illustrates a complex gate structure consisting of several parallel and series connected pulldown transistors. All of the primitive logic gates discussed so far are inverting. Figure 1.22 illustrates a reduced majority logic implementation of a full adder, this time with complex gates [Carr 72, p. 134].

The simplest pass transistor logic network is the selector shown in Fig. 1.16(b). The selector works as follows. When S is high, the input I_1 is connected to the output Z; when S is low, I_0 is connected to Z. Even if I_1 and S are both at V_{DD}, the output would only rise to $V_{DD} - V_T$. But if both inputs, I_1 and I_2, are at $V_{DD} - V_T$, the outputs could still rise to $V_{DD} - V_T$. If S and $\overline{S}$ were somehow both high, we would have the undesirable result of connecting I_1 to I_0 (a sneak path). Figure 1.23 illustrates a pass transistor implementation of a full adder.

We have examined the behavior of two digital MOS circuit forms— the pass gate and the inverting logic gate. We know that we are able to use these forms to build any large digital function because one of these circuit forms includes the two-input NOR gate and that is itself universal. The next step is to examine how these circuit parts can be combined to produce a consistent and reasonable circuit family for building combinational logic. To do this we need to augment (and in some cases change) the composition rules for the parts. We need additional composition rules dealing with the wiring of parts. This exercise of determining the composition rules for circuits is rarely done

FIGURE 1.22   Complex gate implementation of a full adder.

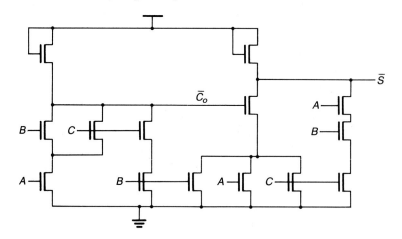

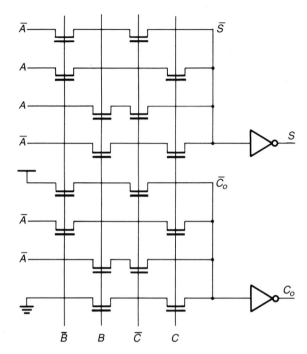

FIGURE 1.23   Pass transistor
implementation of a full adder.

in a formal manner.[7] Nevertheless, understanding how circuits are
constructed and under what conditions and restrictions they can be
wired together is one of the keys to VLSI design. These composition
rules form one of the elements of the circuit-design methodology.

Let us carefully re-examine the enhancement load logic gate with
this more global perspective. There are several composition rules for the
construction of the logic gate. The logic gate consists of a pullup and a
pulldown. The pullup consists of a single enhancement mode transistor
with its drain and gate connected to V_{DD} and its source connected to the
output. The pulldown network consists of enhancement mode transistors
whose sources and drains are connected in a series–parallel two-terminal
network. One of the terminals of this network is connected to ground
while the other is connected to the output node. The gates of these
pulldown transistors are the inputs to the logic gate. The voltage inputs
to the network must be $V_{DD} - V_T$ to represent a high logic level, and V_T
or lower for a low. If there is no path of on transistors to ground, then
the output will be pulled high. The shape factors of the transistors must
be designed so that for every possible path of on transistors between

---

[7] It would be great if it was because then one could prove whether or not a circuit
was correctly constructed. Unfortunately, this is a *very* hard problem.

the output terminal and ground, the output should be pulled below V_T, provided that the gate voltages of the on devices is $V_{DD} - V_T$. In other words, valid input high voltages generate valid output low voltages and vice versa.[8] Let us call these logic gates "Type I" logic gates.

Type I logic gates have no fan-out restrictions (we are ignoring dynamics) to other Type I logic gates because the input impedance of the Type I logic gates is infinite at dc. On the other hand, wiring the outputs of two Type I gates together can cause incorrect operation because two pullups are wired in parallel. A pulldown path whose current-sinking ability is sufficient to pull one pullup low may not be able to pull *two* pullups low.

Therefore we have the first of several composition rules for assembling networks: no two outputs of Type I logic gates may be wired together. The purpose of this restriction is to make the systems design easier. Note that while it is true that a circuit violating this composition rule may not work, there is nothing that *says* it will not work. People have designed logic gates with two pullups that work just fine. Nevertheless, in this, the first and simplest methodology we are developing, we outlaw such logic gates in order to control the complexity of the circuit design task. Having a methodology for the construction of logic gates not only makes them easier to design, but it also makes them easier to debug.

In enhancement-mode pass transistor logic there are two challenges: avoiding sneak paths and keeping track of the threshold drops. Given a pass transistor with high voltages on both the gate v_G and the drain v_D, the voltage at the source v_S is the minimum of v_D and $v_G - V_T$. If v_G is less than V_T, then the output of the device is independent of v_D. Figure 1.24 illustrates a network of pass transistors with the various threshold drops noted on the schematic.

When pass transistors are combined with enhancement-load logic gates, we find a problem with consistency. The highest voltage out of an enhancement-load inverter is $V_{DD} - V_T$. Yet if that voltage is applied to the gate terminal of a pass device, the highest output voltage from the pass device is $V_{DD} - 2V_T$. This voltage is not high enough to be interpreted as a high by a second enhancement-load logic gate. Worse yet, we can have voltages as low as $V_{DD} - 3V_T$ or $V_{DD} - 4V_T$, or even lower, if arbitrary connections of pass transistors are allowed.

There are several fixes one could imagine. The simplest is to outlaw pass transistors and this is sometimes done. Another solution is to design a type of enhancement-load logic gate that has its shape factors adjusted so that the output will pull below V_T even with only $V_{DD} - 2V_T$ on its gate. For pulldown transistors of the same size, this "Type II" gate will have a pullup device with a smaller shape factor. We still

---

[8] This development will be expanded in Chapter 4 to include noise margins.

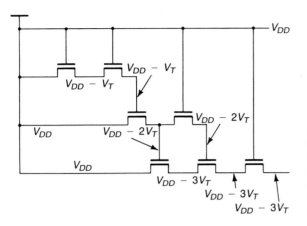

FIGURE 1.24 A network of enhancement-mode n-channel pass transistors with threshold drops noted. A large leakage resistance to ground is assumed at each node.

have the problem of pass transistor output voltages at $V_{DD} - 3V_T$ and below, but this can be avoided by adding yet another restriction on the design methodology: the outputs of pass transistor networks may not be connected to the gates of other pass transistors. The schematic of a logic network containing Type I and Type II enhancement-load logic gates and pass transistor networks constrained to have a minimum high voltage of $V_{DD} - 2V_T$ is illustrated in Fig. 1.25. Table 1.2 lists which logic gate types are legal receivers for which inputs.

We have succeeded in designing a fairly primitive circuit methodology for constructing all enhancement-mode digital MOS combinational circuits. The methodology uses three archetypical circuit forms. A modern VLSI circuit might have fifty such forms. While fifty is much larger than three, it is still orders of magnitude fewer than the number of logic gates in a VLSI chip. Moreover, the circuit methodology could be used for many chips. The key observation is that the complexity of a circuit methodology scales very slowly with the size of the system being designed. There are two forces that tend to make methodologies more complex. The first is that in very large systems new phenomena, such as metal resistance, become important. A more complex methodology may be needed to cope with a more complex situation. The second force is the quest for performance. More complex methodologies can, but do not necessarily, lead to higher performance circuits. Take the example

TABLE 1.2   Methodology specifications

	TYPE I	TYPE II
Logic Gate Input?	Yes	Yes
Pass Transistor Input?	No	Yes
$V_{DD} - 3V_T$ Input?	No	No

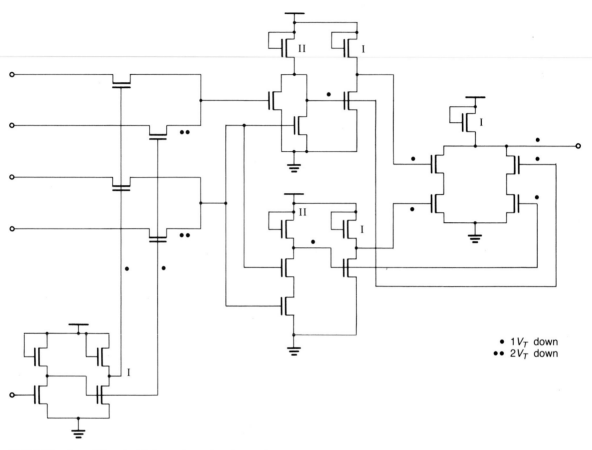

FIGURE 1.25 Threshold drops in a simple methodology.

we have just been investigating. We could simplify the methodology by replacing all the Type I logic gates by Type II logic gates. This reduces the number of circuit forms by 30%, but also slows the circuits down because of the higher pullup resistance. Historically, in the trade-off between simplicity and the performance of low-level building blocks, the engineering and computer science communities have tended to move toward simplicity and designability, but only when the penalty is slight. A full discussion of these forces is outside the scope of this book. It should be noted, however, that the pace of technology is such that products that are behind schedule are being designed in an obsolete technology.

Bootstrapped logic gates

The enhancement-load logic gate wastes the part of the power supply between $V_{DD}-V_T$ and V_{DD} because its output never rises above $V_{DD}-V_T$. As we will see in Chapter 4, it is also fairly slow and has a relatively poor

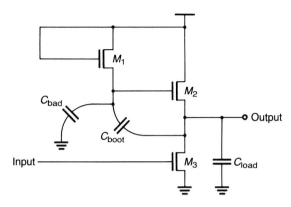

FIGURE 1.26   A bootstrapped inverter.

power-delay product. The reason the enhancement-mode pullup device cannot pull up higher than $V_{DD} - V_T$ is that its gate is tied to V_{DD}. One way to avoid this problem is simply to tie all of the gates of the pullup transistors to a second supply voltage V_{GG} that is larger than $V_{DD} + V_T$. This costs a second power supply and all of the wires to distribute it across the chip, but at least the current drain on this supply is very small and the logic gate's output high voltage increases to the V_{DD} rail.

Another technique is to bootstrap[9] the gate terminal on the pullup transistor above the supply rail [Joynson 72]. Figure 1.26 illustrates a bootstrapped inverter. When the input to the logic gate is high, the output node is pulled down to V_{OL}. V_{OL} is defined as the output logic low voltage and is less than V_T. The source of transistor M_1 is at $V_{DD} - V_T$. The voltage on the capacitor C_{boot} is $V_{DD} - V_T - V_{OL}$. The charge on C_{boot} is $(V_{DD} - V_T - V_{OL})C_{boot}$ and the charge on C_{bad}, the parasitic capacitance on the source of M_1, is $(V_{DD} - V_T)C_{bad}$. When the input voltage falls toward V_T the output voltage rises from V_{OL}. M_2 charges up the capacitor C_{load}. Since M_1 is off, the charge on the boot node is trapped there. The total charge Q_{boot} is

$$Q_{boot} = (V_{DD} - V_T - V_{OL})C_{boot} + (V_{DD} - V_T)C_{bad}. \qquad (1.6)$$

The capacitor C_{boot} will try to keep the voltage across it constant. Thus as v_{out} rises, so does v_{boot}. In order for v_{out} to reach V_{DD}, v_{boot} must reach $V_{DD} + V_T$. We can derive the necessary conditions on C_{boot} for this to happen. When v_{boot} reaches $V_{DD} + V_T$, the charge on C_{bad} is $(V_{DD} + V_T)C_{bad}$ and the charge on C_{boot} is $V_T C_{boot}$. The new charge on the boot node is the sum of these, which must still equal Q_{boot}. We have

$$Q_{boot} = C_{bad}(V_{DD} + V_T) + C_{boot}V_T. \qquad (1.7)$$

Solving for the minimum value of C_{boot} necessary to ensure that the

---

[9] As in "to pull oneself up by one's bootstraps."

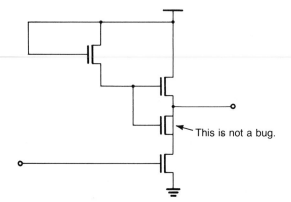

This is not a bug.

FIGURE 1.27   Boot-strapped inverter with MOS capacitor.

output rises to V_{DD}, we have

$$C_{\text{boot}} = \frac{2V_T}{V_{DD} - 2V_T - V_{OL}} C_{\text{bad}}. \qquad (1.8)$$

C_{boot} can be built explicitly using a transistor with its source and drain shorted together as shown in Fig. 1.27. Often the intrinsic gate capacitance of the transistor is sufficient. Layout and transistor capacitances contribute to C_{bad}. Note that in the limit of $C_{\text{bad}} = 0$ the voltage across C_{boot} is constant.

Bootstrapping is an extremely useful technique and will be used in many guises throughout the book. In this case it enabled us to eliminate the V_{GG} power supply and wiring at the expense of an extra transistor. Trading transistors for wire is frequently worthwhile in VLSI.

There are limitations of the bootstrapped load that sometimes make it awkward to use. One limitation is time. After some number of milliseconds, the charge on the boot node will leak off because the off resistance of an MOS transistor is not quite infinite and the output will drift back to $V_{DD} - 2V_T$. If the chip must be stopped periodically or if there are portions of the chip not exercised for long periods of time, bootstrapping may not be appropriate.

The depletion load

Bootstrapped loads pulled all the way to V_{DD} because the v_{GS} voltage was kept approximately constant. Recall from Section 1.1 that a depletion-mode transistor may be thought of as an enhancement-mode device with a built-in battery between the gate and source. To construct a depletion load, one simply connects the gate and source terminals of a depletion-mode transistor [Masuhara 72]. The cost is a slightly more complex and expensive fabrication process. Figure 1.28 illustrates the mask specifications for a pair of depletion-load inverters. Note the buried

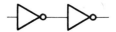

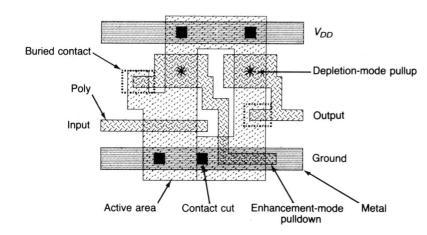

FIGURE 1.28   Mask specifications for a pair of depletion-load inverters.

contact that attaches the gate to the source on the depletion-mode pullup. As a shorthand notation for the depletion threshold adjust mask, we often put a star ($\star$) on the poly gate material of a depletion-mode transistor. A hash mark (#) is used to denote a zero threshold device. If the circuit in Fig. 1.25 were redrawn with depletion-mode gates, the minimum high voltage would be increased from $V_{DD} - 2V_T$ to $V_{DD} - V_T$.

Figure 1.29 illustrates the transient response of a pair of cascaded inverters. Note that the pullup transient is much slower than the

FIGURE 1.29   Transient response of a string of depletion load inverters: (a) the schematic; (b) the waveforms. Note the slowness of the rise times.

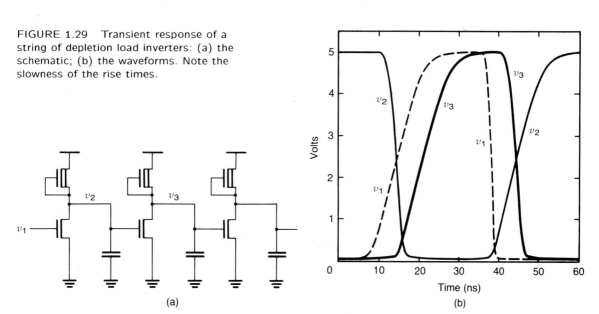

(a)

(b)

pulldown transient. This is because the pulldown transistor must be more conductive than the pullup device in order to pull the output below V_T when the input is high. This constraint on the static behavior of the pulldown is translated into an asymmetry in the dynamic response.

One can always increase the pullup speed of a logic gate that is driving a fixed capacitive load by increasing the width of all the transistors in that gate. There are two problems with this. The first is that the power dissipation of the ratioed logic circuits also increases.[10] The second problem is that the input capacitance (which is the load capacitance on the previous logic gate) also increases. Through clever circuit design, one can get around these problems, up to a point.

One way to speed up the driving of large loads is to use a superbuffer [Mead 80, pp. 17–18]. Figure 1.30 illustrates an inverting superbuffer. This device uses four transistors rather than two. Assuming properly sized transistors, the superbuffer can increase the speed of a depletion-load inverter that is driving a large capacitance at no additional cost in static power. The superbuffer works as follows. Assume that the input is low and the output and internal node voltage v_K are both at V_{DD}. A rising input transition will turn on both pulldown transistors. Internal node voltage v_K will fall first since its load, which consists of some layout parasitics and the gate of M_3, is small compared with C_{load}. This will cause the current out of M_3 to decrease slightly compared with the case when the gate of M_3 is connected to its source. This is not a strong effect and the pulldown speed of a superbuffer is about the same as that for a regular inverter because the conductance of the pulldown transistor determines the discharge dynamics. This is all right since the pulldown transient is the fast transient anyhow. For the pullup case the input to the inverter goes from V_{DD} to V_{OL}. This turns off both pulldown devices. Initially v_{GS} of both pullup transistors is zero, but as v_K quickly rises, the gate-to-source voltage on M_3 becomes large, approaching $V_{DD} - V_{OL}$ for very large load capacitances. Because of this overdrive, M_3 is turned on very hard during the important initial nanoseconds of the output rise time. This can result in pullup speeds two to six times faster, for the same dc power and input capacitance, than in the case of a simple inverter. From our description of the dynamics, it should be clear that this advantage is gained only when driving large loads. Two layouts of nMOS superbuffers are illustrated in Figs. 1.31 and 1.32.

The superbuffer illustrates a mode of thinking often used when designing VLSI circuits. In the superbuffer we were able to partially partition the issues of dynamics and statics and thus achieve a more effective design. We did this at the cost of circuit complexity.[11] The

FIGURE 1.30   An inverting superbuffer.

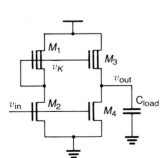

---

[10] Power dissipation is proportional to the current times the power supply voltage V_{DD}.

[11] Compare this with the transition from the bootstrapped enhancement load to the depletion load, where circuit complexity was reduced at the expense of processing complexity.

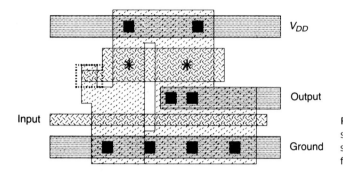

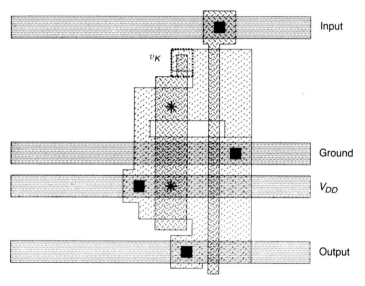

FIGURE 1.31   Mask specifications for an nMOS superbuffer with depletion-follower output pullup.

FIGURE 1.32   Mask specifications for another nMOS superbuffer with the same schematic as the one in Fig. 1.31, but with a different layout idiom.

FIGURE 1.33   A superbuffer with a zero threshold pullup on the output stage; a lower power, but often slower, alternative to the depletion follower superbuffer.

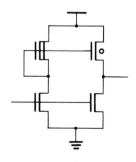

circuit designer's bag of tricks contains many techniques of which the superbuffer is only the first.

In many nMOS processes a "zero" threshold transistor is available in addition to enhancement- and depletion-mode devices. This device is extremely useful for buffers and small fan-in selectors. Figure 1.33 illustrates a buffer that uses the zero threshold device. Compared with a superbuffer implemented with transistors of the same size, the zero threshold buffer uses much less power, is considerably slower, and does not pull the output as high (the maximum output voltage will be $V_{DD}-V_{TZ}$). On the other hand, the zero threshold pullup can be widened with a negligible static power penalty because it is on only during a switching transient. For larger loads the speed of the zero threshold buffer can approach that of the superbuffer, yet still draw less power.

The zero threshold transistor is leaky with $v_{GS} = 0$. Thus we have to be careful that many off zero threshold devices do not overpower one on device. This is the reason only small fan-in selectors can be built with the zero threshold devices. For large fan-in situations, such as buses, enhancement-mode transistors must be used.

The zero threshold device further complicates the design methodology. We typically replace the bootstrapped enhancement loads with depletion loads if depletion transistors are available, but we still have potentially two types of ratioed gates even without the zero threshold devices. They are the gates whose inputs are the outputs of other depletion-load logic gates (which rise to V_{DD}) and logic gates whose inputs come through pass transistors. With zero threshold devices we could either use a third type of ratioed gate (to account for inputs at V_{DD}, $V_{DD} - V_{TE}$, and $V_{DD} - V_{TZ}$) or combine two types. For instance, we could use one type of logic gate for inputs that are at V_{DD} or down one zero threshold threshold, and another type of gate for inputs down an enhancement-mode threshold.

CMOS static logic forms

All of the logic gates we have discussed so far dissipate dc power when the output is low. Because this heat is very hard to remove from the chip and because the performance of MOS transistors decreases as the temperature of the chip increases, the design of nMOS VLSI circuits becomes quite complex. Special circuit forms must be used to keep the power dissipation down. Several of these special nMOS circuit forms will be discussed in the next section.

As usual, we can make the circuit design job easier by making the fabrication more complex and expensive. So far we have introduced three types of nMOS transistors. All have the characteristic that as the voltage on the gate of the transistor rises, the conductivity of the transistor increases. The p-channel transistor works in just the reverse. As the voltage on the gate increases, the pMOS transistor becomes less conductive. This expedites building push–pull type structures, which dissipate power only while switching.

Figure 1.34 illustrates a CMOS inverter, NOR gate, and NAND gate [Wanlass 63]. Note that the pulldown network is exactly the same as it was in the nMOS ratioed logic case. The pullup network is the dual of the pulldown network. Every two subnetworks that are connected in series in the pulldown are connected in parallel in the pullup, and every two subnetworks that are connected in parallel in the pulldown are connected

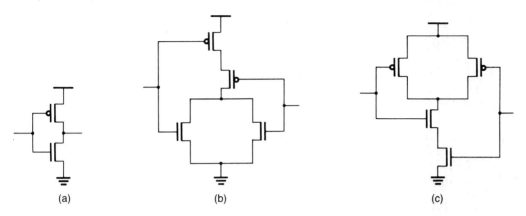

FIGURE 1.34   CMOS logic gates illustrating the schematics for (a) an inverter, (b) a NOR gate, and (c) a NAND gate.

in series in the pullup. The pullup transistors are all enhancement-mode p-channel devices, while the pulldown devices are all enhancement-mode n-channel devices. In a properly constructed CMOS logic gate of this sort, the pulldown network is off when the pullup network is on, and vice versa. The pulldown and pullup networks are on simultaneously only during a switching transient.

The CMOS inverter works as follows. When the input is high (V_{DD}), the pulldown transistor is on but v_{GS} of the p-channel device is zero and hence the pullup is off. Therefore the output pulls all the way to ground. When the input node is low (0 volts) v_{GS} of the p-channel device is $-V_{DD}$ and hence it is on. The pulldown is off. The output rises to V_{DD}.

The dual nature of the pullup and pulldown networks can be understood in terms of the boolean function the gate is performing. Take, for example, the boolean logic function describing the inverse Z of the carry output C_o from a full adder. We have

$$\overline{Z} = AB + BC + AC, \qquad (\mathbf{1.9})$$

where

$$Z = \overline{C}_o. \qquad (\mathbf{1.10})$$

The pulldown network can be read directly from Eq. (1.9) with AND being interpreted as a series operator and OR as a parallel operator. Using DeMorgan's theorem, we have

$$Z = (\overline{A} + \overline{B})(\overline{B} + \overline{C})(\overline{A} + \overline{C}). \qquad (\mathbf{1.11})$$

The topology of the pullup can be directly read from this equation.

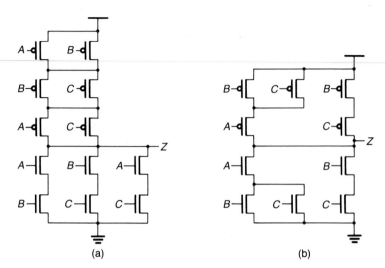

FIGURE 1.35 (a) Schematic of a CMOS circuit for the carry out of a full adder; (b) optimized version of circuit in (a) with fewer transistors and a lower pullup resistance.

(a)   (b)

Equations (1.9) and (1.11) are duals of one another. Every OR is replaced by an AND and every AND by an OR. Every symbol is inverted.

The equations can be simplified and these simplifications can also be interpreted in terms of the topology. We have

$$\overline{Z} = A(B+C) + BC \tag{1.12}$$

and

$$Z = \overline{A}(\overline{B}+\overline{C}) + \overline{B}\,\overline{C}. \tag{1.13}$$

The schematics corresponding to Eqs. (1.9) through (1.13) are shown in Fig. 1.35. (Note that Eqs. (1.12) and (1.13) are *not* duals.)

FIGURE 1.36 A four-to-one selector with two constant inputs.

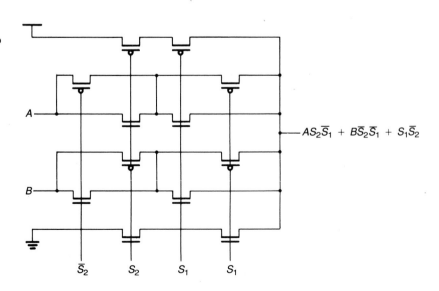

$$AS_2\overline{S}_1 + B\overline{S}_2\overline{S}_1 + S_1\overline{S}_2$$

In CMOS we can use pass transistor circuits without incurring any threshold drops. This is done by using parallel p- and n-type pass transistors whenever the pass gate must transmit both high and low signals. Figure 1.36 illustrates a four-input selector implemented in CMOS. One of the inputs is tied to V_{DD} and the other to ground. For these legs of the circuit, the use of all n-type (if the circuit only pulls low) or all p-type (if the circuit only pulls high) transistors is possible.

Many circuits use both pass transistors and restoring gates very synergistically. Figure 1.37 illustrates several implementations of XOR and XNOR gates.

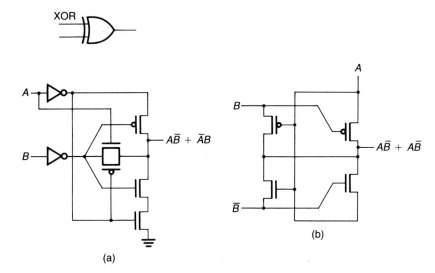

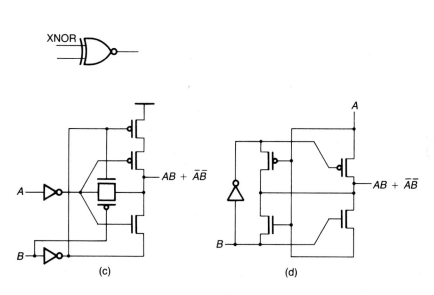

FIGURE 1.37   XOR and XNOR (equivalence) circuits: (a) and (b) are XOR circuits; (c) and (d) are XNOR circuits. These circuits combine aspects of restoring logic gates and pass transistor logic.

Duality

If

$$F = AB + \overline{C} + D,$$

then

$$\overline{F} = (\overline{A} + \overline{B})C\overline{D}. \quad ■$$

■■■■■■■■■■■■■ EXAMPLE 1.3

The dynamic response of several inverter types

Figures 1.38 through 1.42 illustrate the dynamic response of several different types of inverters including an enhancement-load inverter, a bootstrapped inverter, a depletion-load inverter, a depletion follower superbuffer, and a classical CMOS inverter.   ■

FIGURE 1.38   The schematic and waveforms of an all enhancement-mode inverter.

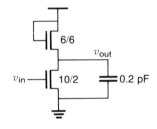

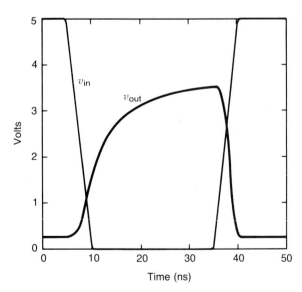

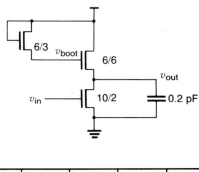

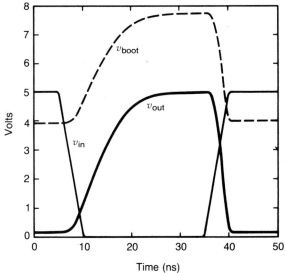

FIGURE 1.39 The schematic and waveforms of an all enhancement-mode bootstrapped inverter.

FIGURE 1.40 The schematic and waveforms of a depletion load inverter.

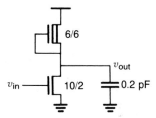

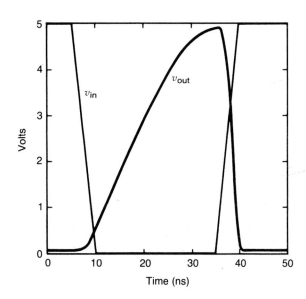

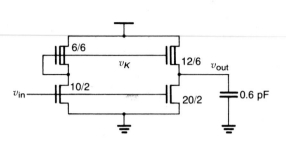

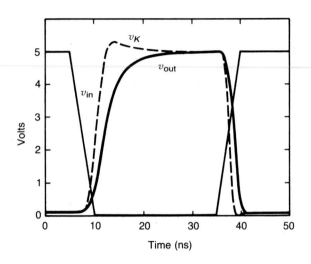

FIGURE 1.41   The schematic and waveforms of a depletion load superbuffer with depletion-mode output pullup.

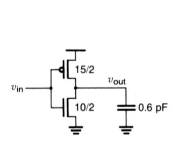

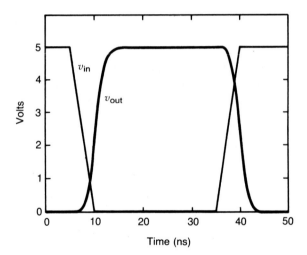

FIGURE 1.42   The schematic and waveforms of a CMOS inverter.

1.3   Storage elements and sequential circuits

In Section 1.2 we examined a number of circuit forms useful for the implementation of combinational logic. In combinational logic circuits, the output is a function only of present inputs. While combinational logic is necessary for the construction of large digital systems, it becomes most useful when combined with memory. A typical digital system must sequence through a number of operations where the results of

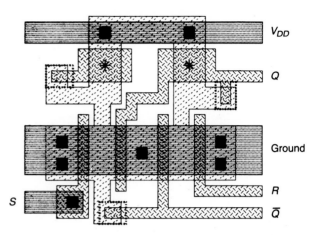

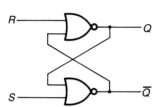

FIGURE 1.43   Cross-coupled NOR gates.

FIGURE 1.44   Layout (mask specifications) of the circuit shown in Fig. 1.43.

each operation depend on the results of previous operations. For such sequential systems, the output depends on the history of the inputs.

Sequential systems are built out of storage elements and combinational logic [Hill 74]. In MOS, there are several types of storage elements. Static storage elements are built from logic gates in a feedback configuration. Figure 1.43 illustrates the schematic for simple SR latch constructed out of cross-coupled NOR gates. Two layouts are illustrated in Figs. 1.44 and 1.45. The latch has two stable states when both S and R are low.

In addition to static storage elements, MOS technology offers the option of building dynamic memory elements. A static memory remembers its state as long as the power supply is turned on;[12] a dynamic memory remembers its state long enough to be useful, but not forever. Dynamic memory elements provide an advantage in that they can be implemented much more compactly than static memory elements can. In MOS technology a dynamic memory cell can retain its information about one millisecond, depending on the temperature. Thus a cell can remember for a period of time on the order of 10^6 times as long as the minimum delay through an inverter pair. In its purest form, the MOS dynamic memory element consists of a switch and a capacitor. A good analogy between restoring logic and static memories, and pass transistor logic and dynamic memory is that the former two actively counteract the effects of noise while the latter two do not.

---

[12] In the case of bubble memories and EPROMs, the state is remembered even after the power is turned off. Memories that remember after the power has been turned off are referred to as nonvolatile. All of the MOS memories discussed in this text are volatile.

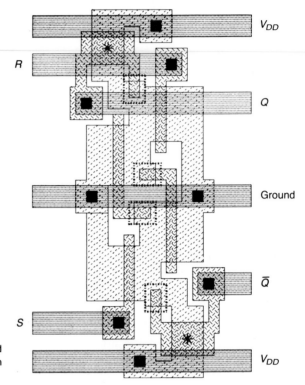

FIGURE 1.45 A second
layout of the circuit from
Fig. 1.43.

A sequential system can be implemented in many ways. The most common and straightforward way is to employ a central clock to synchronize the sequencing of operations. A central clock provides a global time sense and expedites the orderly movement of data about the chip.

The clock period is typically divided into a number of subperiods, called phases, which provide for finer timing granularity. The clocking waveforms are useful both for synchronization and can be exploited in the circuit domain. Clocks provide a good low impedance source of dynamic power.

Shift registers and latches

The shift register is an excellent vehicle for studying the implementation of clocked circuits. The shift register illustrated schematically in Fig. 1.46 is implemented using classical circuit forms of the type found in any elementary book on logic design. Indeed, the whole array of clocked and unclocked latches, registers, and flip-flops familiar to TTL [TI 76]

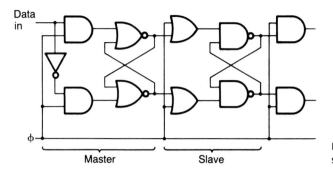

Data in

FIGURE 1.46   A master-slave shift register.

Master   Slave

designers are readily implemented in MOS technology. An example of such a circuit is illustrated in Fig. 1.47.

Figure 1.48 shows both CMOS and depletion load nMOS implementations of the shift register of Fig. 1.46, using dynamic techniques. The data is stored on the capacitances associated with the gates of the inverter transistors and the layout capacitance. The pass transistors act as switches that let charge flow in and out of the capacitors when on and trap the charge when off. The clock is broken into two nonoverlapping

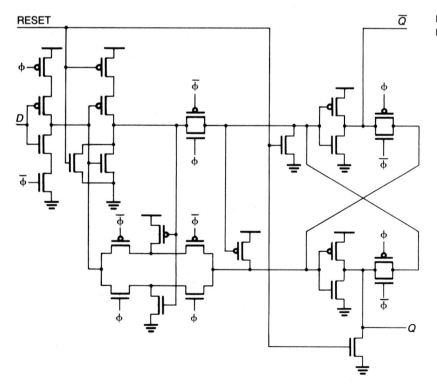

FIGURE 1.47   A CMOS D-type flip-flop with reset.

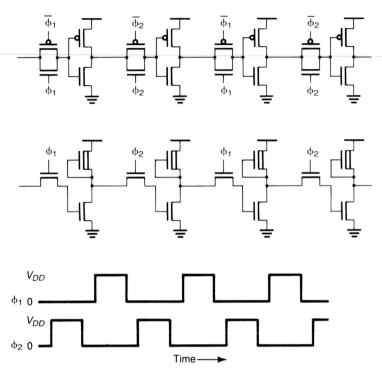

FIGURE 1.48 CMOS and depletion load nMOS dynamic shift register implementations. The nonoverlapping clock waveforms are also shown.

phases. In this methodology [Mead 80, p. 65], the clock waveforms ϕ_1 and ϕ_2 must never be high at the same time or else data can zip from one end of the shift register to the other. This is analogous to the rule that forbids one from opening all of the locks in a canal simultaneously. In our discussions of clocked circuits, we often wish to treat CMOS and nMOS circuits in a unified way. We therefore introduce a notation for the pass gate that transcends the specifics of the technological implementation, much as the standard notation for an inverter is technology-independent. Figure 1.49 illustrates both normally open and normally closed pass gates, together with their MOS implementations. Figure 1.50 illustrates the dynamic shift registers of Fig. 1.48 using this generic MOS notation. Another useful notation is illustrated in Fig. 1.51. It represents a pulldown transistor with its source connected to ground. We also show a NOR gate that uses this new symbol. This notation is most useful for arrays.

Figure 1.52 illustrates another common CMOS shift register circuit. Here the clocked transistors are in series with the sources of the inverter transistors. This circuit has layout advantages over the pass transistor

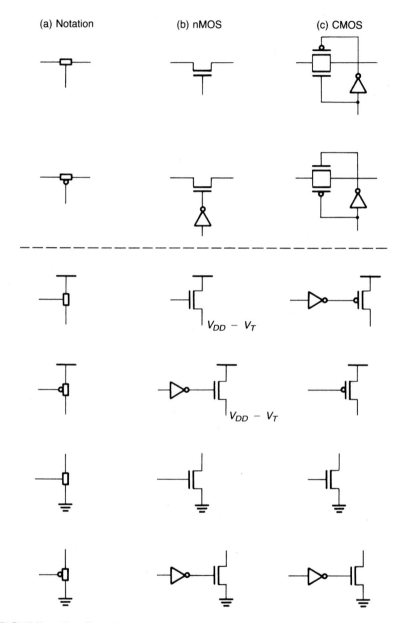

FIGURE 1.49   Generic pass-gate notation.

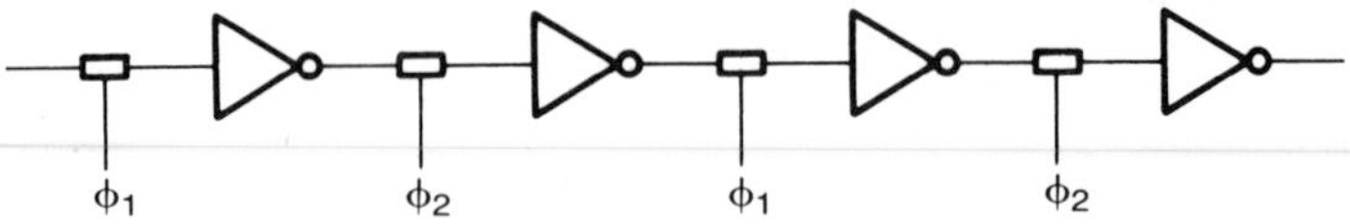

FIGURE 1.50 Generic notation for a MOS shift register.

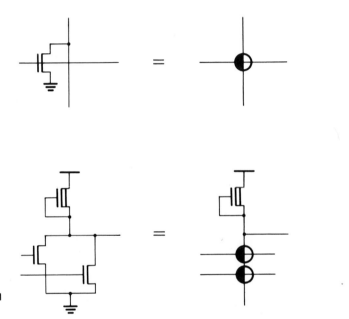

FIGURE 1.51 Pulldown notation, especially useful for arrays.

FIGURE 1.52 Another CMOS shift register cell.

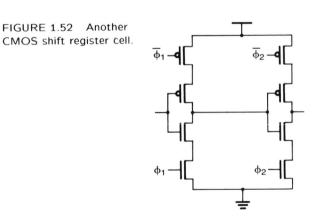

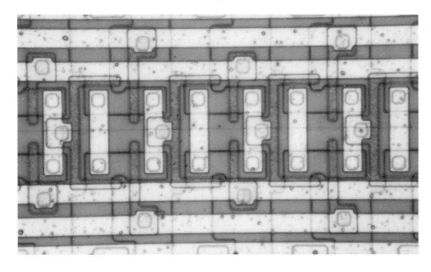

FIGURE 1.53 Photomicrograph of an implementation of the CMOS shift register from Fig. 1.48. ("Circuit fabricated under DARPA sponsorship," IEEE Standard, vol. 581, p. 16, 1978.)

implementation shown in Fig. 1.48. Photomicrographs of these shift registers are illustrated in Figs. 1.53 and 1.54.

With every circuit innovation come new opportunities for error. Clocked circuit techniques introduce the "charge-sharing bug." Charge-sharing problems occur when two capacitors with different voltages are connected together through a pass transistor. When the pass

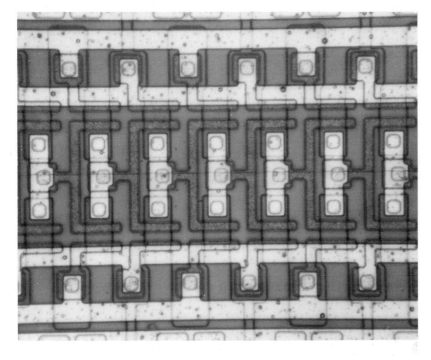

FIGURE 1.54 Photomicrograph of an implementation of the CMOS shift register from Fig. 1.52. ("Circuit fabricated under DARPA sponsorship," IEEE Standard, vol. 581, p. 16, 1978.)

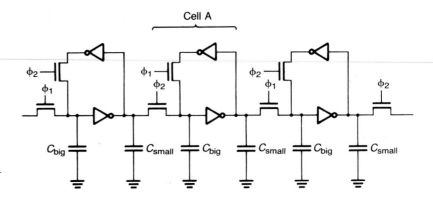

FIGURE 1.55 A shift register that can shift backward due to a charge-sharing bug.

device is turned on, the voltages on the capacitors equilibrate to some intermediate voltage. To see how this can lead to unexpected results, we take a somewhat contrived shift register example. The nMOS shift register in Fig. 1.55 consists of a string of inverters separated by pass transistors clocked either by ϕ_1 or ϕ_2 depending on their position in the string. In order to make the shift register so that it can be stopped for an arbitrary period of time with either of the clocks halted high, clocked feedback paths were added to the design. Through unfortunate coincidence, the layout parasitics of this design were such that the capacitance on the input to a shift register cell was much larger than the output capacitance. These capacitances are labeled on the schematics as C_{big} and C_{small}, respectively. Observe cell A. When ϕ_1 goes high, the feedback path around the cell is closed, making it static. The input to the next cell is also driven by the output of cell A because the pass transistor is also on. However, if the pass transistor is extremely wide, the first thing that happens as ϕ_1 goes high is charge sharing between C_{big} and C_{small}. Because of the size inequality, the voltage on C_{small} assumes the value of the voltage on C_{big}. The output of cell A is now being driven from a very low impedance and can potentially flip the cell. With a little bit of detailed circuit design, the shift register in Fig. 1.55 can even be made to shift backward! We will see many more examples of charge sharing, both helpful and harmful. For instance, the CMOS shift register shown in Fig. 1.52 exhibits charge sharing while the less compact one in Fig. 1.48 does not.

■■■■■■■■■■■ EXAMPLE 1.4

Design of a synchronous finite-state machine to output a pulse every N clock periods
There are many possible solutions to this problem. If N is small, we can use a resettable shift register that passes a token around a shift register loop. This is illustrated in Fig. 1.56. If N is large, this approach is inefficient because the length of the shift register is N.

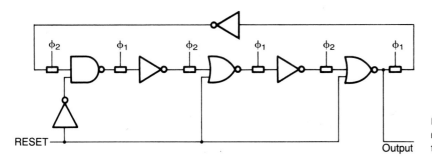

FIGURE 1.56   A shift register implementation of the finite-state machine.

A second solution is to use a binary counter which needs only $\log_2 N$ cells. A MOS implementation of a counter cell is illustrated in Fig. 1.57. If T is true, the cell will toggle. Otherwise it will not. The dynamic nodes are refreshed each cycle. To count up, we note that a cell should toggle if and only if all of the less significant bits are 1. The "all ones" line passes this information about the LSBs. A NOR gate detects the correct pattern to reset the counter after N ticks of the clock. This circuit is illustrated in Fig. 1.58. In practical use, the NOR gate would

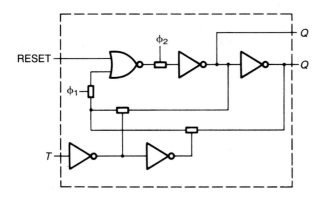

FIGURE 1.57   A toggle cell.

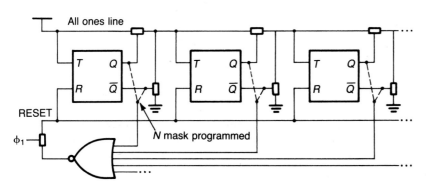

FIGURE 1.58   Binary counter.

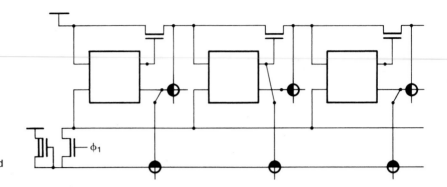

FIGURE 1.59 Binary counter with distributed NOR gate.

be distributed throughout the structure as illustrated for the nMOS circuit in Fig. 1.59.

The binary counter has the nice property that it counts in a recognizable pattern. But it has a very long critical path. In the worst case each cell must toggle in sequence. Another type of counter is the maximal length or pseudorandom sequence counter as shown in Fig. 1.60. An M bit maximal-length counter will sequence through a pattern which

FIGURE 1.60 Maximal length sequence counter.

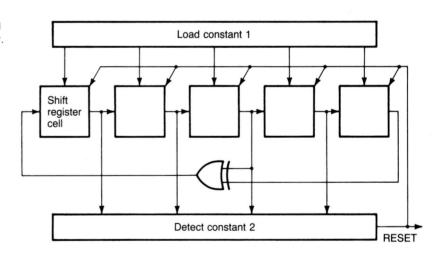

FIGURE 1.61 A precharged bus.

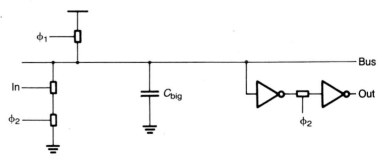

is $2^M - 1$ bits long [Dixon 76]. The pattern is not recognizable but that does not matter in this application. We only need to know what the pattern is. As in the case of the binary counter, we need a method for loading the counter, which now consists only of a shift register and an XOR gate. The maximal-length sequence counter is potentially capable of much higher speeds than the binary counter because its critical path is much shorter. It is also more compact.   ∎

Precharging

The most common application of ratioless nMOS circuits is precharging. In a precharged logic gate or bus, there is still a pullup and a pulldown network. However, they are activated at different times. Figure 1.61 illustrates a MOS bus that is precharged on ϕ_1 and conditionally discharged on ϕ_2. This arrangement has two advantages. First, because there is no direct current path from V_{DD} to ground, the power dissipation of this network is very low. Second, the speed of the circuit is considerably enhanced because the transistors can be made wider without dissipating more power and because the input never causes the bus to pull up. In other words, in a ratioed nMOS gate the high-going transient is slow while the low-going one is relatively fast. In the precharged circuit, if the output should be high, the bus is already high and therefore does nothing. This action is quite fast. When the output should be low, the bus is pulled down by a large transistor. This action is also very fast. All of this assumes that the system architecture allows for a time (in this case, during ϕ_1) when the bus is not being used. The input must also be made glitch free. Note that if the input is high for any time during ϕ_2, the bus will discharge.

Precharged circuits are also extremely useful in CMOS design, not for power reasons but for speed and layout compactness. Consider the case of a NOR gate with a fan-in of q. Without precharging, the NOR gate will take $2q$ transistors with a series connection of q p-type devices. Not only will the device take a lot of area but it will also be slow because the load capacitance must be charged through a high impedance path. The precharged implementation only uses $q + 1$ transistors and has at most one n-channel device limiting the speed. Both of these circuits are shown in Fig. 1.62 for $q = 5$.

FIGURE 1.62   Comparison of (a) a static CMOS bus with (b) a precharged version.

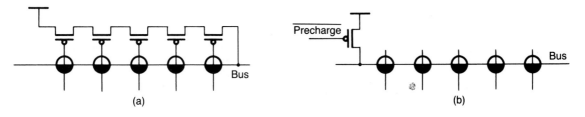

1.4 Regular structures

It would become quite tedious, after the first hundred thousand or so transistors, to design a VLSI circuit if the largest element block one had to work with was the logic gate. Happily, a number of large regular logic structures have been invented that can be made in integrated form. One advantage of these structures is that they enable one to increase the regularity factor [Lattin 79] of the design. The regularity factor is the number of physical transistors in a chip divided by the number of individually designed transistors. Clearly, for design productivity, it pays to have a large regularity factor. It also pays from a layout viewpoint. Experience has shown that the ratio of area taken up per transistor in random logic to regular structures to read-only memory is approximately $10 : 4 : 1$. Unfortunately it is not always simple to decide which, if any, regular structure is most appropriate to use to implement a particular logic function. For instance, despite the fact that read-only memory (ROM) is the most compact structure available on an integrated circuit, it would take a ROM over a meter in both dimensions to compute parity on a 32 bit number. There are better ways to do parity, some of which are also regular structures. Part of the circuit designer's art is choosing the correct structure for the task at hand. Typical regular structures are read-only memories, random-access memories (RAM), programmable logic arrays (PLA), decoders, and shift registers. There are many others. Figs. 1.63 through 1.66 illustrate several forms of MOS decoder circuits.

FIGURE 1.63   NOR-form decoder.

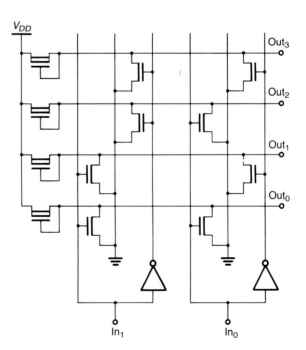

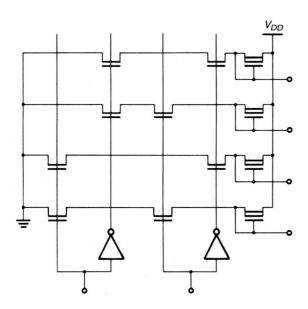

FIGURE 1.64   NAND-form decoder.

FIGURE 1.66   CMOS NAND-form decoder.

FIGURE 1.65   Tree-form decoder.

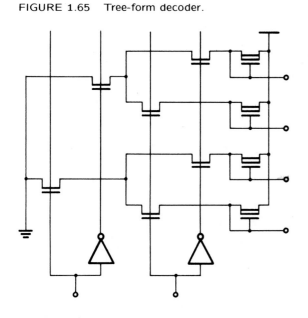

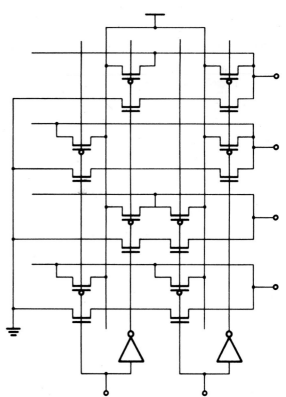

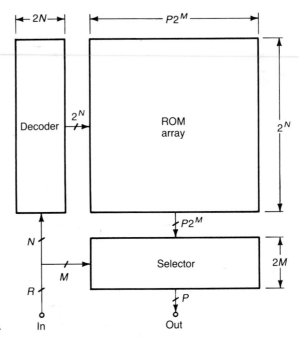

FIGURE 1.67   ROM array.

The first implementation is based on the NOR gate and the second and fourth are based on the NAND gate. The third is a complex gate implementation.

A ROM can be constructed out of a decoder, a NOR array, and a selector as shown in Fig. 1.67. The outputs of the NOR array are interleaved as they enter the selector to minimize wiring. In VLSI, wire costs as much, if not more, than transistors.

The ROM in Fig. 1.67 has R inputs and P outputs. The R inputs are divided between the decoder and the selector. A rough estimate of the area of the array can be obtained by counting wires. If N out of the R inputs go to the decoder, the area A of the array is roughly proportional to

$$A \propto N2^{N+1} + P2^{N+M} + PM2^{M+1}, \tag{1.14}$$

where

$$N + M = R. \tag{1.15}$$

The minimum area occurs at

$$1 + N\ln 2 = P\left(1 + (R - N)\ln 2\right)2^{R-2N} \tag{1.16}$$

In the special case of $P = 1$, we obtain $N = M = R/2$ and the optimum array is square. In most other cases, it is almost square because of the extreme sensitivity of the shape to N. If N is increased by one, the array becomes twice as high and half as wide. Another consideration is that the allocation of N and M may be determined by what will fit on the chip. The M selector signals are needed much later than the N decoder signals. In Chapter 8 we will see how to use this fact to advantage when designing a control store.

Figure 1.68 illustrates a simple nMOS implementation of a ROM. Note that the "personalization" of the ROM occurs only by the placement of transistors in the NOR plane. This is very useful, allowing the programming of the ROM to be bound late in the design effort. It also makes engineering changes easier.

A special case of the ROM occurs when $M = 0$. In this case the selector can be eliminated entirely. The resulting structure is generally referred to as a programmable logic array. The decoder is the "AND plane," and the NOR array is the "OR plane." While PLAs are larger than ROMs when the inputs are fully decoded, PLAs have the advantage that "don't care" conditions in the logic translate into saving PLA area.

FIGURE 1.68   An nMOS ROM implementation.

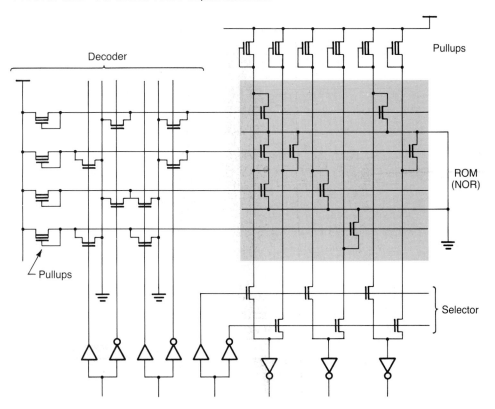

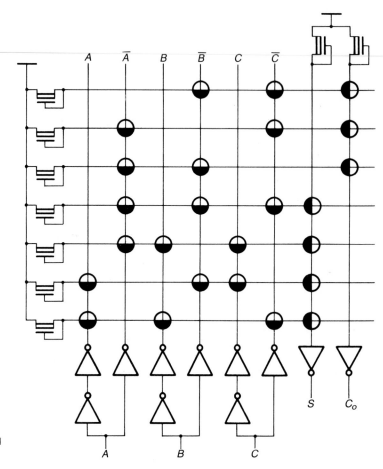

FIGURE 1.69   An nMOS
PLA implementation of a full
adder.

PLAs can be thought of as arrays that implement boolean functions
in sum of products form. Equations (1.4) and (1.5) are examples of this
standard form for expressing logic equations. Because most modest-sized
boolean functions have many don't cares, PLAs are very common. An
nMOS PLA implementation of a full adder is illustrated in Fig. 1.69. In
this case, seven out of the eight possible product terms were required;
usually the area saved is much larger. It is perhaps best to think of PLAs
and ROMs as special cases of a general technique. PLAs and ROMs are
often mixed in creative ways. Adding spare rows to PLAs in order to
expedite engineering changes is standard practice.

Precharging is a technique extremely important to the implementation of large regular structures. Figure 1.70 illustrates a simple
precharged PLA implemented with the NAND form of decoder. When ϕ
is high, the NAND plane output is precharged high, causing the output
of the interplane inverter to be low. Because the inputs to the NOR plane
are low, the NOR plane can be precharged high. When ϕ goes low, the

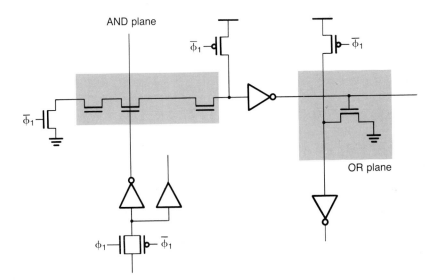

FIGURE 1.70 Single-phase PLA with series-connected AND plane; slow but compact.

NAND plane evaluates. The outputs of the NAND plane ripple into the NOR plane causing it to evaluate in turn. Note that not all of the nodes in the NAND plane will necessarily be precharged. Nevertheless, there are no charge-sharing problems in this circuit. In practice, almost all large regular arrays used for VLSI are precharged.

Memory arrays are ubiquitous in VLSI design. At the very least, a memory array consists of an address decoder and an array of register cells. The register file in Fig. 1.71 is the type one finds in a typical microprocessor. Figure 1.72 shows one of the many possible register cell

FIGURE 1.71   An M-word register array.

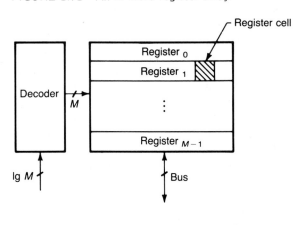

FIGURE 1.72   A register cell.

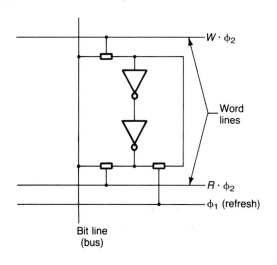

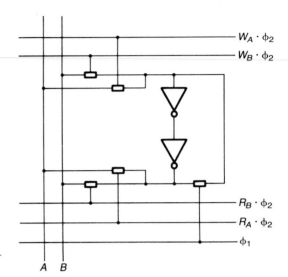

FIGURE 1.73 A multi-ported register cell.

configurations. This register file makes effective use of a precharged bus to expedite reading a word out of the memory. A read is accomplished by precharging the bus high with ϕ_1 and connecting the appropriate word to the bus on ϕ_2. If a cell contains a high value, the bus will be discharged. Otherwise it remains precharged. Writing also occurs on ϕ_2. On ϕ_1 the cell is refreshed by cross coupling the two inverters in the cell. Thus on ϕ_1 the cell is static, but when ϕ_1 is low the cell is dynamic. The system can therefore be stopped indefinitely only when ϕ_1 is held high. Such a configuration is called semistatic. By the addition of more transistors and one or more buses, the cell can be made multiported for reading, writing, or both. Figure 1.73 illustrates a multiported register cell.

Logic and memory can be intermixed to build large regular structures. This mixing is particularly easy in the VLSI circuit domain. Examples include smart memories, pipelined multipliers, convolvers, systolic arrays [Kung 79], content addressable memories, and multiprocessors. There are great opportunities for creativity in this area.

■■■■■ EXAMPLE 1.5

A simple multiplier array
Array multipliers [Dadda 65] are particularly well-suited to MOS implementation. Given two M bit numbers A and B, we can represent the product $P = A \times B$ by a series of partial products X_i. We have

$$X_{i+1} = X_i + a_i B 2^i,$$

where $P = X_M$. In the $i + 1$st row of the array, we add B shifted by i bits to the partial result X_i if $a_i = 1$. Otherwise the results pass

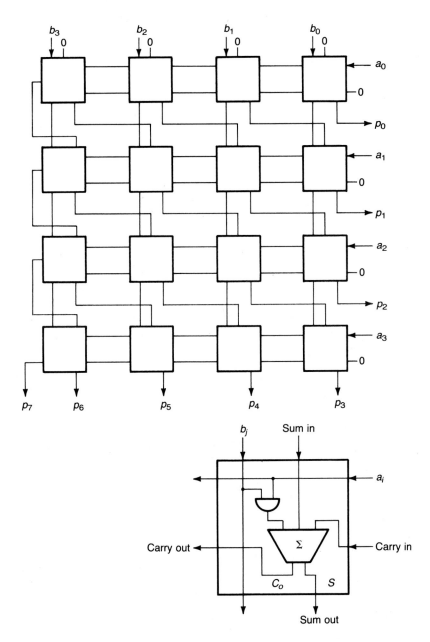

FIGURE 1.74   Array multiplier.

through. The block diagram of a MOS implementation is illustrated in Fig. 1.74. This implementation uses carry propagate adders in each row. This means that the carry might need to propagate $2M$ bits for each multiplication. M carries can occur in any row. Figure 1.75 illustrates a pipelined version of this multiplier. Unfortunately, the maximum carry in any row is M and therefore the clock period would need to be long.

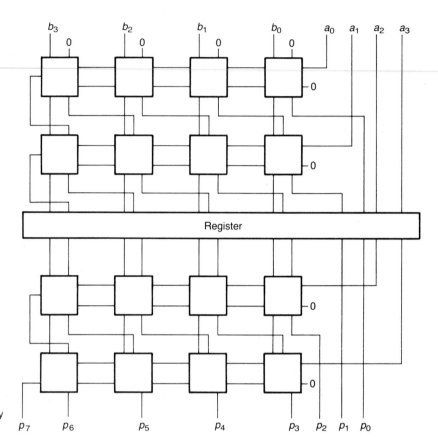

FIGURE 1.75   Pipelined array multiplier with carry propagate adder rows.

Alternatively we could add more logic to each row to speed up the carry, using a technique such as look-ahead, but this costs a great deal of area because every row is affected.

Using carry save adders in all rows but the last is a better technique. A carry save adder reduces the sum of three numbers to the sum of two numbers in the time it takes a signal to propagate through one full adder cell. Contrast this to a carry propagate adder, which reduces the sum of two numbers to a single number, but with delay time proportional to the number of bits to be added unless look-ahead or some other hardware-intensive scheme is used. Figure 1.76 illustrates the pipelined multiplier array implemented with carry save adders. The same basic cell is reused. The maximum carry propagation is now a problem only in the last stage, where a carry look-ahead adder or similar scheme would typically be used to open up this bottleneck. Note that since it is the bottom row of the array that needs the look-ahead circuit, the layout implications are slight.

A number of additional optimizations can be performed. For instance, all the cells in the arrays were illustrated as being identical.

This makes design and layout easy, but quite a bit of space can be saved if one is willing to do a little customization. For instance, the first row of cells in the carry save implementation can be reduced to AND gates. In a practical implementation modified Booth recoding [Rubenfield 75] would be used to halve the number of rows at the expense of some additional hardware. ■

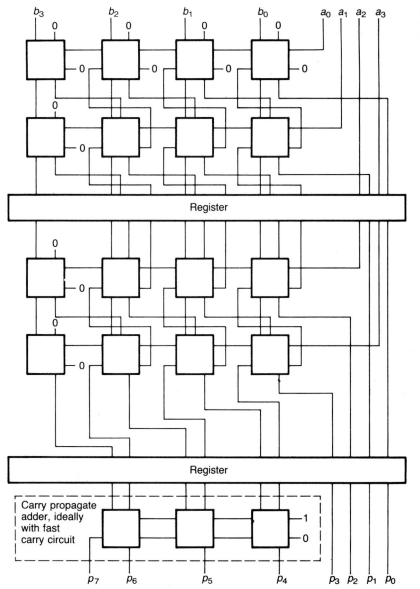

FIGURE 1.76 Pipelined array multiplier with carry save adder rows.

Data in

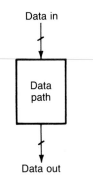

Data out

FIGURE 1.77   A data path.

1.5   Microsystem design

Logic gates and registers are the building blocks with which we engineer VLSI systems. Abstractly, the purpose of a VLSI system is to transform data. Figure 1.77 illustrates a VLSI machine that takes in data from one port, operates on it, and transmits it out the other port. What we have drawn is called a data path.

The implementation of the data path is constrained by the data types, the operations required, and the performance specifications. The multiplier array we saw in Example 1.5 is an instance of a simple data path. Data, encoded as binary words, can generally represent numbers, arrays, alphanumeric strings, pictures, or, indeed, any object whatsoever.

When arithmetic and logical operations are to be performed on the data, an arithmetic logic unit (ALU) is frequently employed. A typical ALU has two input ports and one output port. For arithmetic operations, any one of the adder implementations we have seen could be employed. Logic operations typically consist of bitwise AND, OR, XOR, and so on. Very few primitive operations are needed to make a machine perfectly general. For performance reasons, specialized functions are added to the data path. Typical special functions include shifting and table look-up. Shifting is the sort of function that can be efficiently realized with pass transistors, while table look-up suggests a PLA or ROM implementation. For complex operations one often needs temporary variable storage. Registers are usually employed in this capacity. When the need for several registers is anticipated the registers are often placed in an array called a register file.

Unless the data-path operations are extremely simple and repetitive, as in the multiplication example, an explicit control machine will be required to control the data path. If the control sequence is not too complex, a simple finite-state machine[13] can be used to implement the control path. PLAs are a common idiom in control path implementations. Large PLAs and ROMs in the control path are called control stores. Often we want some degree of programmability. One possibility is to send instructions to the control machine. Instructions coming into the control machine are often stored in a temporary register called the instruction register (IR), as illustrated in Fig. 1.78.

To complete the basic programmed machine we add memory, as illustrated in Fig. 1.79. The output of the control machine now consists not only of control signals for the data path but also of memory addresses for the two memories. A special register in the control unit, called a

FIGURE 1.78   A simple machine.

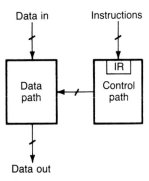

Data out

---

[13] A finite-state machine is a digital machine with a register and combinational logic feedback path. For each clock cycle, the inputs and feedback terms are evaluated and a new output is produced.

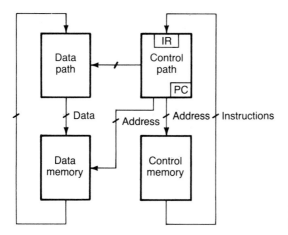

FIGURE 1.79   A programmable machine.

program counter (PC), keeps track of the address of the instruction to be executed next. Given a memory address, the memory returns the data stored at that address if the control inputs to the memory indicate that it is a memory read, or it writes data into that address if the control inputs indicate a write. We have already seen some simple implementations of memories in the discussion of RAM in the last section.

To construct a Von Neumann computer we need a way for the outcome of data operations to influence the control-path sequencing. The memory is shared. A classical Von Neumann computer is illustrated in Fig. 1.80.

Because large memories are typically slow in relation to VLSI logic, it is often economical to multiplex data and address lines onto a single bus. This is a recurrence of the theme that wire is expensive in VLSI. Note that we have two types of "data." One type is data that goes to and from the data path, while the second type of data is actually instructions. The ability to operate on programs as if they were data is

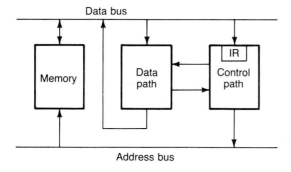

FIGURE 1.80   A computer.

one of the fundamental strengths of the Von Neumann computer. It is sometimes convenient to turn around our thinking and consider all data as instructions. In this case, those instructions that really are simple data are called literals (data to be taken literally, not interpreted as instructions).

We now have a simple machine that fetches instructions and operands (data) from memory. The instructions are interpreted by the control path that dictates the operations that the data path will perform on the operands. The instructions are typically fetched sequentially from the memory except when the results of a data-path computation are required to influence the control flow. At this point we would like to perform a conditional branch in the control flow. The most primitive of conditional branches might take the form

> If the ALU output is zero, do not fetch the next
> instruction stored in the PC, but rather add 77 to
> the contents of the PC, and fetch the instruction
> at that address in memory instead.

Note that the control path now needs the capability of doing arithmetic. Sometimes a second ALU is used, but often the ALU in the data path is used for address computations. When this latter choice is made, the data and control paths become intimately intertwined.

The program branch is one example of a context switch. Another example is the subroutine call. The subroutine call is unique in that context must be kept so the program can return from the subroutine and resume the normal control flow. This requires additional control path registers. In a general register machine, the registers in the register file can be used either by the control machine or the data machine.

When the complexity of the operations required by a computer is large, the control-path finite-state machine becomes very difficult to design. One way out of this predicament is to build all complex operations out of a limited set of simple operations. Thus the computer might need to take several internal cycles to process each incoming instruction, but at least the internal operations would be easily comprehended and implemented. What we have done is to reinvent programming and apply it to the microlevel. Thus we might program the macromachine (which consists of the data path, the control path, and the memory system) and "microprogram" the control path by making it a computer in its own right [Wilkes 53]. Often the microprogram is stored in a ROM or PLA (the control store). The microprogram interprets the complex requirements of the macroprogram and reduces it to a simple sequence of actions that the VLSI chip can readily perform.

1.6 Perspective

The domain of the VLSI circuit designer stretches from device physics to computer architecture. It is a microcosm in which is reflected large portions of electrical engineering and computer science. Indeed, if one were constrained to study just one subject in these areas, VLSI would be the logical choice. In this chapter we have looked, qualitatively, at a number of VLSI topics. We have introduced concepts and examples that will recur throughout the text. We also defined some notation.

We have seen that wire and communication dominate VLSI system planning. To design a VLSI circuit, we have seen that one must develop self-imposed rules for composing a VLSI system out of a limited number of parts—that is, a methodology. Regular structures are an important part of any VLSI methodology.

We have introduced a 2 μm process that will be used again and again in problems and examples. The electrical and physical properties of this process are specified in appendixes. Examples that will recur throughout the book involve adders, ROMs, and microprocessors.

In Chapter 2 we examine the physics of MOS devices. In an ideal world we would be able to abstract these details from the higher levels of representation such as logic design, microarchitecture, and clocking methodologies. Unfortunately this abstraction process has never been done with total success. It is therefore necessary to understand the physics of MOS devices in order to build quality VLSI systems. Other important issues we examine in Chapter 2 are modeling and computer-aided design. The modeling and simulation of a large ROM is a key example. Key equations are denoted by a star ($\star$).

In Chapter 3 the fabrication of VLSI chips is examined with a focus toward understanding how manufacturing uncertainties influence the finished product. Design rules that govern the physical layout of circuit geometries are the key concept.

Noise and fabrication uncertainties must be taken into account in the design of VLSI machines. Reliable digital machines can be built in the face of these real-world concerns. How to do this is the subject of Chapter 4. We also look at a number of primitive circuit forms in quantitative detail. We will see second-order phenomena, such as drain-induced barrier lowering, which we study in Chapter 2, appear with first-order importance in the context of fundamental circuits.

In Chapter 5 we build on the fundamentals in Chapters 2 through 4 to teach a collection of specialized circuit techniques, such as output buffers and sense amplifiers design. Again we will see the interplay of issues such as layout, device physics, noise coupling, processing variations, and communications concerns.

In Chapter 6 we visit the subject of clocking methodologies. Clocks play a central role in controlling the orderly movement of signals about a VLSI system.

In Chapter 7 we build upon the circuits of Chapter 5 and the clocking strategies of Chapter 6 to design large regular structures such as RAMs, ROMs, and systolic arrays. This chapter is a steppingstone to the design of large systems.

In Chapter 8 we look at the microarchitecture of VLSI systems. Architecture has long been driven, for better or for worse, by technology. Chapter 8 examines how a requirement for harmony among circuit concerns, communications issues, and macro-architecture influences the performance of a VLSI machine. The critical issues in data paths and control paths are examined in the context of the microprocessor. In a well-designed VLSI circuit there is great synergy among the various levels of abstraction.

Problems

1.1   Design an up–down counter with reset. Draw a schematic for it in CMOS technology.

1.2   Using generic MOS notation, draw the schematic for a pass-transistor network that takes a number with a single set bit, such as 00010000, and turns it into its thermometer code representation, that is, 00011111.

1.3   It is well known that any logic function can be constructed solely out of NOR gates. Prove it is possible to build any logic function out of only pass transistors and inverters.

1.4   Using nMOS technology, invent a combinational network that computes the parity of an N bit number, using no more than $6N$ transistors.

1.5   What are the logical functions C, G, Z, and V performed by the gates in Figs. 1.81 through 1.84?

FIGURE 1.81   FIGURE 1.82

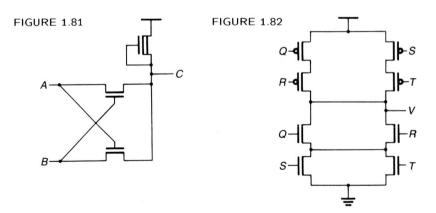

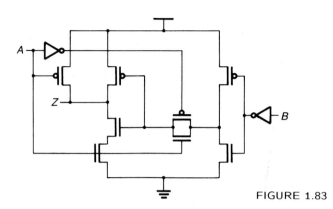

FIGURE 1.83

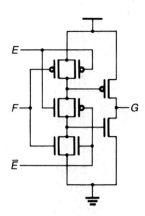

FIGURE 1.84

1.6 A ROM needs to hold 2^{19} bits, which are accessed one at a time, but the place on the chip where it must fit is square, with room for no more than 775 bits on a side. First, what is the problem, and second, how can adding an extra column to one of the decoders solve it? △

1.7 All of the shift register cells in Figs. 1.85 through 1.92 work, given the right assumptions, and all have been used at various times in real chips. Describe how each works, draw rough waveforms, and discuss its

FIGURE 1.85

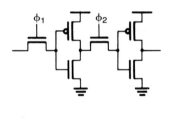

FIGURE 1.86

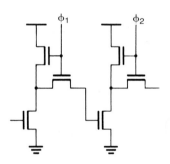

FIGURE 1.87

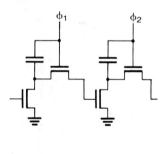

FIGURE 1.88

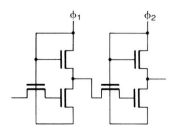

FIGURE 1.89

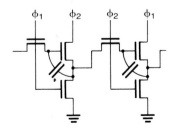

FIGURE 1.90

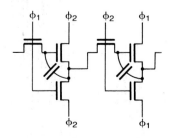

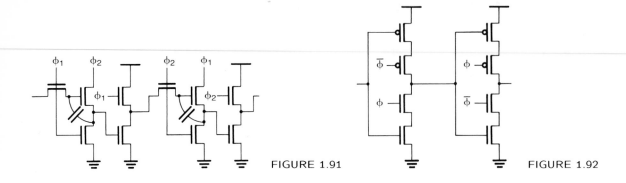

FIGURE 1.91   FIGURE 1.92

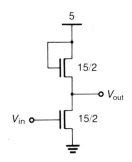

FIGURE 1.93

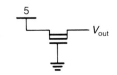

FIGURE 1.94

relative advantages and disadvantages. The clock waveforms ϕ_1 and ϕ_2 are nonoverlapping.

1.8 For the circuit in Fig. 1.93, what (approximately) is V_{out} for $V_{in} = 0$ and $V_{in} = 5$ volts?

1.9 What is the maximum voltage V_{out} in Fig. 1.94, assuming that it starts at ground?

1.10 A generalization of the ROM in Fig. 1.68 uses $2^N + P2^M$ pullups. (See Fig. 1.67 for notation.) Invent a design that uses only $2^N + 2^M$ pullups.

1.11 Design a syntax checker for floating-point numbers.[14] That is, given coded inputs representing character types, determine whether or not the characters make up a valid floating-point number. The checker has three data inputs (A_1, A_0, and start), two clock signals (ϕ_1 and ϕ_2), and one output (bad). The data inputs specify the character types, serially, in their order of appearance from left to right. The character type information is encoded as shown in Fig. 1.95.

A_0	A_1	Type of character
0	0	numeral ("0" through "9")
0	1	decimal point (".")
1	0	plus or minus ("+" or "–")
1	1	exponent character ("E" or "e")

FIGURE 1.95   Input coding.

Figure 1.96 shows examples of valid and invalid floating-point numbers for this checker. There is no restriction on the precision of the numbers.

The timing specification is shown in Figure 1.97. The third input, start, goes high for at least one clock cycle after the last character of a number and before the first character of the next number. During ϕ_1 of the clock cycle after each complete number, the output bad is

---

[14] This problem is due to Steve McCormick.

Valid	Invalid
42	.4
36.	+.72
0.7	+
2E5	1.6E
2.E0	5E+
4.1E+6	+.E7
−50	2.4−E3
5E007	2..5
−5.2E2	5E3.4

FIGURE 1.96   Examples of floating point numbers.

NUMBER STREAM:	...	−	3	.	5		+	

Inputs (on ϕ_1):

start:	1	0	0	0	0	1	0	1
A_0:	X	1	0	0	0	X	1	X
A_1:	X	0	0	1	0	X	0	X

Output (on ϕ_2):

bad:	X	X	X	X	X	0	X	1

FIGURE 1.97   Timing for inputs and outputs for "−3.5" and "+".

sampled—a logical "1" indicating a syntactically incorrect number. Note that all inputs are valid during ϕ_1.

Draw a state diagram for the syntax checker and construct an encoded state transition table from the state diagram. Derive minimized logic equations for each of the next state and output variables.

Define the structure of a PLA that realizes the syntax checker function in tabular format. The following example illustrates the tabular format for the boolean equations $X = A\overline{B} + \overline{A}D$ and $Y = B\overline{C}D + \overline{A} + A\overline{B}$:

A	B	C	D	X	Y
1	0			1	1
0			1	1	
	1	0	1		1
0					1

FIGURE 1.98   Example of tabular representation of PLA.

In Fig. 1.98, a "0" or a "1" shows the position of a transistor, and a blank space indicates the absence of a transistor. Show how the PLA should be clocked. △

1.12 In nMOS technology, three transistors are required to implement a NOR gate. In classical CMOS, due to the pullup structure that is the dual of the pulldown structure, four transistors are required. However, if the inputs and their compliments are available, it is possible to make a three-transistor CMOS NOR gate without using precharging or ratioed logic, and incurring no threshold losses. Show how to make a three-transistor CMOS NOR gate. $\triangle$

1.13 Design a regular combinational logic structure that decodes an N bit number to find the least significant zero. Draw a transistor level schematic in either nMOS or CMOS. There are N inputs and N outputs. A high on the ith output wire means that input i contains the least significant 0. For instance, the input 10010111 produces the output 00001000.

References

[Anderson 80] Anderson and Bogert, *LSI Opportunities: The Integrated Circuit Service Industry*, Los Altos, Calif., 1980. (Or its predecessor *Vertical Dis-Integration*)

[Antognetti 81] P. Antognetti, D. O. Pederson, and H. De man, *Computer Design Aids for VLSI Circuits*, Sijthoff and Noordhoff, Rockville, Md., 1981.

[Boraiko 82] A. A. Boraiko and C. O'Rear, "The Chip," *National Geographic* **162**: 420–457, 1982.

[Bryant 83] R. Bryant (ed.), *Third Caltech Conference on Very Large Scale Integration*, Computer Science Press, Pasadena, Calif., 1983.

[Carr 72] W. N. Carr and J. P. Mize, *MOS/LSI Design and Applications*, McGraw-Hill, New York, 1972.

[Dadda 65] L. Dadda, "Some Schemes for Parallel Multipliers," *Alta Frequenza* **34**: 349–356, 1965.

[Dixon 76] R. Dixon, *Spread Spectrum Techniques*, Wiley, New York, 1976. (Shift register taps of pseudorandom sequence generators.)

[Gray 81] J. P. Gray (ed.), *VLSI 81*, Academic Press, New York, 1981. (Proceedings of the first International Conference on Very Large Scale Integration, Edinburgh.)

[Harris n.d.] *HL Cell Library User's Manual*, Harris Semiconductor, Melbourne, Fla., unpublished.

[Hill 74] F. J. Hill and G. R. Peterson, *Introduction to Switching Theory and Logical Design*, Wiley, New York, 1974.

[Hodges 83] D. A. Hodges and H. G. Jackson, *Analysis and Design of Digital Integrated Circuits*, McGraw-Hill, New York, 1983.

[Hon 80] R. W. Hon and C. H. Sequin, *A Guide to LSI Implementation*, SSL-79-7, Xerox, Palo Alto, Calif., 1980.

[IEEE 78] *IEEE Standard Definitions, Symbols, and Characterization of Metal-Nitride-Oxide Field-Effect Transistors*, IEEE Std. 581–1978, 16, April 28, 1978.

[Jespers 82] P. G. Jespers, C. H. Sequin, and F. van de Wiele (eds.), *Design Methodologies for VLSI Circuits*, Sijthoff and Noordhoff, Rockville, Md., 1982.

[Johnston 82] M. Johnston and C. O'Rear, "Silicon Valley," *National Geographic* **162**: 458–477, 1982.

[Joynson 72] R. E. Joynson, J. L. Mundy, J. F. Burgess, and C. Neugebauer, "Eliminating Threshold Losses in MOS Circuits by Bootstrapping Using Varactor Coupling," *IEEE J. Solid-State Circuits* **SC-7**: 217–224, 1972.

[Kung 79] H. T. Kung, "Let's Design Algorithms for VLSI Systems," *Proceedings of the Caltech Conference on Very Large Scale Integration*, C. L. Seitz (ed.), Pasadena, Calif., 55–90, January 1979.

[Kung 81] H. T. Kung, B. Sproull, and G. Steele, *VLSI Systems and Computations*, Computer Science Press, Pittsburgh, Pa., 1981.

[Lattin 79] B. Lattin, "VLSI Design Methodology: The Problem of the 80's for Microprocessor Design," *Proc. of the Caltech Conference on Very Large Scale Integration*, C. L. Seitz, Pasadena, Calif., 248–252, 1979.

[McCarthy 82] O. J. McCarthy, *MOS Device and Circuit Design*, Wiley, New York, 1982.

[Masuhara 72] T. Masuhara, M. Nagata, and N. Hashimoto, "A High-Performance n-Channel MOS LSI Using Depletion-Type Load Elements," *IEEE J. Solid-State Circuits* **SC-7**: 224–231, 1972.

[Mavor 83] J. Mavor, M. A. Jack, and P. B. Denyer, *Introduction to MOS LSI Design*, Addison-Wesley, Reading, Mass., 1983.

[Mead 80] C. A. Mead and L. Conway, *Introduction to VLSI Systems*, Addison-Wesley, Reading, Mass., 1980.

[Muroga 82] S. Muroga, *VLSI System Design*, Wiley, New York, 1982.

[Olsen 81] R. E. Olsen and D. W. Dobberpuhl, "A 13,000 Transistor NMOS Microprocessor," *IEEE International Solid-State Circuits Conf.*: 108–109, 1981.

[Penfield 82] P. Penfield, Jr. (ed.), *Proc. Conf. on Advanced Research in VLSI*, Cambridge, Mass., 1982.

[Rubenfield 75] L. P. Rubenfield, "A Proof of the Modified Booth's Algorithm for Multiplication," *IEEE Trans. Computers* **C-24**: 1014–1015, 1975.

[Rubinfeld 82] P. I. Rubinfeld, "Two-Chip Supermicrocomputer Outperforms PDP-11 Minicomputers," *Electronics*, pp. 131–136, McGraw-Hill, New York, December 15, 1982.

[Scientific 77] *Scientific American* **237**: (3), September 1977.

[Seitz 79] C. L. Seitz (ed.), *Proc. of the Caltech Conference on Very Large Scale Integration*, Pasadena, Calif., 1979.

[Seitz 81] C. L. Seitz (ed.), *Proc. of the Second Caltech Conference on Very Large Scale Integration*, Pasadena, Calif., 1981.

[Shannon 37] C. E. Shannon, *A Symbolic Analysis of Relay and Switching Circuits*, M. S. Thesis, Massachusetts Institute Technology, Department of Electrical Engineering, 1937.

[TI 76] *The TTL Data Book for Design Engineers*, (2nd ed.), Texas Instruments, Dallas, Tex., 1976.

[Vladimirescu 80] A. Vladimirescu and S. Liu, "The Simulation of MOS Integrated Circuits using SPICE2," Memo. UCB/ERL M80/7, University of California, Berkeley, October 1980.

[Vladimirescu 81] A. Vladimirescu, K. Zhang, A. R. Newton, D. O. Pederson, and A. Sangiovanni-Vincentelli, *SPICE Version 2G User's Guide*, University of California, Berkeley, August 10, 1981.

[Wanlass 63] F. M. Wanlass and C. T. Sah, "Nanowatt logic using field-effect metal-oxide-semiconductor triodes," *IEEE International Solid-State Circuits Conf.*: 32–33, Philadelphia, Pa., 1963. (Early CMOS reference.)

[Wilkes 53] M. V. Wilkes and J. B. Stringer, "Microprogramming and the Design of Control Circuits in an Electronic Digital Computer," *Proc. of the Cambridge Philosophical Society* **49**: 230–238, 1953.

MOS DEVICE ELECTRONICS

2

This chapter examines the physics of MOS devices from a circuit designer's perspective. We show how the various physical mechanisms in MOS capacitors and transistors are manifested in the terminal characteristics of these devices. Because this analysis becomes extremely confusing if one tries to simultaneously analyze p-channel and n-channel devices, we limit most of the analysis to n-channel devices. The results for both device types will be summarized. We also examine the physical properties of the labyrinth of interconnect that dominates VLSI chip area and often chip performance.

While it has always been helpful to have a good fundamental knowledge of device physics when designing circuits, in the fast-moving field of MOS VLSI such knowledge is imperative for the design of high-performance or high-reliability systems. Emerging technologies are always poorly characterized and only by appealing to the fundamental properties of these new devices can we design with a technology before it becomes obsolete.

This is not to say that no place exists for simplified methodologies that abstract the device electronics. But the designer of such a methodology must understand what gremlins lurk beneath the abstractions.

2.1   Capacitors and diodes

The two most fundamental devices in a MOS VLSI circuit are the parallel plate capacitor and the reverse-biased p–n junction diode. Unlike other devices we will discuss later, the theory of these two-terminal devices is well understood and presented in a variety of texts [Grove 67, Muller 77]. We begin by invoking a number of results from the theory of solid-state physics.

Some results from solid-state physics

The concentration of electrons and holes in a semiconductor can be written as

$$n = N_C e^{-(E_C - E_F)/kT} \qquad (2.1)$$

and

$$p = N_V e^{-(E_F - E_V)/kT}, \qquad (2.2)$$

where N_C is the density of states available to electrons at the bottom of the conduction band and N_V is the density of states available to holes at the top of the valence band. N_V and N_C each depend on absolute temperature to the 3/2 power. Equations (2.1) and (2.2) are valid when the magnitude of the exponent is roughly three or greater.[1] Absolute temperature is represented by T and k is Boltzmann's constant. E_C and E_V are the energy levels of the conduction and valence bands, respectively, and E_F is the Fermi energy. E_F and temperature are the fundamental parameters describing the statistical distribution of electron and hole energies. For an object to be in equilibrium, the temperature and Fermi level must be constant throughout. From Eqs. (2.1) and (2.2) we will derive a number of parameters more useful for directly describing the electrical properties of MOS devices.[2] For pure silicon at equilibrium, the number of holes equals the number of electrons. This is because the excitation of one electron into the conduction band leaves behind one hole in the valence band.

In silicon, N_V and N_C are approximately the same. Referring to Eqs. (2.1) and (2.2), we see that E_F must be approximately halfway between E_V and E_C for undoped (intrinsic) silicon. We define this special Fermi level as E_i. Equating Eqs. (2.1) and (2.2) we have

$$E_i = \frac{E_C + E_V}{2} + \frac{kT}{2} \ln \frac{N_V}{N_C}. \qquad (2.3)$$

By substituting impurities into the silicon lattice (doping) the ratio of holes to electrons can be changed. The impurities typically used to

---

[1] This is because Boltzmann statistics are used to approximate what is really a Fermi-Dirac distribution. Also the density-of-states function becomes modified by high carrier concentrations.

[2] There are alternative ways of presenting this material that do not invoke band theory and Fermi levels. The derivation of Gray and Searle [Gray 69] may appeal more to some readers.

dope the silicon are boron for generating holes and either phosphorus or arsenic for generating excess electrons. Boron is an acceptor atom because it accepts (covalently bonds) an extra electron from the silicon lattice. Phosphorus and arsenic are donors because they donate their extra outer shell electron to the lattice. Note that the crystal remains electrically neutral. For example, in phosphorus the fifth electron in the outer shell is bound to the nucleus of the atom by only a few dozen millielectron volts. At room temperature it is easily ionized, becoming free to roam the lattice and carry current. The case of boron is harder to picture. It has three, rather than five, outer electrons and therefore can easily accept an extra electron from the surrounding silicon atoms, each of which have four outer electrons. Of course, if the boron accepts an electron it leaves a silicon atom one short. This hole (much like a bubble in the ocean) propagates in the direction opposite to the electron movement.

If the silicon has more holes than electrons, the Fermi level moves closer to the valence band; if it has more electrons than holes, it moves closer to the conduction band. We define the intrinsic carrier concentration as n_i, which is, by definition, the same for holes and electrons. Rewriting Eqs. (2.1) and (2.2) we have

$$n = n_i e^{(E_F - E_i)/kT} \qquad (2.4)$$

and

$$p = n_i e^{(E_i - E_F)/kT}. \qquad (2.5)$$

Multiplying Eqs. (2.4) and (2.5) we obtain

$$np = n_i^2, \qquad (2.6)$$

where

$$n_i^2 = N_C N_V e^{-E_G/kT}, \qquad (2.7)$$

and where

$$E_G \equiv E_C - E_V. \qquad (2.8)$$

E_G is the band gap energy. Equation (2.6) is true for all semiconductors, doped or undoped, under equilibrium conditions. Typical charge densities in a p-type substrate used for building nMOS devices would be $p = N_A = 2 \times 10^{15} \text{ cm}^{-3}$, $n_i = 1.45 \times 10^{10} \text{ cm}^{-3}$ and $n = 10^5 \text{ cm}^{-3}$. E_G is a weak function of temperature.

We define N_A as the density of acceptor impurity atoms and N_D as the donor density. At room temperature virtually all impurity atoms are ionized.[3] For $N_A \gg n_i$, and $N_A \gg N_D$, we have $p \approx N_A$. For $N_D \gg n_i$, and $N_D \gg N_A$, we have $n \approx N_D$.

---

[3] This ceases to be the case at, or below, liquid nitrogen temperatures or for extremely high impurity concentrations.

From a circuits standpoint it is much more convenient to talk in terms of potentials than energies. We define the Fermi potential

$$\phi_F \equiv \frac{E_F - E_i}{q},$$ (2.9)

where q is the charge on an electron. Note that we are now referring the Fermi potential to its intrinsic value. This is the first of several reference potential shifts we will encounter. We place an additional subscript after the "F" to indicate whether the potential is measured in a p-type material ($N_A > N_D$) or an n-type material ($N_D > N_A$). Assuming that either N_D or N_A is dominant, we have

$$\phi_{Fn} = \frac{kT}{q} \ln \frac{N_D}{n_i}$$ (2.10)

and

$$\phi_{Fp} = -\frac{kT}{q} \ln \frac{N_A}{n_i}.$$ (2.11)

Rewriting Eqs. (2.1) and (2.2) for a final time, we have

$$n = n_i e^{q\phi_F/kT}$$ (2.12)

and

$$p = n_i e^{-q\phi_F/kT}.$$ (2.13)

Fig. 2.1 illustrates the energy band diagrams for intrinsic, p-type, and n-type silicon.

The p–n diode

The p–n diode is ubiquitous in MOS circuits. Its capacitance can be a major factor in the system performance.

The results we have derived so far are valid for a volume of uniformly doped semiconductor. These results are also valid for a classical p–n junction diode for regions far from the junction. Thus deep in the p-type half of the diode the potential is ϕ_{Fp} and deep in the n-type side the potential is ϕ_{Fn}. In equilibrium (no applied voltage) the Fermi levels in the n- and p-type materials must be equal. As one travels from one region to another, one observes a total change in potential of

$$\phi_T = \phi_{Fn} - \phi_{Fp}.$$ (2.14)

Solving, we have

$$\phi_T = \frac{kT}{q} \ln \frac{N_D N_A}{n_i^2}.$$ (2.15)

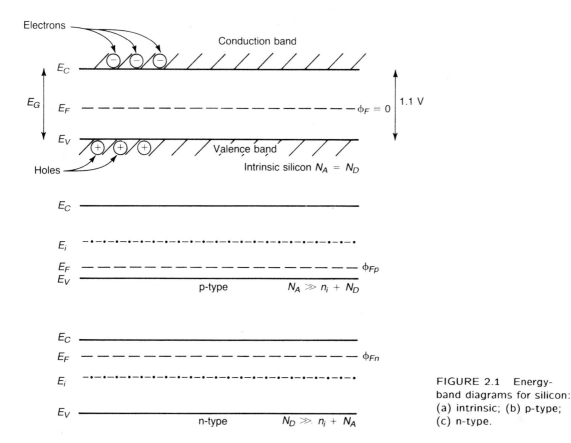

FIGURE 2.1   Energy-band diagrams for silicon: (a) intrinsic; (b) p-type; (c) n-type.

The built-in potential ϕ_T is due to the differences in chemical energy between the n- and p-type materials. While overall the semiconductor is not charged, the discontinuity of the p–n junction causes local volumes of charged material near the interface. For the one dimensional p–n junction shown in Fig. 2.2, Poisson's equation demands that the integral

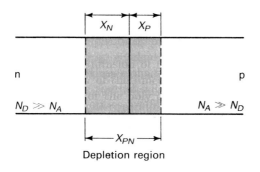

FIGURE 2.2   A p–n junction diode.

of the charge equals ϕ_T. We have

$$\phi_T = -\frac{q}{\epsilon_{Si}} \int \int n(x), \tag{2.16}$$

where ϵ_{Si} is the dielectric constant of silicon, and $n(x)$ is the one-dimensional density of charges.

Physically, at the junction there is a balance between the forces of diffusion and charge neutrality. Holes on the p-side of the interface diffuse to the n-side, while electrons on the n-side diffuse to the p-side. This causes a charge inequality with its accompanying electric field. The electric field is in the direction opposing the flow of charge. At equilibrium these forces exactly cancel. Obtaining exact expressions for the charge distribution near the interface is very diffcult. Excellent results, however, can be obtained by using the depletion region approximation. A depletion region is a portion of semiconductor material that is depleted of mobile charge carriers.

Invoking the depletion region approximation, we assume that the depletion region extends for a distance X_N on the n-type side of the depletion region and X_P on the p-type side. We have

$$q N_D X_N = q N_A X_P \tag{2.17}$$

by charge neutrality. We are assuming an abrupt p–n junction model where the density of donors on the n-side and acceptors on the p-side are constant. The charge density $Q(x)$ is given by

$$Q = \begin{cases} q N_D & -X_N \leq x < 0 \\ -q N_A & 0 < x \leq X_P \\ 0 & \text{otherwise} \end{cases} \tag{2.18}$$

where $x = 0$ defines the position of the p–n interface. Using

$$\phi_T = \frac{1}{\epsilon_{Si}} \int_{-X_P}^{X_N} \int_{-\infty}^{x'} Q(x) dx dx' \tag{2.19}$$

we find

$$\phi_T = \frac{q N_D X_N^2 + q N_A X_P^2}{2\epsilon_{Si}} \tag{2.20}$$

which, when combined with (2.17), reduces to

$$X_{PN} = X_N + X_P = \sqrt{\frac{2\epsilon_{Si}}{q} \frac{(N_A + N_D)}{N_A N_D} \phi_T}, \tag{2.21}$$

where X_{PN} is the total width of the depletion region. We also have

$$\frac{X_P}{X_N} = \frac{N_D}{N_A}. \qquad (2.22)$$

It is very common to have what is called a one-sided step junction in which one of the sides of the junction is heavily doped while the other is lightly doped. In these cases Eq. (2.21) reduces to

$$X_{PN} = \sqrt{\frac{2\epsilon_{Si}}{q}\frac{\phi_T}{N}}, \qquad (2.23)$$

where N is N_D if $N_D \ll N_A$ and N_A if $N_A \ll N_D$. Note that the depletion width is dominated by the effects of the more lightly doped semiconductor.

The equations in this section were derived under the conditions of zero applied voltage. Under reverse-bias conditions ϕ_T must include not only the built-in voltage of Eq. (2.15) but also the applied voltage. Rewriting Eq. (2.23) to include the applied voltage v we have[4]

$$X_{PN} = \sqrt{\frac{2\epsilon_{Si}(\phi_T + v)}{qN}}. \qquad (2.24)\star$$

Examining Eq. (2.24) we see that a large reverse bias v will increase the size of the depletion region. (Reverse bias v is defined such that a positive value of v implies the potential applied to the p-type material is negative with respect to the potential applied to the n-type material.)

We write the capacitance of a reverse-biased one-sided step p–n junction as

$$C \equiv \frac{dQ}{dv} = \frac{d}{dv} qNX_{PN}$$

$$= \sqrt{\frac{\epsilon_{Si}qN}{2(\phi_T + V)}}. \qquad (2.25)\star$$

By convention, we will use uppercase variables (V) when referring to dc quantities and lowercase variables (v) when referring to large signal or dynamic quantities.

The voltage across a wide depletion region is limited by avalanche breakdown. When the electric field in a semiconductor is sufficiently strong, mobile electrons can gain an energy between collisions that significantly exceeds that of the band gap energy E_G. When a collision between one of these highly energetic electrons and a silicon atom does occur, the silicon atom may be ionized. Since each electron can liberate two additional carriers (one electron and one hole), the current begins to multiply. If the carrier generation rate exceeds the recombination rate,

---

[4] Equations with a $\star$ are of central importance.

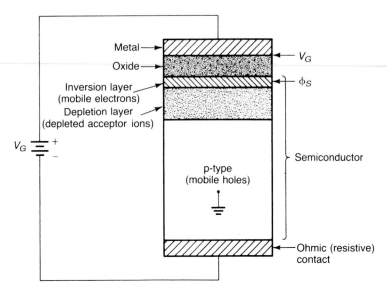

FIGURE 2.3   A MOS capacitor.

avalanche breakdown occurs. The breakdown voltage V_M for a one-sided junction is given by

$$V_M = \frac{\mathcal{E}_M X_{PN}}{2} = \frac{\epsilon_{\text{Si}} \mathcal{E}_M^2}{2qN}, \tag{2.26}$$

where $\mathcal{E}_M$ is the electric field at which avalanche multiplication occurs. An empirical expression for $\mathcal{E}_M$ is

$$\mathcal{E}_M = \frac{4 \times 10^5}{1 - \frac{1}{3} \log_{10}\left(N/10^{16}\right)} \text{ V/cm}, \tag{2.27}$$

where N is measured in units of cm^{-3} [Sze 81, p. 102].

The MOS capacitor

The MOS capacitor is not only the basis of the MOS transistor, but is a fascinating and useful circuit element in its own right. It is an extremely nonlinear device and this nonlinearity is fundamental to a number of circuit techniques.

The MOS capacitor combines the characteristics of the parallel plate capacitor and the p–n junction diode. It consists of a sandwich of semiconductor, oxide, and metal, as illustrated in Fig. 2.3. In a typical nMOS case, the semiconductor would consist of lightly doped p-type silicon, the oxide of thermally grown silicon dioxide, and the metal of either aluminum or degenerately doped[5] n-type silicon. As long as the

---

[5] When a semiconductor is degenerately doped, the doping is so high that the Fermi level reaches either the valence (p-type) or conduction (n-type) band. The semiconductor then acts almost exactly as a metal.

semiconductor is sufficiently conductive that it can support displacement currents, this configuration results in a parallel plate capacitor with, in this example, the aluminum as one electrode, the p-type silicon as the other electrode, and the silicon dioxide as the dielectric. We define C_{OX} as the capacitance per unit area of this structure. We have

$$C_{OX} \equiv \frac{\epsilon_{\text{SiO}_2}}{T_{OX}},$$

(2.28)⋆

where T_{OX} is the thickness of the oxide and ϵ_{SiO_2} is the dielectric constant of silicon dioxide. It is important to note that C_{OX} is defined per unit area.

We know from our analysis of the p–n junction that the chemical potential of silicon depends on its doping level. It should not be surprising that aluminum, silicon dioxide, and p-type silicon also each have different chemical potentials and that these potentials can set up built-in electric fields and nonuniform charge distributions. This has the effect of bending the bands in the energy band diagram. For most simple geometries and doping profiles, we can compensate for the band bending by applying an external voltage, making the semiconductor neutrally charged. The voltage needed to achieve this result is the flat-band voltage. It is an important reference potential. Fig. 2.4 illustrates a MOS capacitor at flat-band.

For an applied voltage $v_G < V_{FB}$, the bands bend as shown in Fig. 2.5(a). In this regime, called accumulation, the lower applied voltage causes holes to be drawn up from the substrate and accumulate at the silicon/silicon dioxide interface. The capacitance does not change.

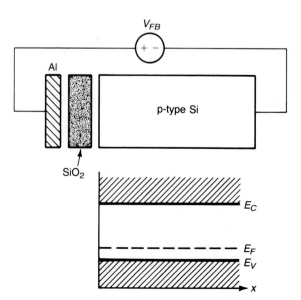

FIGURE 2.4   The energy band diagram of a MOS capacitor at flat-band.

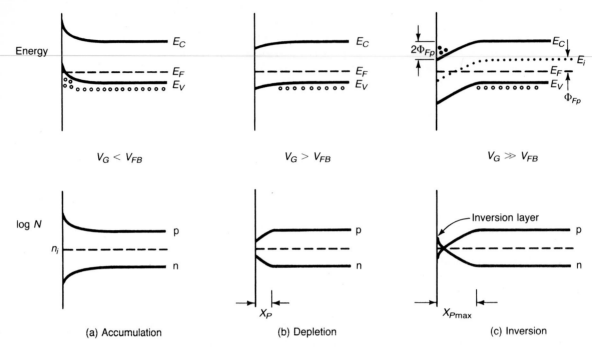

FIGURE 2.5 Diagrams showing the energy bands and carrier concentrations near the surfaces of a p-type semiconductor under conditions of (a) accumulation, (b) depletion, and (c) inversion.

When the applied voltage v_G is slightly greater than V_{FB} the device is in the depletion regime. In this case the bands bend downward as shown in Fig. 2.5(b). Because the potential at the semiconductor surface is greater than it is in the bulk (ϕ_{Fp}), the number of holes begins to decrease according to Eq. (2.13). Thus the region becomes depleted of holes. The quantity np, however, remains at its constant value of n_i^2. This depletion region acts as a capacitor just as it does in the case of a reverse-biased p–n junction diode. The result is that the effective plate of the MOS capacitor moves down from the oxide interface into the semiconductor. We now have two dielectrics in series, ϵ_{SiO_2} and ϵ_{Si}. This results in a capacitance per unit area that is less than C_{OX}, and is the series connection of these two capacitors. We have

$$C_{MOS} = \left(\frac{1}{C_{OX}} + \frac{1}{C_{Si}} \right)^{-1}$$

$$= \left(\frac{T_{OX}}{\epsilon_{SiO_2}} + \frac{X_P}{\epsilon_{Si}} \right)^{-1}, \tag{2.29}$$

where X_P is the width of the induced depletion region.

If we continue to increase v_G, pushing the bands down, we can make the concentration of electrons at the surface greater than the concentration of holes at the surface, as illustrated in Fig. 2.5(c). When this happens we say that we have entered the weak inversion regime. If we further increase v_G, the concentration of electrons at the surface will equal, and then exceed, the concentration of holes in the substrate. This regime, called strong inversion, is very important from a circuits standpoint. The voltage at which inversion occurs is the threshold voltage V_{TX}. A convenient definition of the voltage at which (strong) inversion[6] takes place is that the surface is inverted at the voltage at which the electron density at the surface is the same as the hole density deep in the substrate. We know that with reference to an intrinsic semiconductor, in the substrate the potential ϕ_F is ϕ_{Fp}. The potential at the surface must therefore be $-\phi_{Fp}$ (a positive number since ϕ_{Fp} is negative) for the charge densities to be equal. The difference in potentials is $-2\phi_{Fp}$. This is a very important quantity. To keep our sanity in later manipulations of these equations, we define

$$\Phi_{Fp} \equiv |\phi_{Fp}|. \tag{2.30}$$

At this point it is advantageous to shift reference potentials again. We want to consider the substrate as being at ground. That is, instead of referencing the potential deep in the bulk to the Fermi potential of intrinsic silicon, we offset the bulk potential to zero. This moves the intrinsic potential to $-\phi_{Fp}$ and the potential of the inverted surface to $2\Phi_{Fp}$, a positive quantity. Out of the $2\Phi_{Fp}$, one Φ_{Fp} was required to get from p-type to intrinsic and the other Φ_{Fp} was needed to get from intrinsic to n-type. The potential at the inverted surface[7] is $2\Phi_{Fp}$.

The capacitance of a MOS capacitor in inversion is much more complicated than it was in the other two regimes. To understand why this is so we must ask where the electrons at the inverted surface come from. In one sense they come from Eq. (2.6) in which we observed that in equilibrium the quantity np was constant. The key is "in equilibrium." How long does it take the semiconductor to reach equilibrium? The only source of electrons at the silicon/silicon dioxide interface is the thermal generation of carriers. This process is slow (on the order of milliseconds) under normal operating conditions. Thus if we measure the capacitance of the MOS structure in inversion *very slowly*, we will observe a capacitance value of C_{OX}. This is because charges on the metal plate will be mirrored by charges in the inversion layer. The thermal generation of carriers will short out the depletion layer capacitance. This

---

[6]   From now on, inverted will mean strongly inverted.

[7]   We will need to modify this result somewhat in Section 2.3 because doping levels at the surface of the silicon are different than doping levels in the bulk.

is shown in Fig. 2.6. On the other hand, if we perform the measurement at high frequency, we find that the capacitance does not return to its maximum value of $C_{OX}WL$ as v_G increases past the threshold value at which inversion occurs, but neither does it continue decreasing. This is a subtle but important point. As the voltage v_G increases slowly past the voltage at which inversion occurs, all further increases in v_G are seen across the oxide while the potential ϕ_S at the silicon surface remains pinned near $2\Phi_{Fp}$. To understand this effect we can do a sensitivity analysis on the effect of a change in the charge density at the interface.

A 10% change in the surface charge density will result in a 10% change in the voltage across the oxide capacitor because $\Delta v = q\Delta n/C_{OX}$. On the other hand, the surface potential ϕ_S is related to the electron charge density by Eq. (2.12). At inversion we have

$$n = n_i e^{q\Phi_{Fp}/kT} = N_A. \tag{2.31}$$

At this operating point,

$$\frac{\partial n}{\partial \phi_S} = \frac{q}{kT}N_A. \tag{2.32}$$

A 10% change in n results in only a 2.5 mV change in ϕ_S. This is because the same exponential mechanism that makes it difficult to drop more than ϕ_T volts across a forward-biased p–n diode makes it difficult to raise the surface potential of a MOS capacitor above $2\Phi_{Fp}$ by increasing the voltage on the gate electrode. A small change in voltage at the surface of the silicon causes a flood of electrons.

We can relate X_P to ϕ_S using Eq. (2.23). We have

$$X_P = \sqrt{\frac{2\epsilon_{Si}}{q}\frac{\phi_S}{N_A}}, \tag{2.33}$$

where X_P has a maximum value of

$$X_{P\max} = \sqrt{\frac{2\epsilon_{Si}}{q}\frac{2\Phi_{Fp}}{N_A}}. \tag{2.34}\star$$

The ratio of the maximum to the minimum capacitance of a MOS capacitor is given by[8]

$$\frac{C_{\max}}{C_{\min}} = 1 + C_{OX}\sqrt{\frac{4\Phi_{Fp}}{\epsilon_{Si}qN_A}}, \tag{2.35}$$

where $C_{\max}$ is $C_{OX}WL$. This equation can be used to determine N_A from capacitance measurements. Figure 2.6 illustrates ideal C–V curves

---

[8] This value is actually slightly high because the device enters the "weak inversion" regime before $\phi_S = 2\Phi_{Fp}$.

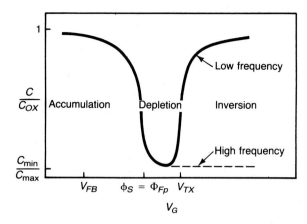

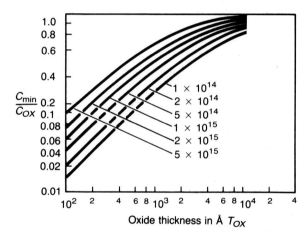

FIGURE 2.6 Capacitance characteristics of a MOS capacitor.

FIGURE 2.7 C_{min}/C_{OX} versus T_{OX} as a function of doping density in units of cm^{-3}. [Goetzberger 66, p. 1121] Reprinted with permission from The Bell System Technical Journal. Copyright 1966, AT&T.

of a MOS capacitor. Figure 2.7 plots C_{min}/C_{max} as a function of oxide thickness and doping level [Goetzberger 66]. If N_A is nonuniform the capacitance versus voltage characteristics can be deconvolved to obtain the doping profile.

Figure 2.8 illustrates a common circuit configuration with a side connection to the MOS capacitor. When this capacitor is in inversion the n+ connection on the side can provide electrons to the inversion region without needing to rely on the slow process of thermal generation. Thus the C–V characteristics of this device follow the low frequency branch of the characteristics up to very high frequencies.

Below threshold the inversion layer does not exist and thus there is almost no capacitance between the gate and the side contact. The capacitance between the gate and the body was already seen in Fig. 2.6.

FIGURE 2.8 MOS capacitor circuit configuration.

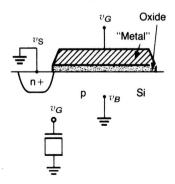

EXAMPLE 2.1

Lowering the capacitance per unit area of a MOS capacitor below $C_{\min}$

The capacitance can be lowered below $C_{\min}$ by applying a large voltage pulse to the gate. This will increase the size of the depletion region by increasing ϕ_S above $2\Phi_{Fp}$ before carriers can be generated at the surface to restore equilibrium. This effect is called "deep depletion." ■

EXAMPLE 2.2

Level restoring a MOS bus using varactor pumping

The purpose of the bus booster circuit in Fig. 2.9 is to boost the bus voltage to V_{DD} when the bus is high and not to boost it when the bus is low. On ϕ_1 the bus is precharged to $V_{DD} - V_T$ through transistor M_1. On ϕ_2 the inputs are evaluated and the bus either remains at $V_{DD} - V_T$ or is discharged to ground, depending on the state of the inputs. On ϕ_3 the bus boost circuit comes into play. If the bus is high, then M_3 is acting in a diode configuration and v_{boost} is at $V_{DD} - V_T$. This means that M_4 is above threshold and a channel exists under the gate. Therefore the capacitance to ϕ_3 is large, and when ϕ_3 rises v_{boost} will be pumped to over $V_{DD} + V_T$, enabling the bus to rise to V_{DD} (assuming we have correctly sized M_4).

If the bus is low, then M_3 is on and v_{boost} is at ground. This means that M_4 is below threshold and no channel exists. The capacitance is thus extremely low, and when ϕ_3 goes high v_{boost} will not rise above V_T (again assuming correctly sized devices). The bus will stay low. ■

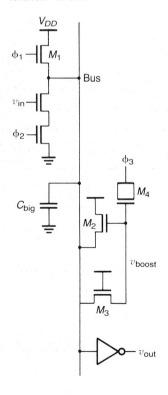

FIGURE 2.9 A bus booster circuit that exploits the varactor effect.

The flat-band voltage

We have seen the importance of the flat-band voltage in the analysis of MOS devices. The flat-band voltage is influenced in major ways by three phenomena. The first is work function differences, the second is charges in the insulator, and the third is surface states at the silicon/silicon dioxide interface.

When conductive materials of different chemical composition are connected together their Fermi levels must equalize. This causes a relative displacement of the conduction bands in the two materials, just as we saw in the case of the p–n diode.

The chemical potentials of materials are, by convention, measured in reference to vacuum. This potential, called the work function, is the energy necessary to remove one electron from the Fermi level of the material and bring it to infinity. This is a somewhat tricky measurement to perform and has only limited applicability to the design of solid-state devices. Chemical potentials have another property, however, that makes work functions useful. Chemical potentials obey Kirchhoff's voltage law. Thus the work function difference between, say, a metal and a semiconductor would be $\Phi_{MS} = \Phi_M - \Phi_S$, where Φ_M is the work function of the metal and Φ_S is the work function of the semiconductor. Consider the case of a metal connected to an oxide connected to a semiconductor connected back to the metal. Because the sum of the potentials around the loop must be zero, we have

$$0 = \Phi_{MO} + \Phi_{OS} + \Phi_{SM} \qquad (2.36)$$

or

$$\Phi_{MS} = \Phi_{MO} + \Phi_{OS}. \qquad (2.37)$$

Note that in some sense, the fact that the oxide is between the metal and the semiconductor does not matter. For a MOS device we have a couple of situations. If the gate is a classical metal like aluminum, then we are concerned with the chemical potential difference between the aluminum and the silicon. For intrinsic silicon the difference in chemical potentials Φ_{MS} is -0.61 volts. For gold it would be -0.11 volts because the work functions of aluminum and gold differ by 0.5 volts [Glaser 77, p. 92]. As the silicon is doped, ϕ_F changes and Φ_{MS} changes with it. As the silicon becomes more n-type Φ_{MS} increases, and as it becomes more p-type Φ_{MS} decreases. For an aluminum gate, we have

$$\Phi_{MS} = -0.61 + \phi_F. \qquad (2.38)$$

For a degenerately doped n-type poly gate the situation is numerically different but qualitatively the same, providing we accept the fact that it is legal to use energy band diagrams for the poly. This becomes more reasonable when we realize that leaky diodes can in fact be built on poly and that the recrystallization of poly is an active research area [Gibbons 80]. For the n+ poly gate, we have

$$\Phi_{MS} = -0.55 + \phi_F, \qquad (2.39)$$

where 0.55 is about half of qE_G.

Charges in the oxide will also affect the flat-band voltage. A sheet of charge Q buried in the oxide will induce an opposite charge in the gate and the substrate, which will have to be compensated for by the flat-band voltage. For this case, the solution of Poisson's equation yields

$$\Delta V_{FB} = -\frac{x}{T_{OX}} \frac{Q}{C_{OX}}, \qquad (2.40)$$

where x is the distance from the gate into the oxide toward the substrate. In the general case, we have

$$\Delta V_{FB} = -\frac{1}{C_{OX}} \int_0^{T_{OX}} \frac{x}{T_{OX}} \rho(x) dx, \qquad (2.41)$$

where $\rho(x)$ is the charge density per unit area. The voltage shift depends on the placement of the charge. If the ions are mobile they will move under the influence of both built-in and applied electric fields. This will change V_{FB} and hence the device thresholds. Mobile ionic charge Q_m typically consists of the alkali ions Na+, K+, and Li+. In the early days of MOS, sodium contamination was a major cause of threshold instability in MOS devices precisely because of this mechanism. The other type of charge that can exist in the oxide is called oxide trapped charge Q_{ot}. Oxide trapped charge can be created by X-rays or highly energetic (hot) electrons.

All MOS structures on silicon have a sheet of charge located near the silicon/silicon dioxide interface. The exact nature of this interface is still not fully understood. It is believed that as the silicon becomes silicon dioxide it goes through a transition monolayer of SiO_y, where $1 < y < 2$, and then through about 10 to 40 Å of strained SiO_2. There are two types of charge at this interface. One is a fixed charge[9] Q_f located within about 30 Å of the interface. The other is the interface trapped charge Q_{it}. This charge is roughly proportional to the number of bonds seen at the silicon surface. For instance, $\langle 100 \rangle$ material has a generally lower value of Q_{it} than $\langle 111 \rangle$ material. These interface charges have donor or acceptor energy states in the silicon forbidden band gap. N_{it} (Q_{it}/q) can be reduced as low as 10^{10} cm^{-2}. Both Q_{it} and Q_f are very dependent on processing; a typical final value for $N_f + N_{it}$ is 1.5×10^{11} ions/cm^2.

The surface charge contribution to V_{FB} can be calculated using Eq. (2.40) with $x = T_{OX}$. We have

$$\Delta V_{FB} = -\frac{Q_f + Q_{it}}{C_{OX}}. \tag{2.42}$$

Summing the three components of the flat-band voltage, we have

$$V_{FB} = \Phi_{MS} - \frac{Q_f + Q_{it}}{C_{OX}} - \frac{1}{C_{OX}} \int_0^{T_{OX}} \frac{x}{T_{OX}} \rho(x)dx. \tag{2.43}$$

All of these parameters are controlled by fabrication; therefore only V_{FB} is important to the circuit designer once a technology has been chosen.

2.2   First-order phenomena in MOS transistors

The MOS capacitor can be converted into a transistor by adding conductors to the opposite ends of the structure, as shown in Fig. 2.10. These conductors are the source and drain terminals. As in the case of

FIGURE 2.10   A MOS transistor.

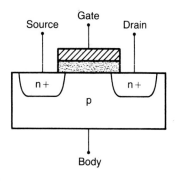

Source Gate Drain

n+   n+

p

Body

---

[9] Q_f was originally denoted Q_{SS}.

the capacitor, the channel exhibits three major regimes of operation. They are

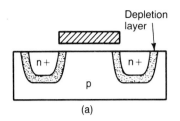

- **Accumulation:** $v_G < V_{FB}$. (Fig. 2.11a) The surface is strongly accumulated with substrate majority carriers. A reverse-biased p–n junction exists at both source and drain, between which no significant current flows.
- **Depletion:** $v_G > V_{FB}$. (Fig. 2.11b) The surface is depleted of majority carriers. There are no free carriers and no significant current flows. This is the regime in which the gate capacitance drops.
- **Inversion:** $v_G \gg V_{FB}$. (Fig. 2.11c) Minority carriers are at the surface. They form a conduction path between the source and the drain. The higher the gate voltage, the more the carriers and thus the larger the current.

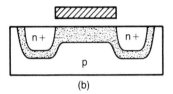

Each of these regions are important from a circuit design viewpoint. The most important parameter is the threshold voltage, which divides the regimes of depletion and inversion (off from on). For simplicity, we will first examine the case when the source and body voltages are at the same potential.

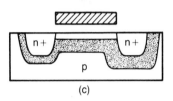

FIGURE 2.11   Three regimes of operation of a MOS transistor: (a) accumulation, (b) depletion, and (c) inversion.

Threshold voltage

We have seen that by applying a positive voltage v_G to the MOS structure, an inversion layer and depletion layer can be induced under the gate. We will now reexamine this structure with the objective of enumerating the charges in each region.

We define Q_S as the total charge per unit area induced in the silicon and Q_G as the total charge per unit area on the gate electrode. Charge neutrality requires that

$$Q_G + Q_S = 0. \tag{2.44}$$

The electric potential at the surface of the silicon is ϕ_S. Relating the charge on the gate oxide capacitor to the voltage across it, we have

$$Q_G = (v_G - V_{FB} - \phi_S)C_{OX}. \tag{2.45}$$

Note that when $v_G = V_{FB}$, $\phi_S = 0$ and therefore $Q_G = 0$. We can partition the charge in the silicon into two components. Q_B is the fixed charge in the depletion region and Q_N is the mobile charge in the inversion layer. We have

$$Q_S = Q_B + Q_N \tag{2.46}$$

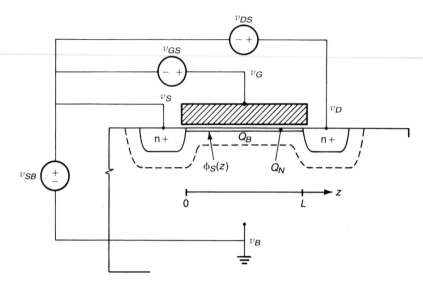

FIGURE 2.12   MOS transistor potentials and charges.

and, using Eq. (2.45), we have

$$v_G - V_{FB} = \phi_S - \frac{Q_B + Q_N}{C_{OX}}. \qquad \textbf{(2.47)}$$

Fig. 2.12 illustrates the various charges and voltages in a MOS transistor. Q_N versus the surface potential ϕ_S is illustrated in Fig. 2.13 [Sze 81, p. 369].

FIGURE 2.13   Variation of charge density in the semiconductor as a function of the surface potential ϕ_S with $N_A = 4 \times 10^{15}$ cm^{-3} at room temperature. [Sze 81, p. 369] (S. M. Sze, Physics of Semiconductor Devices, ©1981. Reprinted by permission of John Wiley & Sons, Inc.)

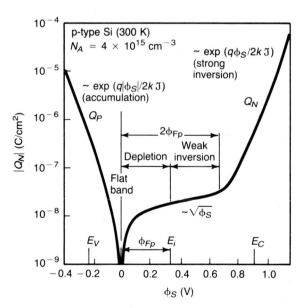

We can now develop an equation for the threshold of a MOS transistor. The onset of conduction in a MOS transistor occurs as the transistor enters the inversion regime. At the edge of this regime $Q_N = 0$. We define the threshold voltage V_{TX} as the gate voltage at which the channel is formed. From Eq. (2.47) we have

$$V_{TX} = -\frac{Q_B}{C_{OX}} + \phi_S + V_{FB}. \qquad (\mathbf{2.48})$$

At the onset of inversion the surface potential in the silicon must be such that electrons can be supported in the channel. From the analysis of the MOS capacitor we argued that inversion occurs approximately when the potential at the surface is $2\Phi_{Fp}$. This analysis assumes an abrupt transition to the conducting state. The exact transition is exponential and occurs over about a third of a volt. Despite this qualification, V_{TX} is an extremely useful parameter for describing the characteristics of MOS transistors as they are used in digital circuits. At inversion $\phi_S = 2\Phi_{Fp}$ and Q_B is defined as Q_{B0}. We have

$$V_{TX} = -\frac{Q_{B0}}{C_{OX}} + 2\Phi_{Fp} + V_{FB}. \qquad (\mathbf{2.49})$$

Q_B, the charge in the depletion region, can be found from the analysis of the p–n diode. V_{TX} becomes lower with increasing temperature. The temperature coefficient is typically between -1 and -3 mV/$^\circ$ C for a silicon gate process. The dominant temperature effect comes through the temperature dependence of n_i which doubles every 11° C. [Sze 81, p. 20]. For p-channel transistors the threshold becomes more positive with temperature, at about the same rate.

Voltage-controlled resistance

The drain-to-source current in a MOS transistor is a function of the quantity of charge in the channel that is available for conduction Q_N and the charge transport relations.

Free charge in the channel of a MOS transistor is propelled by the electric field applied through the drain and source electrodes. The average velocity of this charge is given by

$$\mathcal{V} = \mu \mathcal{E}, \qquad (\mathbf{2.50})$$

where μ is the mobility of the carriers and $\mathcal{E}$ is the electric field. The mobility is determined by the effective mass of the carriers, collisions of the carriers with lattice atoms (phonon scattering), coulomb scattering off charge centers (oxide charge, interface charge, and charge due to impurity atoms), and surface roughness at the oxide interface. The relative importance of these parameters is a function of the applied and built-in voltages as well as the fabrication methods. The physics is quite complicated and most models are somewhat empirical. We will be discussing mobility in more detail later. For now we assume that the mobility is constant.

To gain a feel for the physics we first analyze the MOS transistor in the linear region where all fields and charge distributions are assumed uniform. Figure 1.6 shows cross-sectional and top views of a MOS transistor. Assume that both the source and the body are grounded. We refer to the distance between drain and source as the channel length L. The width of the channel W is shown in the top view. Assuming a uniform surface charge density n, integrated over the depth of the channel, the current flow is given by

$$i_D = qnW\mathcal{V} = -\mu Q_N W \mathcal{E}, \quad (2.51)$$

where the electric field is assumed to point from drain to source.

For a uniform lateral field we have

$$\mathcal{E} = \frac{v_D}{L} \quad (2.52)$$

and

$$i_D = -\mu Q_N \frac{W}{L} v_D, \quad (2.53)$$

where v_D is the drain-to-source voltage. The incremental conductance g_D is given by

$$g_D = \frac{\partial i_D}{\partial v_D} = -\mu \frac{W}{L} Q_N = -\mu S Q_N, \quad (2.54)$$

where S is the shape factor defined in Eq. (1.3). From Eq. (2.47) we have

$$Q_N = -\left(v_G - V_{FB} + \frac{Q_B}{C_{OX}} - \phi_S\right) C_{OX}. \quad (2.55)$$

If we assume v_D to be very small, then the potential variation from source to drain is small; therefore we have

$$\phi_S \approx 2\Phi_{Fp}. \quad (2.56)$$

Equation (2.55) then becomes

$$Q_N = -\left(v_G - V_{FB} - 2\Phi_{Fp} + \frac{Q_{B0}}{C_{OX}}\right) C_{OX}. \quad (2.57)$$

Substituting V_{TX} from Eq. (2.49) we obtain

$$Q_N = -(v_G - V_{TX}) C_{OX} \quad (2.58)$$

and from Eq. (2.54) we obtain

$$g_D = \mu S C_{OX}(v_G - V_{TX}). \quad (2.59)$$

The current is given by

$$i_D = \mu S C_{OX}(v_G - V_{TX}) v_D \quad \text{for} \quad v_D \ll v_G - V_{TX}. \quad (2.60)$$

Equation (2.60) is valid only for small drain voltages. We will now use the same techniques, with fewer assumptions, to derive a useful result.

If v_D is not small, then the variation in ϕ_S cannot be ignored. For this analysis we will still assume Q_B is constant across the channel. This assumption turns out to be good for contemporary nMOS circuits. Let

$$\phi_S(z) = 2\Phi_{Fp} + \phi_C(z), \qquad (2.61)$$

where z is the distance along the channel, measured from the source, and ϕ_C is the voltage along the channel. From Eq. (2.55) we have

$$Q_N(z) = -\left(v_G - V_{FB} - 2\Phi_{Fp} + \frac{Q_{B0}}{C_{OX}} - \phi_C(z)\right)C_{OX}, \qquad (2.62)$$

which reduces to

$$Q_N(\phi_C) = -(v_G - V_{TX} - \phi_C)C_{OX}. \qquad (2.63)$$

Each incremental region of the channel is resistive. Given a current i_D, the voltage drop across an increment is given by

$$d\phi_C = i_D dr, \qquad (2.64)$$

as seen in Fig. 2.14. Using Eq. (2.54), we obtain

$$dr = dg_D^{-1} = -\frac{dz}{W\mu Q_N}. \qquad (2.65)$$

Rearranging Eq. (2.65) and integrating from source to drain, we obtain

$$\int_0^{v_D} W\mu Q_N d\phi_C = -i_D \int_0^L dz \qquad (2.66)$$

or

$$i_D = -\mu S \int_0^{v_D} Q_N d\phi_C. \qquad (2.67)$$

Substituting for Q_N and evaluating, we obtain

$$i_D = -\mu S C_{OX} \int_0^{v_D} -(v_G - V_{TX} - \phi_C)d\phi_C$$

or

$$i_D = \mu S C_{OX}\left((v_G - V_{TX})v_D - \frac{1}{2}v_D^2\right). \qquad (2.68)$$

Note the similarity to the previous result of Eq. (2.60). The equations are the same with the exception of the term $v_D^2/2$, which for small v_D approaches zero.

Figure 2.15 shows a plot of I_D vs V_D for a particular gate voltage using Eq. (2.68). As shown by the solid line on the plot, the current initially increases with increasing V_D. It reaches a peak and then begins to decline. This decline does not seem reasonable and it is not observed in real devices. Instead, the measured current follows the equation closely to the peak and then saturates for further increases in V_D, as shown by the dotted line. Clearly, something has been overlooked in our analysis.

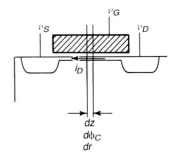

FIGURE 2.14   A differential slice through the channel of a MOS transistor.

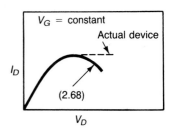

FIGURE 2.15 I_D versus V_D from (Eq. 2.68). The equation predicts unphysical behavior past the point of saturation.

Examining Eq. (2.63) closely, we see that $Q_N(\phi_C)$ goes to zero at $v_G - V_{TX} = \phi_C$. The maximum value of ϕ_C will be at the drain end of the channel, where we have $v_G - V_{TX} = v_D$. At this point, Eq. (2.62) is no longer valid since the channel vanishes.

Using Eq. (2.68), we compute the peak current by taking the derivative of i_D with respect to v_D and setting it equal to zero. At this peak we suspect that our model may cease to be valid. We have

$$\frac{\partial i_D}{\partial v_D} = \mu S C_{OX}(v_G - V_{TX} - v_D) = 0. \tag{2.69}$$

We define $V_{D\text{sat}}$ as the voltage at which the current saturates. We have

$$V_{D\text{sat}} = v_G - V_{TX}. \tag{2.70}$$

At this voltage $Q_N = 0$ at the drain end of the channel. The problem shows up where predicted. At this point we must look for a new physical model of what happens when the channel vanishes at the drain end. We know that a depletion region can exist in this region and that free charges will be propelled to the edge of the depletion region by the field in the channel. When they get to the depletion region, they will be swept across to the drain by the potential that exists between the end of the channel and the drain. Any drain voltage higher than $V_{D\text{sat}}$ will appear across the depletion region but not across the inversion region, which will have an apparent drain voltage limited to $V_{D\text{sat}}$. Thus we can rationalize that the current should saturate at a value determined by $V_{D\text{sat}}$. Unlike the case of the depletion region in a reverse-biased p–n junction diode, no potential barrier is at the drain end of the channel and the electric field is pointing in a direction to accelerate the charges, not repel them. We define the saturation current as

$$I_{D\text{sat}} \equiv \frac{\mu S C_{OX}}{2}(v_G - V_{TX})^2. \tag{2.71}$$

In actual devices, there is some dependence of i_D on v_D above $V_{D\text{sat}}$. Several second-order effects must be considered in order to model this dependency. We will treat these effects in Section 2.3.

While the previous development was good for nMOS devices with $v_S = 0$, we also need to know what happens when $v_S \neq 0$. The next step considers the effect of a source-to-substrate bias. Consider the configuration of Fig. 2.12 with four independent terminals. We will let v_B be the datum (that is, $V_B \equiv 0$). At the source end of the channel, the potential is $2\Phi_{Fp} + v_S$. This can be accounted for explicitly in setting the limits for integration in Eq. (2.67). However, we have used this potential implicitly in computing the value of Q_B and Q_N. The new equation for Q_N at the source end of the channel becomes

$$Q_N = -\left(v_G - V_{FB} - 2\Phi_{Fp} + \frac{Q_B}{C_{OX}}\right)C_{OX}. \tag{2.72}$$

V_{TE} is defined as

$$V_{TE} \equiv V_{FB} + 2\Phi_{Fp} - \frac{Q_B}{C_{OX}}. \qquad (2.73)$$

To compute Q_B, we use the formula developed for the one-sided step junction, given as

$$Q_B = -\sqrt{2\epsilon_{Si}qN_A(v_S + 2\Phi_{Fp})}. \qquad (2.74)$$

The threshold voltage becomes

$$V_{TE} = V_{FB} + 2\Phi_{Fp} + \frac{\sqrt{2\epsilon_{Si}qN_A(v_S + 2\Phi_{Fp})}}{C_{OX}}. \qquad (2.75)$$

We define γ, the body or back gate factor, as

$$\gamma \equiv \frac{\sqrt{2\epsilon_{Si}qN_A}}{C_{OX}}. \qquad (2.76)\star$$

Note the intimate relationship between γ and Q_B. Re-expressing the threshold voltage as a function of γ, we obtain

$$V_{TX} = V_{FB} + 2\Phi_{Fp} + \gamma\sqrt{2\Phi_{Fp}} \qquad (2.77)\star$$

and

$$V_{TE} = V_{TX} + \gamma\left(\sqrt{v_S + 2\Phi_{Fp}} - \sqrt{2\Phi_{Fp}}\right). \qquad (2.78)\star$$

The body factor γ determines the threshold shift in a device as a function of source-to-substrate voltage. It is typically larger in CMOS circuits than in nMOS circuits because most nMOS circuits use a substrate-bias voltage V_{BB} in the range of -2 to -3 volts, while in CMOS the V_{BB} voltage on nMOS transistors is usually zero. This allows the substrate doping to be lower in the nMOS case.[10] From Eq. (2.76) we see that a lower value of substrate doping lowers the body effect. The body effect parameter is fairly technology-dependent, with older technologies having larger values of γ. One can encounter values of γ ranging from less than 0.1 $V^{1/2}$ to over 1 $V^{1/2}$.

To determine i_D, we integrate Q_N with the new limits and obtain

$$i_D = -\mu S \int_{v_S}^{v_D} Q_N(\phi_C)d\phi_C. \qquad (2.79)$$

Substituting for Q_N, we obtain

$$i_D = -\mu S C_{OX} \int_{v_S}^{v_D} -(v_G - V_{TE} - \phi_C)d\phi_C, \qquad (2.80)$$

which evaluates to

$$i_D = \mu S C_{OX}\left((v_{GS} - V_{TE})v_{DS} - \frac{v_{DS}^2}{2}\right), \qquad (2.81)$$

---

[10] The limiting phenomenon is drain-to-source punch-through. See Section 2.3.

where

$$v_{DS} \equiv v_D - v_S \qquad (2.82)$$

and

$$v_{GS} \equiv v_G - v_S. \qquad (2.83)$$

The transconductance parameter[11] is defined as

$$K' \equiv \frac{\mu C_{OX}}{2}. \qquad (2.84)\star$$

The following is a summary of these results for the three regions of operation.

1. **Cutoff:** $(v_{GS} - V_{TE}) \leq 0$,

$$i_D = 0. \qquad (2.85)\star$$

2. **Nonsaturation:** $(v_{GS} - V_{TE}) \geq v_{DS}$,

$$i_D = K'S\left(2(v_{GS} - V_{TE})v_{DS} - v_{DS}^2\right). \qquad (2.86)$$

3. **Saturation:** $(v_{GS} - V_{TE}) \leq v_{DS}$,

$$i_D = K'S(v_{GS} - V_{TE})^2. \qquad (2.87)$$

Because we have ignored the variation in Q_B along the channel, these results are accurate only for processes in which γ is below about $0.2 \ \mathrm{V}^{1/2}$.

Charge distribution in the channel region

Using the equations we have developed so far, we can find the charge distribution along the channel. This analysis is very important in developing an intuitive understanding of MOS device operation.

Qualitatively, we know that for the nonsaturation region of operation (see Eq. 2.86) the number of charge carriers decreases as we move from source to drain due to the reduction in voltage from gate to channel across the MOS capacitor. Since the current is the same at the source and drain ends, the carriers must move faster at the drain to compensate for their reduced numbers. To quantify this relationship, we start with Eq. (2.86). At any point $0 \leq z \leq L$ along the channel, we can apply Eq. (2.86) to obtain

$$i_D = \frac{W}{z}K'\left(2(v_{GS} - V_{TE})\phi_{CS} - \phi_{CS}^2\right), \qquad (2.88)$$

where ϕ_{CS} is the voltage from point z to the source. Combining Eqs.

---

[11] K' is *not* the same as KP of SPICE. KP= $2K'$.

(2.86) and (2.88), we obtain

$$\frac{z}{L} = \frac{2(v_{GS} - V_{TE})\phi_{CS} - \phi_{CS}^2}{2(v_{GS} - V_{TE})v_{DS} - v_{DS}^2}. \tag{2.89}$$

Solving for ϕ_{CS}, we obtain

$$\phi_{CS} = v_{GS} - V_{TE} - \sqrt{(v_{GS} - V_{TE})^2 - (2(v_{GS} - V_{TE})v_{DS} - v_{DS}^2)\frac{z}{L}}. \tag{2.90}$$

We define the saturation ratio X as

$$X \equiv \frac{v_{DS}}{v_{GS} - V_{TE}}. \tag{2.91}$$

Re-expressing Eq. (2.90), we obtain

$$\phi_{CS} = (v_{GS} - V_{TE})\left(1 - \sqrt{1 - (2X - X^2)\frac{z}{L}}\right). \tag{2.92}$$

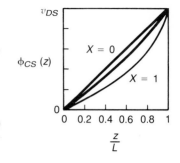

FIGURE 2.16 ϕ_{CS} versus z/L.

Figure 2.16 illustrates ϕ_{CS} versus z for several different saturation ratios. The slope of the curves at z represents the electric field at that point in the channel.

We compute the electric field $\mathcal{E}$ by differentiating ϕ_{CS} with respect to z. It is given that

$$\mathcal{E} = -\frac{d\phi_{CS}}{dz}. \tag{2.93}$$

We therefore obtain

$$\mathcal{E} = -\frac{(v_{GS} - V_{TE})}{2L}\frac{(2X - X^2)}{\sqrt{(1 - (2X - X^2)\frac{z}{L})}} \tag{2.94}$$

or, equivalently,

$$\mathcal{E} = -\frac{v_{DS}}{L}\frac{1 - \frac{X}{2}}{\sqrt{1 - (2X - X^2)\frac{z}{L}}}. \tag{2.95}$$

Fig. 2.17 shows the electric field as a function of distance along the channel, parameterized in X. For $X = 0$, the electric field is constant and evaluates to v_{DS}/L. For $X = 1$, the electric field tends to infinity at $z = L$. This is a consequence of our physical model of Q_N approaching zero at this point. Incorporation of the depletion region in our mathematical model would make further calculations very difficult. Instead we will note that at the saturation boundary our mathematical model is not very accurate.

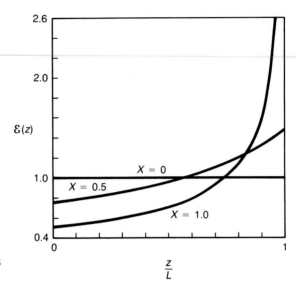

FIGURE 2.17 $\mathcal{E}(z)$ versus z/L, normalized to V_{DS}/L.

We now examine the charge distribution quantitatively. Recall that the charge at any point in the channel is given by

$$Q_N = -(v_{GS} - V_{TE} - \phi_{CS})C_{OX}. \qquad (2.96)$$

At the source $(z = 0)$, we have

$$Q_{\text{source}} = -(v_{GS} - V_{TE})C_{OX} \qquad (2.97)$$

FIGURE 2.18   Mobile charge in the channel, normalized to its value at the drain, as a function of z/L.

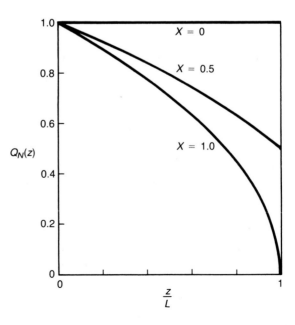

and in general

$$\frac{Q_{\text{channel}}}{Q_{\text{source}}} = \frac{v_{GS} - V_{TE} - \phi_{CS}}{v_{GS} - V_{TE}}.$$

(2.98)

Expressed in terms of the saturation ratio, Eq. (2.98) can be re-expressed as

$$\frac{Q_{\text{channel}}}{Q_{\text{source}}} = \sqrt{1 - (2X - X^2)\frac{z}{L}}.$$

(2.99)

This relationship is shown in Fig. 2.18. Figs. 2.16 through 2.18 should be studied carefully to understand the operating characteristics of a MOS device.

Improved models for the drain current

Because the substrate dopings are typically higher in CMOS technologies, γ and Q_B are higher also. This makes errors in the estimation of Q_B more significant. In the next derivation we will improve our models for CMOS technology by developing a better approximation of Q_B.

The derivations of Eqs. (2.85) through (2.87) assumed that the charge Q_B in the depletion region under the channel was constant. Physically, this cannot be true when the channel potential varies from source to drain. The magnitude of error depends on the ratio Q_B/Q_N. When the substrate is heavily doped or the gate oxide is relatively thick, the error is greatest. We will now rederive the current equations allowing for the variation in Q_B. Rather than use the exact value of Q_B, we will approximate the variation with a Taylor expansion about v_S. We do this because the exact equations lead to somewhat impenetrable results and are not that much more accurate.

From Eq. (2.62) we have the basic expression for Q_N without simplifications. Using the equation for a step junction, we have

$$\frac{Q_B}{C_{OX}} = \frac{-1}{C_{OX}}\sqrt{2\epsilon_{\text{Si}}qN_A(v_S + \phi_{CS} + 2\Phi_{Fp})}$$

(2.100)

or

$$\frac{Q_B}{C_{OX}} = -\gamma\sqrt{v_S + \phi_{CS} + 2\Phi_{Fp}}.$$

(2.101)

Using the first two terms of a Taylor expansion about the source voltage v_S, we obtain

$$Q_B \approx Q_B(v_S) + \phi_{CS}\left.\frac{dQ_B}{d\phi_{CS}}\right|_{\phi_{CS}=0},$$

(2.102)

which evaluates to

$$\frac{Q_B}{C_{OX}} = -\gamma\sqrt{v_S + 2\Phi_{Fp}} - \phi_{CS}\frac{\gamma}{2\sqrt{v_S + 2\Phi_{Fp}}}.$$

(2.103)

Substituting into Eq. (2.55) and using Eq. (2.78) for V_{TE}, we obtain

$$Q_N = -\left(v_G - V_{TE} - \frac{\gamma\phi_{CS}}{2\sqrt{v_S + 2\Phi_{Fp}}} - \phi_{CS} - v_S\right)C_{OX}. \qquad \textbf{(2.104)}$$

As before, to find the current we integrate the charge from source to drain using ϕ_{CS}. We have

$$i_{DS} = -\mu S \int_0^{v_{DS}} Q_N(\phi_{CS})d\phi_{CS} \qquad \textbf{(2.105)}$$

or

$$i_{DS} = \mu S C_{OX} \int_0^{v_{DS}} \left(v_{GS} - V_{TE} - (1+\delta)\phi_{CS}\right)d\phi_{CS}, \qquad \textbf{(2.106)}$$

where we have defined

$$\delta \equiv \frac{\gamma}{2\sqrt{v_{SB} + 2\Phi_{Fp}}}. \qquad \textbf{(2.107)}\star$$

The result is

$$i_{DS} = K'S\left(2(v_{GS} - V_{TE})v_{DS} - (1+\delta)v_{DS}^2\right). \qquad \textbf{(2.108)}\star$$

This equation is very similar to Eq. (2.86), with the addition of the correction factor δ. The effect of δ is to reduce the current at higher drain voltages. At higher drain voltages the error in Q_B is largest. The depletion region under the channel is largest at the drain end, but we used the value at the source in all our earlier calculations. This means we underestimated Q_B and hence overestimated Q_N and the current.[12] The correction factor δ is most important in technologies where the substrate doping is high and the gate oxide is thick.

To find the saturation voltage, we find the peak value of Eq. (2.108) in the usual manner. We have

$$\frac{di_{DS}}{dv_{DS}} = 0 = v_{GS} - V_{TE} - (1+\delta)V_{Dsat}, \qquad \textbf{(2.109)}$$

which reduces to

$$V_{Dsat} = \frac{v_{GS} - V_{TE}}{1+\delta}. \qquad \textbf{(2.110)}$$

---

[12] While we have used a Taylor series to derive δ, the first two terms of a Taylor series are not necessarily the best approximation to a function. In this case, δ is actually too large to provide the "best" linear fit to $Q_B(z)$. For this reason, values of δ slightly smaller than that defined in Eq. (2.107) are sometimes used.

We summarize the results of this analysis for the three regions of operation.

1. **Cutoff:** $(v_{GS} - V_{TE}) \leq 0,$

$$i_{DS} = 0. \qquad (2.85)\star$$

2. **Nonsaturation:** $(v_{GS} - V_{TE}) \geq (1 + \delta)v_{DS},$

$$i_{DS} = K'S\left(2(v_{GS} - V_{TE})v_{DS} - (1 + \delta)v_{DS}^2\right). \qquad (2.108)\star$$

3. **Saturation:** $(v_{GS} - V_{TE}) \leq (1 + \delta)v_{DS},$

$$i_{DS} = K'S\frac{(v_{GS} - V_{TE})^2}{1 + \delta}, \qquad (2.111)\star$$

where

$$K' \equiv \mu\frac{C_{OX}}{2}, \qquad (2.84)\star$$

$$V_{TE} = V_{TX} + \gamma\left(\sqrt{v_{SB} + 2\Phi_{Fp}} - \sqrt{2\Phi_{Fp}}\right), \qquad (2.78)\star$$

$$V_{TX} = V_{FB} + 2\Phi_{Fp} + \gamma\sqrt{2\Phi_{Fp}}, \qquad (2.77)\star$$

$$\gamma \equiv \frac{\sqrt{2\epsilon_{Si}qN_A}}{C_{OX}}, \qquad (2.76)\star$$

$$\delta \equiv \frac{\gamma}{2\sqrt{v_{SB} + 2\Phi_{Fp}}}. \qquad (2.107)\star$$

These equations are the principal results of the section. They will be used in hand calculations throughout the book.

<hr>

■ EXAMPLE 2.3

MOS device parameter calculations

A MOS capacitor is built on a 20 $\Omega\cdot$cm p-type substrate with a $\langle 100\rangle$ orientation. The gate material is degenerately doped n+ poly and the oxide is 300 Å thick.

From Fig. 2.19 we see that 20 $\Omega\cdot$cm material corresponds to a doping level of $N_A = 9 \times 10^{14}$ cm^{-3} [Irvin 62]. From Eq. (2.11) we find $\phi_{Fp} = -0.285$ V, or $2\Phi_{Fp} = 0.570$ V.

We also have $C_{OX} = 0.00115$ F/m^2=0.00115 pF/μm^2 from Eq. (2.28). At inversion $X_{P\max}$ is given by Eq. (2.34). We have $X_{P\max} = $

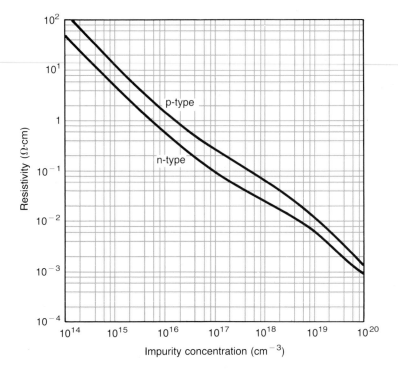

FIGURE 2.19 Resistivity of silicon at 300 K as a function of donor and acceptor impurity concentrations. [Irvin 62] Reprinted with permission from The Bell System Technical Journal. Copyright 1962, AT&T.

0.91 μm. The charge Q_B in the semiconductor at inversion is

$$Q_{B\text{max}} = -qN_A X_{P\text{max}} = -1.3 \times 10^{-4} \text{ C/m}^2 = 814 \text{ ions}/\mu\text{m}^2,$$

and its capacitance is

$$C_{\text{Si}} = \frac{\epsilon_{\text{Si}}}{X_{P\text{max}}} = 1.14 \times 10^{-4} \text{ pF}/\mu\text{m}^2.$$

We see that $C_{\text{min}} = 1.04 \times 10^{-4}$ pF/μm^2 and $C_{\text{max}}/C_{\text{min}} = 11$. Note that from the exact calculations of Goetzberger, $C_{\text{max}}/C_{\text{min}}$ is actually a little under 10 [Goetzberger 66, p. 1121]. Our error is a result of the approximation $\phi_S = 2\Phi_{Fp}$.

Using $N_f = 1.5 \times 10^{11}$ cm^{-2} and $Q_{it} = \rho(x) = 0$, we compute V_{FB}. From Eq. (2.39), $\Phi_{MS} = -0.835$ V. From Eq. (2.43), we have

$$V_{FB} = \Phi_{MS} - \frac{qN_f}{C_{OX}} = -1.04 \text{ V}.$$

We compute $V_{TX} = -0.37$ V from Eq. (2.49), $\gamma = 0.15$ V$^{1/2}$ from Eq. (2.76), and $\delta = 0.1$ from Eq. (2.107).   ∎

Capacitance characteristics of the MOS transistor

While algebraically somewhat complicated, our analysis of the current–voltage characteristics of the four-terminal MOS transistor is nonetheless on quite firm theoretical ground. Things were comparatively simple because, in normal operation, dc current flows only between the source and the drain. Capacitances, on the other hand, exist between each pair of terminals [Meyer 71].

Consider for the moment only incremental capacitance (see Appendix B). We can linearize the transistor characteristics about some operating point, and therefore the concept of impedance makes sense. Taking one terminal of the MOS transistor as the datum, the capacitance matrix of the MOS transistor contains nine independent terms. They include, for instance, C_{GS} as well as C_{SG}. In most familiar problems, these capacitances are the same, that is, the capacitance matrix is reciprocal. There are some knotty theoretical issues here. For instance, it is known that if a network has a nonreciprocal capacitance matrix, it can generate infinite power at infinite frequency. On the other hand it is also known that the best fit to the capacitance characteristics of a MOS transistor results in nonreciprocal capacitances. These nonreciprocal capacitances have been measured [Paulos 82]. Part of the problem is that the transistor is a distributed structure that, if we were going to be perfectly rigorous, must be modeled by transmission lines [Paulos 83]. This is, of course, much more involved and precise than anything we would ever need, so we truncate these higher-order effects—and obtain nonreciprocal capacitances. In this book we will sidestep all of these issues to present only the simplest of models. Be aware, however, that the ice is thin.

When a MOS transistor is in the active region, we can conceptualize the gate capacitance as split between a gate-to-source component and a gate-to-drain component as shown in Fig. 2.20. C_{GS} and C_{GD} are defined as

$$C_{GS} \equiv \frac{\partial Q_G}{\partial v_{GS}}\bigg|_{V_{GD}\,=\,\text{constant}} \qquad (2.112)$$

and

$$C_{GD} \equiv \frac{\partial Q_G}{\partial v_{GD}}\bigg|_{V_{GS}\,=\,\text{constant}} \qquad (2.113)$$

Using Eq. (2.63), we can find the charge density Q_N at any point in the channel. From Eq. (2.46), $Q_S = Q_N + Q_B$, where Q_S is the total charge density induced in the silicon. Q_S must be balanced by an equal charge on the gate, thus $Q_G = -Q_S = -(Q_N + Q_B)$. Using Eqs. (2.63) and (2.99), we obtain

$$Q_N = -(v_{GS} - V_{TE})C_{OX}\sqrt{1 - (2X - X^2)\frac{z}{L}}, \qquad (2.114)$$

FIGURE 2.20   Simple capacitance model of a MOS transistor.

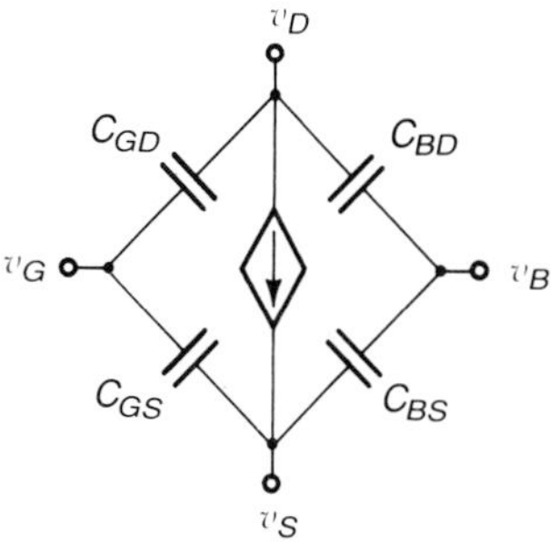

where X is the saturation ratio. Equation (2.114) ignores the body effect. To find the total charge in the channel we must multiply by the channel width and integrate Q_N across the channel length. Thus

$$Q_{TN} = -(v_{GS} - V_{TE})WC_{OX}\int_0^L \sqrt{1 - (2X - X^2)\frac{z}{L}}\, dz \qquad (2.115)$$

or

$$Q_{TN} = (v_{GS} - V_{TE})C_{OX}WL\frac{2(X^2 - 3X + 3)}{3(X - 2)}, \qquad (2.116)$$

where Q_{TN} is the total charge induced in the channel. Using the chain rule, Q_{TN} can be differentiated per Eqs. (2.112) and (2.113), resulting in

$$C_{GS} = -\frac{\partial Q_N}{\partial v_{GS}} - \frac{\partial Q_N}{\partial X}\frac{\partial X}{\partial v_{GS}} \qquad (2.117)$$

and

$$C_{GD} = -\frac{\partial Q_N}{\partial X}\frac{\partial X}{\partial v_{GD}}, \qquad (2.118)$$

where $v_{DS} = v_{GS} - v_{GD}$. Evaluating, we obtain

$$C_{GS} = \frac{2}{3}WLC_{OX}\left(1 - \frac{(1 - X)^2}{(2 - X)^2}\right) \qquad (2.119)$$

and

$$C_{GD} = \frac{2}{3}WLC_{OX}\left(1 - \frac{1}{(2 - X)^2}\right). \qquad (2.120)$$

At the extremes of X, we find

$$C_{GS} = C_{GD} = \frac{1}{2}C_{OX}WL \quad \text{at} \quad X = 0, \qquad (2.121)\star$$

$$C_{GS} = \frac{2}{3}C_{OX}WL \quad \text{at} \quad X = 1, \qquad (2.122)\star$$

and

$$C_{GD} = 0 \quad \text{at} \quad X = 1, \qquad (2.123)\star$$

FIGURE 2.21 MOS transistor capacitances, $C_T \equiv C_{GS} + C_{GD}$.

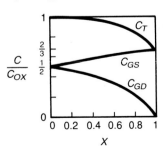

where X is the saturation ratio defined in Eq. (2.91). Equation (2.121) has an easy, intuitive interpretation: since nothing distinguishes the source from the drain at $X = 0$, the capacitances should be equal. At $X = 1$, the device has just saturated and the operation is independent of drain voltage, so Eq. (2.123) is also intuitive. For Eq. (2.122), we just have to trust the mathematics.

Fig. 2.21 plots the curves of Eqs. (2.119) and (2.120). Also included in the figure is a plot of C_T, where $C_T = C_{GS} + C_{GD}$. Although our equations are not valid for $X \geq 1$, we can easily see that C_{GS} and C_{GD}

will be the same as in Eqs. (2.122) and (2.123) beyond the point of saturation.

These equations are sufficient for most digital applications. On occasion, however, one needs to account for the capacitance between the channel and the substrate. As with the gate-to-channel capacitance, we apportion part of this capacitance to the source C_{BS} and part to the drain C_{BD}. In the analysis of C_{GS} and C_{GD}, we assumed Q_B constant. This would lead to $C_{BS} = C_{BD} = 0$. In reality Q_B does change; we saw this change when we analyzed the body effect. The analysis of the body effect tacitly assumed v_{DS} constant. For this condition, we have

$$\frac{1}{C_{OX}} \frac{dQ_B}{dv_S} = \frac{dV_{TE}}{dv_S}\bigg|_{V_{DS}=\text{constant}} \tag{2.124}$$

or

$$\frac{dQ_B}{dv_S} = bC_{OX}, \tag{2.125}$$

where we have defined b as

$$b \equiv \frac{dV_{TE}}{dv_S}. \tag{2.126}$$

This analysis suggests [Tsividis 80] that

$$C_{BS} = bC_{GS} \tag{2.127}$$

and

$$C_{BD} = bC_{GD}. \tag{2.128}$$

Nonlinear capacitances are fairly tricky to implement in a circuit simulation program because of the importance of numerical errors. If the program does not explicitly keep track of charge, integration errors can make the circuit appear not to be conserving charge. This is a problem, not with the model but with the choice of primary and secondary variables in the simulation program. Most early simulators, which were written before this problem was understood, apparently made the wrong choice. This includes SPICE2.G. Thus simulations of dynamic RAMs, switched capacitor filters, and charge coupled devices cannot be counted on to conserve charge and typically lead to quite dubious results in these specialized applications.

■ EXAMPLE 2.4

$Q_B(z)$ as a function of the saturation ratio
Using the complete expression for Q_B from Eq. (2.101) and approximation of $\phi_C(z)$ from Eq. (2.92), we find

$$\frac{Q_B(z)}{Q_B(0)} = \sqrt{1 + \frac{v_{GS} - V_{TE}}{v_{SB} + 2\Phi_{Fp}} \left(1 - \sqrt{1 - (2X - X^2)\frac{z}{L}}\right)},$$

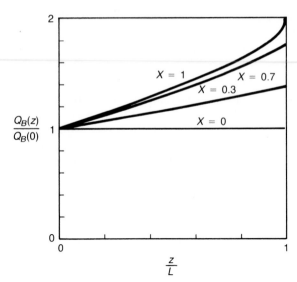

FIGURE 2.22 Depletion layer charge along the channel of a MOS transistor.

where $Q_B(0)$ is the bulk charge at the source. This function is plotted in Fig. 2.22. Note that

$$\frac{Q_B(L)}{Q_B(0)} = \sqrt{1 + \frac{V_{D\text{sat}}}{v_{SB} + 2\Phi_{Fp}}}$$

for $X = 1$. The Taylor expansion of Q_B used to derive the current

TABLE 2.1 Summary of threshold equations

n-CHANNEL	p-CHANNEL
$\phi_{Fp} = -\frac{kT}{q}\ln\frac{N_A}{n_i}$	$\phi_{Fn} = \frac{kT}{q}\ln\frac{N_D}{n_i}$
$X_{P\max} = \sqrt{\frac{2\epsilon_{\text{Si}}\left(2\lvert\phi_{Fp}\rvert+V_{SB}\right)}{qN_A}}$	$X_{N\max} = \sqrt{\frac{2\epsilon_{\text{Si}}\left(2\lvert\phi_{Fn}\rvert+\lvert V_{SB}\rvert\right)}{qN_D}}$
$V_{TX} = V_{FB} + 2\lvert\phi_{Fp}\rvert + \gamma\sqrt{2\lvert\phi_{Fp}\rvert}$	$V_{TX} = V_{FB} - 2\lvert\phi_{Fn}\rvert - \gamma\sqrt{2\lvert\phi_{Fn}\rvert}$
$\gamma = \frac{\sqrt{2\epsilon_{\text{Si}}qN_A}}{C_{ox}}$	$\gamma = \frac{\sqrt{2\epsilon_{\text{Si}}qN_D}}{C_{ox}}$
$\phi_S(\text{at inversion}) = 2\lvert\phi_{Fp}\rvert + V_{SB}$	$\phi_S(\text{at inversion}) = -2\lvert\phi_{Fn}\rvert + V_{SB}$
$V_{TE} = V_{TX} + \gamma\left(\sqrt{V_{SB} + \lvert 2\phi_{Fp}\rvert} - \sqrt{\lvert 2\phi_{Fp}\rvert}\right)$	$V_{TE} = V_{TX} - \gamma\left(\sqrt{\lvert V_{SB}\rvert + \lvert 2\phi_{Fn}\rvert} - \sqrt{\lvert 2\phi_{Fn}\rvert}\right)$
$\phi_{Fp} < 0$	$\phi_{Fn} > 0$
$V_{SB} \geq 0$	$V_{SB} \leq 0$

expression in terms of δ yields

$$\frac{Q_B(L)}{Q_B(0)} = 1 + \frac{V_{D\text{sat}}}{v_{SB} + 2\Phi_{Fp}}$$

under the same conditions, clearly an overestimate for large saturation voltages. ∎

2.3   Second-order phenomena in MOS transistors

In the previous sections, we developed an analytical model for predicting device characteristics from a few basic physical parameters. Unfortunately, real devices have additional important dependencies. The effects we have examined so far are good for modeling processes circa 1975. In more modern processes, the current predicted by the equations in the last section can be off by a factor of two. So before we can apply our models to real circuits we must include these additional effects. Some of these effects can be modeled analytically, while others require an empirical approach.

New models for short and narrow channel devices are being developed daily and more accurate formulas for these effects will surely continue to be published. The material in this section is instructive for reading the future work. The important lessons in this section involve the physics of these second-order phenomena and the general way in which they affect the terminal characteristics of MOS devices. Well understood or not, these phenomena do influence the performance and even the functionality of VLSI systems. We will start with two of the most important effects, which are also the simplest.

Variations in length and width

When designing a circuit, the two parameters most directly under our control are the device width W and device length L. For several reasons, the final value of these parameters will not match the design values exactly. One obvious reason is the manufacturing tolerances. Chapter 4 discusses the effect of manufacturing tolerances on all the design parameters and develops the idea of worst-case models. In addition to the statistical variations, there are some basic physical reasons for a bias on the value of W and L.

Consider the case of device width W. Define

$$W \equiv W_{\text{drawn}} - 2\Delta W. \tag{2.129}\star$$

ΔW is the parameter for describing process width variations. We will assume that the variation is independent of the drawn width or any other basic process parameter (that is, it is an independent process variable). Given this definition, ΔW can be made to accommodate a number of different effects. In a modern silicon gate, locally oxidized, MOS process, ΔW is caused by two major phenomena.

The first effect is a physical movement of the [active area] device edge caused by the field oxidation. In a locally oxidized process, an active area mask is used to define a region to be blocked from field oxidation. Typically this is accomplished by defining a nitride patch prior to oxidation. This patch blocks the flow of oxygen from the surface to the silicon. At the edges of the patch, a diagonal source of oxygen is available. The result is commonly referred to as a "bird beak" for obvious reasons, which can be seen in Figs. 2.23 and 1.5. The exact shape of the bird beak is a function of the oxidation parameters, but in all cases it results in an effective decrease in the width of the transistors. In simple processes, the ΔW encroachment is of the same order as the thickness of the field oxide.

The second effect is seen when a field implant is used to control surface leakage. Because we have not yet addressed ion implantation, we will make a slight digression to examine those points pertinent to this discussion.

Ion implant technology provides a very convenient means for adjusting impurity concentrations in silicon. Basically the ion implanter is a linear accelerator for impurity ions. It consists of an ion source, an accelerating column (by static field adjustable in the range of 30k–300k volts), and a target area on which silicon wafers are placed. By adjusting

FIGURE 2.23   Device width definition.

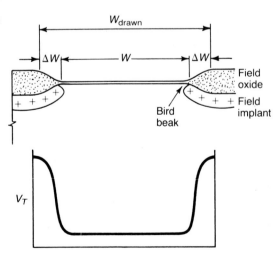

the acceleration voltage, the depth of penetration can be precisely controlled. By adjusting the beam current and implant time, the total impurity dose can be precisely controlled. Several techniques are used in various stages of processing to mask the implant into selected areas.

In MOS technology, oxide thickness is the primary parameter used to distinguish active (transistor) areas from inactive (field) areas. However, any interconnect passing over a thick field area defines the basic structure of a MOS transistor. Whether the device is active or not depends on its threshold. Recall from Eq. (2.75) that

$$V_{TE} = V_{FB} + 2\Phi_{Fp} + \frac{1}{C_{OX}}\sqrt{2\epsilon_{\text{Si}}qN_A(v_S + 2\Phi_{Fp})}. \qquad (\mathbf{2.130})$$

Since C_{OX} is inversely proportional to the oxide thickness, the thick field oxide will cause an increase in V_{TE}. The magnitude of the increase also depends on the substrate doping concentration N_A. Using a combination of thick oxide and ion implant control of N_A, process technologists maintain V_{TF} (thick field threshold) well above the operating voltages of most circuits.

The field implant normally occurs before field oxidation using the nitride patch as an implant mask. During the high-temperature field oxidation, the impurity atoms diffuse out into the silicon (and into the field oxide) due to the concentration gradient. The result, shown in Fig. 1.5, is a band of more heavily doped silicon that encroaches on the active transistor region. As N_A in the field region is made larger, not only does this encroachment increase but the side wall capacitance of diffused lines also increases and the breakdown voltage of the n+ diffused wires decreases. Thus there are trade-offs that limit the magnitude of V_{TF}.

The combination of graded oxide thickness and impurity profile results in a continuum of threshold voltages at the edge of the transistor, as illustrated in Fig. 2.23. If the curve is steep enough, we can approximate it as a cliff and use ΔW as a catchall parameter. As W_{drawn} becomes large with respect to ΔW, any errors in our approximation become third order. Note however that very narrow devices (in the range $W \leq 2\Delta W$) will behave very peculiarly.

We define the effective channel length L as

$$L \equiv L_{\text{drawn}} - 2L_{\text{diff}} - 2\Delta L_{\text{poly}}. \qquad (\mathbf{2.131})\star$$

L_{diff} represents the lateral diffusion of the drain and source junctions under the gate (typically 0.7 of the junction depth X_J). The difference between L_{drawn} and L_{final} for the poly gate is ΔL_{poly}:

$$\Delta L_{\text{poly}} = \frac{L_{\text{drawn}} - L_{\text{final}}}{2}. \qquad (\mathbf{2.132})$$

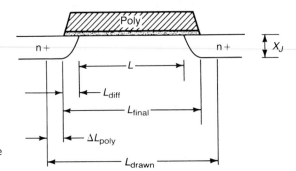

FIGURE 2.24 Device length definition.

These definitions are illustrated in Fig. 2.24. The reason we break down L into two components is that we need both parameters to calculate the gate capacitance correctly. The previous discussion of MOS capacitance included the transistor channel region only (in other words only the intrinsic part of the transistor). A real transistor has two additional fixed capacitors between gate and source, and gate and drain, due to the thin oxide overlap of the lateral diffusion area. These capacitors are calculated to be

$$C_{GSO} = WL_{\text{diff}}C_{OX} \qquad (2.133)$$

and

$$C_{GDO} = WL_{\text{diff}}C_{OX}, \qquad (2.134)$$

where C_{GSO} is the gate-to-source overlap capacitance and C_{GDO} is the gate-to-drain overlap capacitance. The corrected values of device width and length replace W and L in all device equations.

Mobility degradation

We have reached the end of the models that are used in hand calculations. Though there are exceptions, most of the models we will discuss in the remainder of the section are used principally as part of computer-aided design (CAD) programs. Unfortunately this does not mean we do not need to study these models. We must for several basic reasons. First, these effects can make a difference between a circuit that works and one that does not. This is not a rare occurrence; it happens all the time. To be able to do design one must understand where and how these effects will be important. One must also realize that CAD tools are not completely accurate. Even the best CAD tools have built-in assumptions of how they will be used. An inevitable outgrowth of creative effort is that one will want to apply the program to applications with characteristics never imagined by the program's authors. How does one know the program is doing the right thing? First, one must be an expert with the tool to get the most from it; second, one must know what the answer should be or whether the assumptions are violated.

When we introduced mobility, we assumed it had a constant value. We also indicated that this is an oversimplification. Although several underlying physical mechanisms modify mobility, when viewed from the device terminals there are three observable effects. The first is the characteristic mobility μ_o. This is a function of the fabrication process and depends on the interface charge density, substrate doping, and the wafer surface crystal orientation. The characteristic mobility μ_o is a strong function of device temperature and accounts for most of the temperature variation in MOS device characteristics. It is degraded in several ways. We partition these effects into vertical and horizontal field mobility degradation components which are usually characterized using one or more empirical parameters.

The velocity saturation of the carriers causes horizontal field mobility degradation. When the carriers are accelerated to a large horizontal velocity, the horizontal velocity saturates. This effect is usually characterized by the measured value of the saturated velocity $V_{\max}$. While μ_o for n-type silicon is much larger than μ_o for p-type silicon, both types of carriers saturate at about the same velocity. For high-performance MOS devices in which the carriers travel at saturated velocity through most of the channel, the performance of p-type transistors and n-type transistors are much more closely matched than for long channel devices. It is not that the p-type devices have improved, but that n-type devices have degraded.

To determine μ_o, one usually measures devices under conditions of low applied vertical and horizontal electric fields. From Eq. (2.108), K', which is proportional to mobility, can be determined under various operating conditions if given W and L. To determine μ, one must make an independent evaluation of C_{OX}. This can be done from capacitance measurement or by physical measurement of oxide thickness during processing.

We can express μ as

$$\mu = \mu_o(T)f_v(v_G, v_S, v_D)f_h(v_G, v_S, v_D), \tag{2.135}$$

where f_v is the vertical field degradation function, f_h is the horizontal field degradation function, and T is temperature. The variation of mobility with temperature can be characterized by the equation

$$\mu_o(T_2) = \mu_o(T_1)\left(\frac{T_2}{T_1}\right)^{-M}, \tag{2.136}$$

where T_1 and T_2 are absolute temperature. M is an empirical constant[13]

---

[13] SPICE uses $M = 1.5$, which is the correct value in the bulk, while Sabnis and Clemens [Sabnis 79] report $M = 2$ in the inversion layer of enhancement mode n-channel MOS transistors. Our experience has been $M \approx 1.6$.

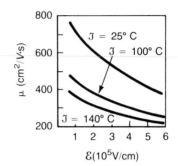

FIGURE 2.25   An n-channel inversion layer mobility as a function of the vertical electric field. [Sabnis 79, p. 21] (©1979 IEEE)

normally taken to be between about 1.5 and 2. Typical values for μ_o at room temperature are 600 cm^2/(V· s) for n-channel and 250 cm^2/(V· s) for p-channel MOS devices.

An equation often used for the vertical field dependency is

$$f_v = \begin{cases} 1 & \text{for } v_v \leq V_c \\ (V_c/v_v)^U & \text{for } v_v \geq V_c, \end{cases} \qquad (2.137)$$

where the vertical voltage v_v is taken as the average voltage drop between the gate and channel. In Eq. (2.137)

$$v_v = v_{GS} - V_{TE} - \frac{v_{DS}}{2} \qquad (2.138)$$

and V_c is the critical voltage, equal to $\mathcal{E}_c T_{OX}$. The critical electric field, $\mathcal{E}_c$, is on the order of 2×10^5 V/cm and U is approximately 0.25 for n-channel and 0.15 for p-channel devices. Measured values of μ as a function of the average vertical field and temperature are illustrated in Fig. 2.25 for an n-channel MOS transistor [Sabnis 79].

A simplified equation is sometimes used (notably in the SPICE2 circuit simulation program) to express f_v. It is

$$f_v = \frac{1}{1 + \theta(v_{GS} - V_{TE})}, \qquad (2.139)$$

where θ is an empirical constant. Although usually determined by measurement, θ can be estimated for a particular process using Eqs. (2.137) and (2.138).[14] Vertical field mobility degradation can lower $I_{D\text{sat}}$ by 25 to 50% in typical devices for large gate voltages.

---

[14] There are basically two ways of curve fitting to obtain device parameters. One way is to find a region of the curves in which a parameter is dominant, and to extract that parameter from data local to that region. V_T is an obvious example of a parameter that can be extracted this way. The extracted parameter is then assumed to be known and one may use it in the extraction of the next parameter. This technique generally produces parameter values that have obvious physical meaning. The other technique is to consider all of the parameters and all of the data simultaneously and to solve for the best fit of the model to the data [Ward 82]. This technique produces a better fit to the data, but at the sacrifice of some physical insight.

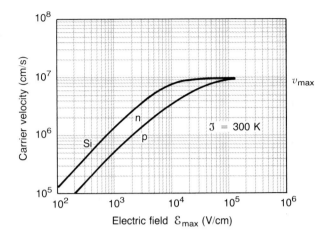

FIGURE 2.26   Carrier velocity versus lateral electric field for high resistivity silicon. The saturated velocity is fairly independent of doping. [Smith 80, p. 797]

The lateral field mobility degradation has a more significant impact on our device current equations than does the vertical field. Fig. 2.26 illustrates the velocity of both n- and p-type carriers as a function of applied electric field. As can be seen from the figure, one can characterize the velocity saturation either by the maximum carrier velocity $\mathcal{V}_{max}$ or by the critical field $\mathcal{E}_{max}$ at which velocity saturation occurs. The temperature dependence of the saturated velocity of electrons is shown in Fig. 2.27 [Jacoboni 77].

In Section 2.2, we saw that the basic long channel model assumptions lead to an infinite lateral electric field at the point of saturation. Qualitatively, we would expect that once the carriers reach $\mathcal{V}_{max}$, further increases in drain voltage will have no effect. From our previous analysis we know that the maximum field occurs at the drain. To compute the

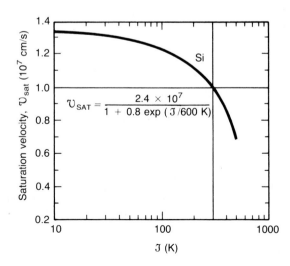

FIGURE 2.27   Saturated velocity of electrons versus temperature in silicon. [Jacoboni 77, p. 84]

point where saturation occurs, we use the continuity equation at the drain. We have

$$i_{DS} = -WQ_{N(z=L)}\mathcal{V}_{\max}. \tag{2.140}$$

Using Eq. (2.108) and (2.140) and the condition at the drain that $v_{DS} = \phi_{CS} = V_{Dsat}$, where V_{Dsat} is the new value due to velocity saturation, we obtain

$$\frac{\mu S C_{OX}}{2}\left(2(v_{GS} - V_{TE})V_{Dsat} - (1+\delta)V_{Dsat}^2\right) =$$
$$WC_{OX}\left(v_{GS} - V_{TE} - (1+\delta)V_{Dsat}\right)\mathcal{V}_{\max}. \tag{2.141}$$

Solving for V_{Dsat}, we obtain

$$V_{Dsat} = \frac{v_{GS} - V_{TE}}{1+\delta} + \frac{\mathcal{V}_{\max}L}{\mu} - \sqrt{\left(\frac{v_{GS} - V_{TE}}{1+\delta}\right)^2 + \left(\frac{\mathcal{V}_{\max}L}{\mu}\right)^2}. \tag{2.142}$$

The vertical field mobility degradation should be taken into account in Eq. (2.141), so $\mu_o f_v$ is used for μ. Note that when $\mathcal{V}_{\max}$ approaches infinity, the equation approaches our previous results in Eq. (2.110). Equation (2.142) gives a value for V_{Dsat} that is somewhat less than Eq. (2.110). To account for the gradual reduction in mobility that leads to the point of V_{Dsat}, we write the horizontal field degradation factor as

$$f_h = \begin{cases} \frac{1}{1+\mu v_{DS}/(\mathcal{V}_{\max}L)} & \text{for } v_{DS} \le V_{Dsat} \\[2mm] \frac{1}{1+\mu V_{Dsat}/(\mathcal{V}_{\max}L)} & \text{for } v_{DS} > V_{Dsat}. \end{cases} \tag{2.143}$$

Again, $\mu_o f_v$ is used for μ, where μ_o has the temperature dependence given in Eq. (2.136). For very short channel devices at high voltages, velocity saturation is a dominant effect. From Eq. (2.51), we have

$$i_{DS} = -Q_N W\mathcal{V}. \tag{2.144}$$

The quadratic characteristics of the current versus $v_{GS} - V_{TE}$ in saturation came from the fact that as $v_{GS} - V_{TE}$ increased, both Q_N and $\mathcal{V}$ increased. When velocity saturation dominates, $\mathcal{V}$ approximately equals $\mathcal{V}_{\max}$, and the only component of the current that varies with $v_{GS} - V_{TE}$ is Q_N. Therefore, for high lateral field devices, we have $I_{DS} \propto v_{GS} - V_{TE}$ rather than $(v_{GS} - V_{TE})^2$. Note also that the device length L drops out of the current equation.

One reliability problem of short channel devices, which is related to carrier velocity, is the hot electron effect. Hot electrons are electrons that are lucky enough to have enough energy to leave the channel region. For instance, electrons that cross the silicon/silicon dioxide barrier and get trapped in the oxide can cause long-term shifts in the flat-band voltages (through Q_{ot}). This effect is worst at low temperatures because annealing of the traps is minimized. Other hot electrons contribute to the substrate current. To reduce the hot electron effect, and also because the electric fields can become high enough to cause avalanche breakdown near the drain, graded diffusions (which limit the maximum field) are sometimes used on processes with channel lengths shorter than 2 μm [Ogura 82].

Channel length modulation

Thus far, although we have calculated V_{Dsat} several ways, all our models predict i_{DS} to be constant for $v_{DS} > V_{Dsat}$. Since this is not the case with real devices, our next topic will develop a modification to account for this effect.

Physically, a shortening of the channel length can account for the increase in current beyond V_{Dsat}. It is reasonable to assume that the depletion width between the end of the inversion region and the drain should increase as the voltage across it increases. It then follows that although the saturation voltage at the end of the inversion layer does not change, the distance across which the voltage is developed is reduced. This makes the driving electric field larger at all points in the inversion layer, thus increasing the current. In other words, we increase electrically the effective shape factor of the transistor. This is corroborated with the observation that everything else being equal, a short channel device shows more incremental conductance (di_{DS}/dv_{DS}) than a long one.

We represent the corrected value of the channel length by L'. We have

$$L' = L - \Lambda, \tag{2.145}$$

where Λ is the width of the depletion region at the drain. From Eq. (2.24), we have

$$\Lambda = \sqrt{\frac{2\epsilon_{Si}}{qN}(v_{DS} - V'_{Dsat})}, \tag{2.146}$$

where $v_{DS} - V'_{Dsat}$ is the voltage across the depletion region. Note that this effect is lower in devices where the substrate doping is higher.

Equation (2.146) predicts too much curvature in i_{DS} above I_{Dsat}. This is principally due to an insufficient accounting of velocity saturation. Another problem is that Eq. (2.146) does not account for the electrons in the depletion region, which causes N to be effectively larger than predicted from the doping level. A more accurate model is presented in Example 2.5.

When analytical solutions to a problem are very difficult, another approach is to find an empirical formula that matches the behavior. An empirical relation for the drain current in saturation is

$$i'_D = i_D \left(1 + \frac{v_{DS} - V_{Dsat}}{V_A + V_{Dsat}}\right), \qquad (2.147)$$

where

$$V_A \approx \mathcal{E}_A L \sqrt{\frac{N_A}{N_T}} \qquad (2.148)$$

and where $\mathcal{E}_A = 5$ V/μm and $N_T = 10^{15}$ cm^{-3} [Merkel 77].

There are two main limitations of the empirical approach. The first is that it gives no physical insight, and the second is that it is difficult to determine where the formula is valid. Nevertheless, practical models tend to combine aspects of both analytical and empirical modeling.

■■■■■■■■■■■■■■ EXAMPLE 2.5

Circuit simulation models for channel length modulation

While the discussion in Section 2.3 was adequate for obtaining some intuition for channel length modulation, we acknowledged that it was not very accurate. An accurate model is needed for circuit simulation. Examine what is done to model this phenomenon in SPICE2G. Baum and Beneking [Baum 70] have examined the value of Λ more closely. By solving the one-dimensional Poisson equation, they obtained

$$\Lambda = \sqrt{\left(\frac{\mathcal{E}_{\max}}{2a}\right)^2 + \frac{v_{DS} - V'_{Dsat}}{a}} - \frac{\mathcal{E}_{\max}}{2a},$$

where

$$a \equiv \frac{qN_A}{2\epsilon_{Si}}.$$

This is more accurate than Eq. (2.146), but it still has two problems. First, Λ must be known to compute V'_{Dsat}, and V'_{Dsat} must be known to compute Λ. That is, we have a pair of simultaneous nonlinear equations in the inner loop of a circuit simulation program. Since this is too costly to solve by iteration, and a closed form solution is impossible, we approximate V'_{Dsat} by the value of V_{Dsat} calculated with $\Lambda = 0$. This turns out to be fairly accurate since $|V'_{Dsat} - V_{Dsat}|$ usually does not exceed a few hundred millivolts. A second problem is that the real geometries at the drain are two-dimensional, but a one-dimensional approximation was used to derive the model equations. Therefore a couple of empirical factors are introduced to allow curve fitting. In SPICE, the substitutions [Vladimirescu 80] are

$$N_A \rightarrow \text{NEFF} N_A$$

and

$$v_{DS} - V'_{Dsat} \rightarrow \text{KAPPA}(v_{DS} - V_{Dsat}).$$

As we can see, modeling for circuit simulation can be quite involved because of the interplay of efficiency, accuracy, and our limited ability to obtain analytical expressions for important phenomena.   ■

Threshold variations due to short channel effects

All analysis thus far has taken a one-dimensional approach, either in the vertical or lateral dimension, implicitly implying that all field effects are orthogonal. This assumption is poor at the edges of the channel region. When the channel is very short or very narrow, these effects can be a significant component in the total device behavior. In this section we examine some of the effects that can be attributed to very short channel lengths.

One of the most important short channel effects is an observed reduction in V_{TE} with respect to channel length. This is primarily attributed to field lines from charges in the depletion region that is under the channel terminating laterally in the source and drain regions, instead of vertically at the surface. The simplest attack on this problem approximates the two-dimensional field problem with an equivalent one-dimensional problem. It does this by subtracting out that portion of the charge associated with the lateral field. We let

$$Q'_B = Q_B - Q_L \qquad (2.149)$$

where Q_L is the equivalent laterally connected charge. Q'_B is a function of the junction depth and shape as well as the voltage on the source and drain. In reviewing the methods used to calculate Q'_B, keep in mind that this entire approach is an approximation to the real two- (or three-) dimensional problem. Also note that as L gets large, $Q_B \gg Q_L$, and Q'_B approaches Q_B as expected. T_{OX} also plays an important role in translating Q_B to V_{TE}.

The first to take this approach was Yau [Yau 74], who approximated the shape of the depletion area as a trapezoid with the result that

$$Q'_B = Q_B \left(1 - \left(\sqrt{1 + \frac{2X_{Pmax}}{X_J}} - 1 \right) \frac{X_J}{L} \right), \qquad (2.150)$$

where Q_B is the charge computed in the conventional manner, X_J is the junction depth, and X_{Pmax} is the width of the nominal channel-to-substrate depletion region. These geometries are illustrated in Fig. 2.28. Under conditions of large V_{SB}, the trapezoid can become a triangle, making these equations no longer valid. Note from this equation that

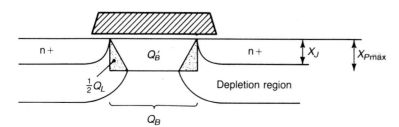

FIGURE 2.28 Charge model of short-channel effects. Some of the depletion layer charge Q_L under the channel is assumed to belong to the source or drain.

the junction depth X_J plays the important role of normalizing the various other distances. Thus whether or not a device is short channel depends not so much on the actual length of the channel, but rather on the ratio of that length to the junction depth. A 5 μm device with phosphorus source/drain diffusions can evidence more severe short channel characteristics than a 3 μm device with arsenic sources and drains. This is because the arsenic junction depth can be more shallow than that of the phosphorus.

Dang [Dang 79] modified Yau's model by assuming a cylindrical shape for the drain and source connected depletion regions. This more accurate expression for the threshold is used in the level-three model of SPICE2.

Recall that the body factor γ is proportional to the charge under the channel. We have

$$Q_B = -\gamma C_{OX}\sqrt{v_{SB} + 2\Phi_{Fp}}. \tag{2.151}$$

As this charge decreases due to short channel effects, the effective body factor also decreases. We define γ' such that

$$\gamma' = \gamma \frac{Q'_B}{Q_B}. \tag{2.152}$$

Thus, everything else being equal, short channel devices have a lower body effect than long channel devices.

Drain-induced barrier lowering and punch-through

Our analysis of short and narrow channel effects on the threshold voltage assumed that both the source and drain were at the same potential. If separate potentials are assumed, the results become very complex. The drain can act as a second gate and modulate the amount of charge Q_B under the channel. This causes an apparent shift in the threshold voltage as a function of drain voltage as the drain modulates the potential barrier in the channel. We introduce an empirical parameter σ, which operates directly on V_T and is given by

$$V'_{TE} = V_{TE} - \sigma v_{DS}, \tag{2.153}$$

where σ is the coefficient of static feedback. Ratnakumar, Meindl, and Scharfetter [Ratnakumar 81] propose that

$$\sigma = \frac{6T_{OX}}{X_{P\max}} \exp\left(\frac{-\pi L}{4X_{P\max}}\right), \tag{2.154}$$

where, in the case of an implanted transistor, we interpret $X_{P\max}$ as the depletion region depth under the channel.[15]

Drain-induced barrier lowering is a very strong effect for short channel devices operated near threshold. For such a device operating in

---

[15] SPICE2G, level 3, uses $\sigma \propto L^{-3}$.

saturation, σ is the principal factor determining the output conductance. We will see, in Chapter 4, two circuit examples in which drain-induced barrier lowering is extremely important. Drain-induced barrier lowering limits the maximum gain of a MOS inverter. It also plays a central role in determining the noise margins of ratioed and precharged circuits.

When v_{DS} is large enough, punch-through can occur. In the punch-through condition a typically subsurface[16] current, minimally controlled by the gate, flows between source and drain. By the same mechanism by which the gate potential inverts the silicon surface, the drain voltage can lower the potential barrier below the channel. When the barrier becomes lowered sufficiently, current can flow through this path. Because the problem is two-dimensional in nature, it is very difficult to obtain an exact equation for the punch-through voltage V_{PT}.

The punch-through voltage increases roughly linearly with the doping density and quadratically with L. As the magnitude of V_{BB} increases, the subsurface punch-through voltage also increases[17] because the potential barrier is raised. Another technique for increasing the subsurface punch-through voltage is to do a deep implant beneath the channel. This will have the undesirable side effect of causing the body effect to increase with v_{SB}.

Subthreshold conduction

In the development of the device equations, transistor turn-on was postulated to occur abruptly when the source-to-channel barrier was reduced by $2\Phi_{Fp}$ or, equivalently, when the channel potential was reduced by $v_{SB} + 2\Phi_{Fp}$. This condition was used to define V_{TE}. In real devices, the turn-on condition is not perfectly sharp. Rather, devices display an exponential i_{DS} versus v_{GS} behavior below V_{TE}.

When the channel potential is between $v_{SB}+\Phi_{Fp}$ and $v_{SB}+2\Phi_{Fp}$ the transistor is in the weak inversion regime. Two phenomena influence the drain-to-source current. The first is that the number of electrons in the channel that have enough energy for conduction is roughly proportional to $\exp(q\phi_S/kT)$. These electrons are the ones that can traverse the source-to-channel potential barrier and contribute to i_{DS}. Thus the exponential I–V characteristics in this regime are analogous to the exponential I–V characteristics of a forward-biased p–n diode.

The second important consideration is the leverage that v_{GS} has on ϕ_S. As v_{GS} increases toward threshold, the ability of v_{GS} to control ϕ_S decreases dramatically. Thus we would expect i_{DS} to have a milder dependence on v_{GS} than on ϕ_S. To first order, we can expect the drain current to be proportional to $\exp(qv_{GS}/nkT)$, where n captures the fact that a 10 mV change in v_{GS} does not result in a 10 mV change in ϕ_S.

---

[16] Punch-through can actually occur at any depth where the potentials are right.
[17] The folklore stating that punch-through occurs when the depletion regions touch is wrong.

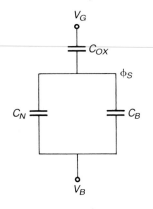

FIGURE 2.29 Capacitance model of a transistor in weak inversion.

As usual, a more careful accounting of the charge will improve the model. Following Swanson and Meindl [Swanson 72], we develop a small signal model for the region. Let C_N represent the capacitance of the inversion layer electrons, and let C_B represent the capacitance of the bulk depletion layer. We have

$$C_N = \frac{\partial Q_N}{\partial \phi_S} \qquad (2.155)$$

and

$$C_B = \frac{\partial Q_B}{\partial \phi_S}. \qquad (2.156)$$

A circuit model using these capacitances is illustrated in Fig. 2.29. In the weak inversion regime, C_N is much less than C_B. From this model, we develop a new threshold voltage that defines the interface between subthreshold conduction and strong inversion. We have

$$V'_{TE} = V_{TE} + \frac{nkT}{q}, \qquad (2.157)$$

where

$$n = \frac{C_B + C_{OX}}{C_{OX}}. \qquad (2.158)$$

Note that this threshold is not a true threshold in a circuit design sense. It acts more as a point at which the behaviors of the strong and weak inversion regimes are pieced together. There can be significant conduction at V'_{TE}. Although n has a physical basis, it is usually determined from experimental data.

It has been shown that for $v_{GS} < V'_{TE}$, the current can be expressed in the form given by

$$i_{DS} = I_0 \exp\left(\frac{q(v_{GS} - V'_{TE})}{nkT}\right). \qquad (2.159)$$

A useful figure of merit for design purposes is $dV_{GS}/d\log I_{DS}$, which predicts the reduction in gate voltage required to reduce the leakage current by one order of magnitude. We have

$$\log I_{DS} = \log I_0 + \frac{q\log e}{nkT}\left(V_{GS} - V_{TE} - \frac{nkT}{q}\right). \qquad (2.160)$$

Differentiating, we obtain

$$d(\log I_{DS}) = \frac{q\log e}{nkT} dV_{GS}. \qquad (2.161)$$

We define α as the subthreshold figure of merit. We have

$$\alpha \equiv \frac{dV_{GS}}{d(\log I_{DS})} = \frac{nkT}{q \log e} = 2.3 \frac{nkT}{q}. \qquad (2.162)$$

Typical values of n range from 1.0 to 2.5. Because kT/q is 0.0257 V at 25° C, α is typically in the range of 60 to 150 mV/decade at room temperature and 75 to 185 mV/decade at 100° C. It can take many hundreds of millivolts to shut off a transistor to the point where the drain-to-source leakage current is of the same order of magnitude as the reverse-biased p–n junction leakage current between the active area and the substrate.

Subthreshold currents are exacerbated by short channel behavior. Brews et al. [Brews 80] have discovered, through extensive experimentation and two-dimensional simulation, that the minimum device length (in μm) for long channel subthreshold behavior is given by the empirical relation

$$L_{\min} = 0.4 \left(X_J T_{OX} \left(X_S + X_D \right)^2 \right)^{1/2}, \qquad (2.163)$$

where X_J is the junction depth (in μm), T_{OX} is measured in angstroms, and X_S and X_D are X_{PN} for the source and drain depletion depths, respectively (in μm). When L is less than $L_{\min}$, the subthreshold currents increase because the effective length of the channel is decreased by the encroaching source and drain depletion regions. Fig. 2.30

FIGURE 2.30 Subthreshold leakage characteristics of transistors with a variety of channel lengths: (a) $N_A = 10^{15}$cm^{-3}; (b) $N_A = 10^{14}$cm^{-3}. [Sze 81, p. 470] (S. M. Sze, Physics of Semiconductor Devices, ©1981. Reprinted by permission of John Wiley & Sons, Inc.)

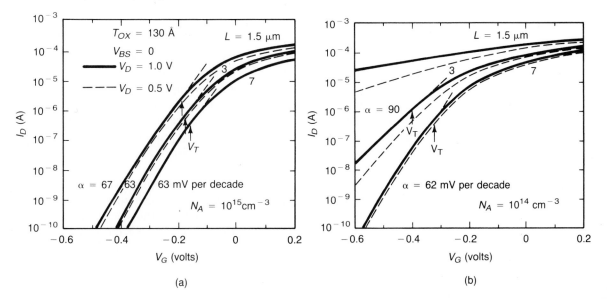

(a)

(b)

illustrates the subthreshold characteristic of two different nMOS devices. The effects of substrate doping and channel length are evident.

Channels fabricated with ion implantation

Ion implantation provides a means of controlling device thresholds that is nearly independent of all other device parameters. The threshold voltage and the body effect parameter depend, among other things, on the substrate doping. Ion implantation technology enables one to selectively control N_A. One can vary the value of N_A in different areas of the chip and control the doping profile with high precision. Since ion implantation is a low-temperature process, many materials, including photoresist, can be used for masking the implant.

The accelerating potential and the characteristics of the impurity species control the impurity penetration depth. The distribution profile can be modeled by a Gaussian, but can be modified by a subsequent high-temperature diffusion if desired. The profile is characterized by the depth of the Gaussian peak R_p and its standard deviation ΔR_p, also called the "straggle." We define the dose N as the total number of atoms per square centimeter implanted into the surface of the wafer. Measured values of R_p and ΔR_p have been tabulated for silicon and silicon dioxide layers as a function of implant energy and atomic species [Gibbons 75]. We define the peak value of the doping concentration, which occurs at R_p, as N_p and the value of the background concentration as N_b. Integrating the impurity distribution under the Gaussian we have

$$N = \sqrt{2\pi}\Delta R_p(N_p - N_b). \tag{2.164}$$

After implantation, high temperature annealing is required. It allows the lattice to relax, repairs damage caused by the ion beam, and allows the impurities to be accommodated into lattice sites.

Implanted transistors are very common in VLSI circuits. A light implant of acceptor impurities is common for nMOS enhancement mode devices, while a heavier implant of donor impurities is common in nMOS depletion mode devices. Implants are generally used for both types of CMOS devices. The implant of the device in the well is often used to lower the effective doping concentration rather than to raise it. For instance, an arsenic implant might be used to slightly compensate the heavy boron well doping.

We first examine the case of an enhancement mode nMOS device. Specifically, we examine the case where we wish to raise the device threshold by adding a light boron implant to the silicon surface. This results in a p-type Gaussian-shaped region sitting on a more lightly doped uniform background, as illustrated in Fig. 2.31. Counting the charges correctly is much more important than perfectly modeling their

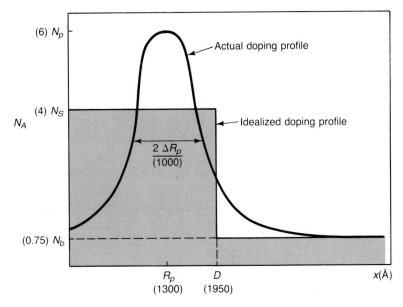

FIGURE 2.31   Implanted enhancement-mode transistor. (Numbers in parentheses are from example.)

distribution. This is because the way that charges get translated into terminal voltages is through integration in Poisson's equation. This integration filters out shape details.[18] Therefore, to simplify the analysis, we approximate the Gaussian by a box. This may seem overly crude, but remember that the tails of a Gaussian fall off very steeply. Fig. 2.31 also illustrates the idealized view of the implanted device. The problem reduces to one in which there are only two different doping levels under the channel.

An important point is that the implant does nothing to affect the flat-band voltage. There are many ways to understand this. By Kirchhoff's voltage laws applied to chemical potentials, we know that if we place any conducting material between the silicon surface and the point deep in the p-type body, which we have taken to be ground, the material's effects will exactly cancel out. V_{FB} will not change. To look at it another way, assume that the silicon surface is more heavily doped than our original reference point. In this case we might conclude that we have incorrectly computed the flat-band voltage because V_{FB} depends on ϕ_F. But this is only true if we measure the flat-band voltage referenced to this more heavily doped material. On the other hand, this apparent error is exactly the voltage encountered when traveling from the more heavily doped material to the the more lightly doped material, where the potential is being referenced. Again everything cancels out.

---

[18] On the other hand, incremental parameters do depend on the detailed shape of the doping profiles.

As the gate bias on the transistor moves from V_{FB} up toward V_T, the region under the silicon/silicon dioxide interface becomes depleted, starting at the surface. For notational convenience, we label the equivalent box surface doping as N_S and its corresponding body effect parameter as γ_S. The box extends into the body to a distance D. Deep in the bulk, the parameters are N_A and γ_A. A critical issue is whether or not D is greater than the depletion width $X_{P\max}$. We can compute the source voltage V_{SB} at which $X_{P\max} = D$. We have

$$V_{SB} = \frac{qN_S D^2}{2\epsilon_{Si}} - 2\Phi_{Fps},\tag{2.165}$$

where Φ_{Fps} is Φ_{Fp} for the surface doping N_S.[19] If V_{SB} is positive, then V_{TX} depends only on N_S. In this case, we have

$$V_{TX} = V_{FB} + 2\Phi_{ps} + \gamma_S\sqrt{2\Phi_{Fps}}.\tag{2.166}$$

Thus we find the fairly intuitive result that if all of the charges that matter are in the box, then the device behaves exactly as if the entire substrate was doped to the N_S level. Nevertheless, the implanted device, even in this case, has two significant advantages. First, since the doping is high only under the transistors, the junction capacitance beneath the source/drain interconnect will be significantly lower than if the substrate was more heavily doped. In addition, in source follower type configurations,[20] the body effect that is seen after the source voltage has exceeded the critical value from Eq. (2.165) will be much lower. This is highly desirable in these situations because the gain of the voltage follower becomes more nearly unity. A plot of the body effect parameter γ versus source voltage is shown in Fig. 2.32. In most implanted devices, D is greater than $X_{P\max}$, as was assumed here. If this is not true, one must break the problem into two parts. Until the depletion region reaches D, we have the problem we have just been investigating. But beyond that point, one must use N_A rather than N_S.

In nMOS technologies that make use of a nonzero V_{BB} voltage, the edge of the depletion region can be made to remain where γ is low. Other situations are possible. For instance, in a typical 2 μm p-well CMOS process, the body effect of the p-channel device goes down as the source moves away from V_{DD}. However, the body effect of the nMOS device in the well goes up as the source voltage increases.

Unfortunately, most circuit simulators, including SPICE2, do not allow for a nonconstant value of γ. To accommodate this deficiency, one must juggle the available parameters to get a reasonable fit to the actual

FIGURE 2.32 Body effect versus source voltage of an implanted device.

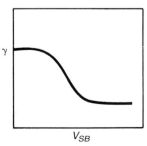

---

[19] It doesn't matter whether we use $2\Phi_{Fps}$ or $\Phi_{Fps} + \Phi_{Fp}$ to define the onset of inversion. At what surface concentration is the surface inverted?

[20] In these configurations, the source voltage is not fixed.

device curves. If D approximately equals $X_{P\max}$, then over the vast majority of the operating range γ takes on the value γ_A, corresponding to N_A. If we use this value without any other adjustments, the simulator will calculate the wrong value for V_{TX}, which is a critical circuit design parameter. To account for this, we can adjust the value of V_{FB} to get a proper V_{TX} result. Juggling these parameters can result in a decent overall fit to the current characteristics. Note that, while V_{FB} is not a critical circuit design parameter, it does have an effect on the MOS capacitance model. This should be considered when modeling circuits that have a high sensitivity to capacitances around V_{FB}.

A very important use of implantation is in the fabrication of depletion-mode MOS devices. With n-channel technology, either arsenic or phosphorus is implanted in the channel region to form a depletion device.

Even more than in the enhancement mode transistor case, major simplifications must be made to enable analysis. To calculate V_{TX}, one must find the gate voltage at which $Q_N = 0$. Just prior to inversion, the channel will consist of a depletion region of negative boron ions in the bulk, with a surface layer of positive arsenic or phosphorus ions. Given that the positive ions are close to the surface, they are exactly analogous to Q_f, as analyzed in Section 2.1. Unlike the enhancement-mode case, we will look for a shift in the flat-band voltage. Actually, in both cases the term "flat-band" is somewhat of an anachronism. The bands cannot be totally flattened for either device. But from a circuits viewpoint, the flat-band voltage is the maximum voltage at which the gate-to-body capacitance is WLC_{OX}. Adding different dopings of p-type material to a p-type substrate does not confuse this issue, but adding n-type dopants does. The reason is that if we used our former definition of flat-band, then at the point where there is no excess charge at the surface, the surface is n-type; not only does a depletion region exist, changing the capacitance, but the transistor is on as well. In a simple analysis it is therefore better to assume that the flat-band voltage of interest does indeed change. The change is given by

$$\Delta V_{FB} = -\frac{Q_S}{C_{OX}}, \qquad (2.167)$$

where Q_S is the implanted surface charge density qN_S.

This analysis is fairly accurate for shallow implants. Deep implants require a much more complicated analysis, including a mobility model that varies with distance from the surface. SPICE has no such feature, so we must juggle available parameters to fit characteristics. Usually this involves adjusting γ in addition to V_{FB}. Note that unlike enhancement implants, there is an actual shift in V_{FB} although it has a more complex behavior than our analysis predicts.

The turning off of a depletion-mode transistor with a deep implant is difficult. As the gate voltage is lowered toward threshold, a depletion

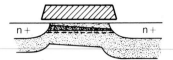

FIGURE 2.33 Depletion device under conditions of reverse gate bias. A channel exists below the surface.

region forms at the top of the silicon surface. Thus the depletion-mode device actually has a channel that is below the surface, sandwiched on top and bottom by depletion regions as illustrated in Fig. 2.33. The effective gate dielectric now consists of the original T_{OX} plus the top depletion layer.[21] This lowers the effectiveness of the gate electrode in controlling Q_N. Depletion devices can be very leaky unless a negative body voltage is used to push the channel to the surface.

■ EXAMPLE 2.6

Threshold calculations for an implanted transistor

This example of an implanted transistor is drawn from the classical paper on scaling theory by Dennard et al. [Dennard 74]. In this example, $T_{OX} = 350$ Å and the substrate is 2 Ω·cm p-type material. Looking up this resistivity in Fig. 2.19, we discover that 2 Ω·cm corresponds to $N_A = 7.5 \times 10^{15}$ cm^{-3}. The device is implanted with boron at 40 keV. The dose is 6.7×10^{11} cm^{-2}. From the range tables we discover that $R_p = 1300$ Å and $\Delta R_p = 500$ Å; 3% of the implant is estimated to be lost in the oxide. We make a rough approximation that the equivalent box depth D is 1950 Å, and compute an equivalent concentration N_S of 4×10^{16} cm^{-3}. Using this doping level, we compute that at inversion $\phi_S = 0.768$ volts. The applied voltage at which the depletion width X_P equals D is $V_S = 0.155$ volts. Note that under zero bias conditions the depletion region is entirely contained in the implant. Assuming $V_{FB} = -1$ V, we can use the standard V_{TX} equation to compute the zero bias threshold. Because the charges Q_B that are important to this calculation are only the ones in the implanted region, we use the γ that corresponds to N_S. We have $\gamma = 1.167$ V$^{1/2}$ inside the implant region and $V_{TX} = 0.79$ volts. We can compare this to the case without the implant, in which ϕ_S at inversion would be 0.681 volts, $\gamma = 0.505$ V$^{1/2}$, and $V_{TX} = 0.097$ volts. Figures 2.31 and 2.34 illustrate the relevant geometries.

Now consider the depletion transistor case with arsenic replacing boron as the implanted species. A dose of 1.2×10^{12} atoms/cm^2 leads to a threshold of -1.9 volts. Remember to use N_A. The body effect is assumed to be unchanged by the implant.   ■

FIGURE 2.34 Implanted enhancement-mode MOS transistor.

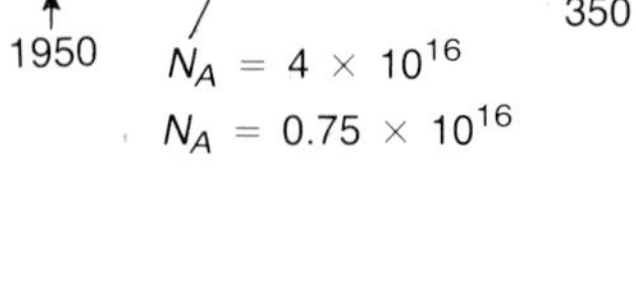

Threshold variations due to narrow channel effects

Recall that we developed a ΔW correction factor to help account for the effects of the bird beak and field implant encroachment on the active area region. This correction factor is not sufficient to model accurately the operations of very narrow devices. ("Narrow" is a function of the

[21] With very thin oxides, the effective T_{OX} is larger than the physical T_{OX} for even unimplanted devices because of the finite thickness of the inversion layer.

process vertical scaling but, roughly, $W < 4~\mu$m would be considered narrow for a 2 μm process.)

The most pronounced narrow width effect is a positive change in the device threshold voltage for narrow width devices. This effect is most prominent in processes with light substrate concentrations and heavy field implants. In this case, the change is primarily due to lateral diffusion of the field implant into the channel region, increasing the effective background concentration. We will concentrate on this mechanism since it predominates in most processes.

Other mechanisms affecting narrow width thresholds are the tapering of the oxide due to the bird beak (which also affects the device gain) and simple fringing of the vertical fields at the edges of the device. The vertical fringing effect also acts to increase the threshold voltage.

The limiting case for lateral diffusion occurs when the field implant extends completely across the active area. In this case, one must add together the field implant, the device threshold adjust implant, and the substrate concentration in order to calculate the worst-case threshold change due to the field implant. We know of no analytical solution to this problem that takes into account the nonuniform concentrations and oxide thicknesses that will exist in the real device.

Yang and Chatterjee [Yang 82] have empirical models for the dependence of γ and V_{TX} on W. They suggest that

$$\gamma' = \gamma + \frac{\gamma_w}{W} \qquad (2.168)$$

and

$$V'_{TX} = V_{TX} + \frac{V_{TXW}}{\sqrt{W}}. \qquad (2.169)$$

Equation (2.169) is suggested by a model that approximates linearly the transition of Q_B from its ideal value under the gate oxide to its value in the field region. Note that as W decreases, γ and V_T increase, and as L decreases, γ and V_T decrease.

■■■■■■■■■■■ EXAMPLE 2.7

Narrow channel effects
Consider the enhancement-device implant in Example 2.6. Typical field implant concentrations are in the range of 3×10^{16} cm^{-3}, thus for a narrow enhancement-mode device, the effective doping concentration N_S consists of three components: a background concentration of 7.5×10^{15}, a device implant of 4×10^{16}, and a field implant of 3×10^{16}. This will cause γ to increase from 1.167 V$^{1/2}$ to 1.49 V$^{1/2}$ in cases in which the field implant diffuses completely across the transistor. This in turn increases the threshold by 238 mV, thus bringing it to 1.073 V. Note also that the field implant extends quite deeply into the substrate since it diffuses

during the long field oxide growth. Thus the higher value of γ will tend to remain constant, even with bias on the source.

For the depletion device, γ increases from 0.505 to 1.13 and V_{TX} increases by 550 mV, going from -1.9 V to -1.35 V. The increase is greater because the effective starting concentration is lower. ∎

Latchup in bulk CMOS processes

Inherent in bulk CMOS processes is a four-layer p+/n/p/n+ path running from V_{DD} to ground. This pnpn path contains three p–n junctions that become a pair of cross-coupled pnp and npn bipolar transistors, shown in Fig. 2.35 for an n-well process. These two transistors form a positive feedback path with each transistor configured so that its collector drives the base of the other, as seen in Fig. 2.36.

Throughout this section, an n-well process will be used for the description. The description for a p-well process corresponds by having supply polarities and n- and p-types reversed.

The pnp transistor M_W is a vertical transistor, which can have values for the current gain, beta, ranging from 50 to several hundred. Transistor M_S is a lateral npn transistor for which the substrate is the base, and the well is the collector. Beta for this transistor is usually in the 0.5 to 10 range, and will fall off to zero exponentially as the spacing between the n+ emitter and the well is increased.

FIGURE 2.35 Latch path inherent in bulk CMOS. (Gates and drains of the n- and p-channel transistors, whose sources are shown, have been omitted for clarity. For some process designs, surface p+ substrate contacts may be optional if packaging methods permit a contact to the backside of the chip.)

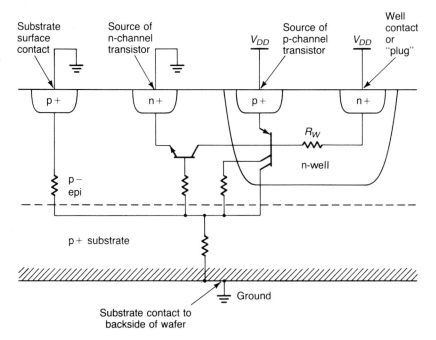

The resistors R_W and R_S, shown in Fig. 2.36, shunt the base and emitter of their respective transistors. R_W is typically in the 1k to 20k ohm range. The substrate resistor R_S strongly depends on whether the substrate is p— throughout, or is made with a p— epi layer grown on a highly doped p+ substrate. In this p—/p+ epi case, the substrate acts as a ground plane, resulting in a value of R_S as low as a few ohms. In the non-epi case, the R_S may typically be as high as 500–700 ohms.

If the current gains of the two transistors and the values of resistors R_W and R_S are sufficiently high, then the circuit can be triggered by an external disturbance into a regenerative condition where each transistor keeps driving the other. The currents in both transistors will increase until they self-limit or until they result in the destruction of the chip or its bonding leads. This situation is called latchup.

If a latched pnpn path does not result in the chip's destruction, it will almost always cause malfunctioning of the logic circuitry. To restore a chip to an unlatched state, the chip would generally have to be powered down (V_{DD} set to zero) and then repowered.

A single latched pnpn path may not in itself be destructive for some process designs or layout configurations since the currents may self-limit to a safe level. However, the disturbance from a latched pair may cause other nearby pnpn paths, which may otherwise have been "safe," to become latched; thus latchup may propagate over large portions of the chip to result in failure in what would otherwise have been a nondestructive situation.

Because the consequences of latchup are so serious (either failure or logic malfunction), it is important to understand which circuit disturbances can potentially cause latchup in order to incorporate into the technology measures that will completely eliminate the possibility of latchup. We list some of the common disturbances and mechanisms that can trigger latchup.

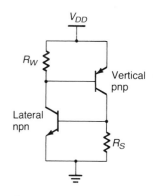

FIGURE 2.36   Basic latch loop of path in Fig. 2.35.

Disturbances through signal I/O pins

- **Transmission-line reflections or ringing:** If ringing or reflections on an output bus or transmission line pull the p+ source or drain of a p-channel transistor more than approximately 0.6 V above V_{DD}, then the vertical pnp transistor that has that p+ region as its emitter will turn on. If that transistor is part of a susceptible latch loop, latching could result. A similar situation holds for the forcing of an n+ source or drain of an n-channel transistor below ground.

- **"Hot plug in" of an unpowered circuit board or module:** If an unpowered circuit board is plugged into a system that is already under power, such as might be done when replacing a defective module, then the signal pins on the chip can see voltages that are more than 0.6 V higher than V_{DD}, which is rising more slowly to its full value. When the chip comes up to its full power, parts of the chip could be latched unless proper suppressive measures are taken or unless, of course, one decides not to permit such "hot" plugging in of modules.

- **Electrostatic discharge (ESD):** The spark from an electrostatic discharge (usually originating from a person's body in low-humidity environments) can trigger latchup. This pulse of charge will usually enter an input pad to be clamped by the ESD protection circuit. The device of the clamping circuit will inject minority carriers into either the substrate or the well, potentially triggering another pnpn path nearby. If the ESD protection circuit is not sufficiently guarded, the circuit will be latched after the charge pulse has passed. The design of ESD protective circuitry is discussed in Section 5.7.

Disturbances generated internal to the chip or through V_{DD} or ground

- V_{DD} slewing during initial turn on can cause capacitive displacement currents due to the well capacitance, in both the well and the substrate. The effect of these displacement currents is worse in the well than in the substrate. This is because the displacement charge must all go out through the small number of well plugs (ohmic connections to V_{DD}) for the well, whereas for the substrate, the charge has a relatively lower resistance path to ground. Lateral voltage drops across the well could result in turn on of one of the vertical bipolar transistors in the well, and consequent latchup.

- Sudden transients on power or ground buses due to internal switching of a large number of drivers might be able to turn on one of the bipolar transistors of a pnpn latch path, resulting in latchup. Whether this can happen in a specific situation should be checked carefully.

- Leakage currents across the well junction could cause large enough lateral voltage drops to turn on the vertical bipolar transistor in the well.

- Radiation—X-ray, cosmic, or alpha rays, for example—could generate hole-electron pairs in both the well and substrate regions that the particle passes through, which could trigger potential pnpn latch paths.

- Hot-electron induced minority carrier currents in the substrate around n-channel devices could, after being collected by a nearby well, turn on a vertical transistor in that well and thereby trigger a pnpn path. Also, majority carrier substrate currents could cause the potential in the substrate to be increased sufficiently to turn on nearby n+ active (source/drain) regions. The use of an epi/p+, as discussed below, should totally eliminate the majority-carrier effect.

Precise necessary and sufficient conditions for latching will not be derived here. There are several different treatments of this subject in the literature [Estreich 82, Gregory 73, Schroeder 80]. A necessary condition for latching that will serve our needs in developing measures for preventing latchup is the following:

> For a pnp/npn pair of transistors to be triggered by some disturbance into a latched state, the triggering disturbance must bias the devices into a condition where the incremental loop gain exceeds unity.

Measures to prevent latchup, either by process design or by appropriate layout or guarding techniques, are usually aimed at preventing the incremental loop gain from exceeding unity for a given disturbance. This can be accomplished either by blocking or attenuating the disturbance itself, or by attenuating the incremental loop gain so that it will not exceed unity even when the transistors are turned on by the disturbance.

Since latchup has such serious consequences, steps must be taken in design of either the underlying process structure or the lateral circuit geometries to suppress latchup so that it will never occur in a given system environment. We next discuss four common suppression measures.

Beta Reduction: In order for the npn/pnp pair of transistors to latch, the product of their (bipolar) betas must exceed unity. This condition follows from, and is weaker than, the main necessary condition stated above. It does suggest a strategy for preventing latchup: use device structures (by suitable vertical process design and horizontal spacings) that will ensure that the product of the two betas will not exceed unity. The problem with this strategy is that for the lateral spacings and vertical device structures encountered in VLSI processes, the beta product is always much greater than unity; to increase lateral spacings enough to reduce the product sufficiently will result in unsatisfactorily low circuit densities. Hence, beta reduction is, at best, only a partial strategy for most processes.

Resistive Shunting: The most effective strategy, sometimes used in conjunction with beta reduction, is that of making the resistors R_S and R_W small enough to ensure that they will bypass would-be base current to the supply rails. By making one or both of these resistors small enough, the base-emitter voltage on at least one of the transistors can be kept safely below the 0.6 volt level at which it begins to turn on. If even one of the transistors is prevented from turning on, the loop gain can never exceed unity, and latchup will have been suppressed. R_W is reduced by using enough well plugs (ohmic connections from the well to V_{DD}, usually through metal). To achieve a much lower value of R_W at some sacrifice in area, an n+ (for an n-well process) collar can be placed around the periphery of a well. This greatly reduces R_W (from several kilohms to less than 100 ohms) and is most useful for I/O circuits, where latching disturbances are most severe.

To reduce R_S in non-epi processes, frequent plugs (ohmic contacts connected to ground) may be used. An appropriate spatial frequency of these contacts is generally ensured by design rule specification. The well and substrate plug spatial positioning is one of the more difficult tasks for CAD tools to manage.

The use of an epitaxial layer on a highly doped substrate strongly lowers R_S, and is considered one of the most effective latchup prevention strategies. Although the use of epi alone (that is, without other measures) will not completely prevent latchup, its use with other measures makes control of latchup a much easier task.

The use of epi should be satisfactory for latchup control in processes with feature sizes at least down to the 1.5 μm region. To go below that (and have the n+ to p+ spacing scale proportionate), other preventive steps must be taken.

If a backside contact is made to the wafer with an epi/highly doped substrate structure, then rarely, if at all, will surface contacts be needed to ground the surface of the epi. Some packaging schemes, however, preclude the use of a backside contact, in which case surface contacts are needed. This is still less of a burden than it would be with a non-epi process. A topside ring of surface contacts (plugs) is almost as effective as a backside contact.

Some strategies involve also lowering the well resistance R_W by making well structures that have a lower sheet resistivity. One approach uses a deep implant to achieve a retrograde well doping profile. This lowers the sheet resistivity of the well, while maintaining an appropriate surface concentration for the transistors at the surface. The use of a buried layer for accomplishing the same goal has also been described [Estreich 80, Walczyk 83]; buried layer approaches are usually considered in conjunction with a bipolar device capability in the same CMOS process.

Other Process Structure Strategies: For processes with feature sizes at sub-micron levels, more drastic steps must be taken. These might include trench isolation [Rung 82], where an oxide trench deep enough to reach the low-resistivity substrate is used to isolate the well from the substrate region; silicon on insulator, where the pnpn path is totally eliminated since the body regions are either on the surface of an insulator or insulated from each other by some insulator (other than a reverse-biased p–n junction); or other dielectric isolation schemes that also eliminate the pnpn path.

Prevention at the Circuit and Layout Design Level: The preceding measures deal with latchup suppression through appropriate design of the basic process structures. Additional measures, applied at the circuit design and layout level, must also be taken. Some of these are now discussed.

The most critical lateral spacing is that between n+ and p+ active regions (sources or drains, as opposed to ohmic plugs). Since the well junction also passes between any pair of active p+ and n+ regions, reducing this spacing is important for achieving the high circuit densities needed for VLSI. In a twin-well type of process, both well edges will pass between any two active p+ and n+ regions. Making the n+ to p+ spacing too small can cause well punch-through. This can cause high leakage currents that can in turn cause latchup. Also, the lateral bipolar transistor's beta will increase sharply as the spacing is reduced. These factors are too complex and process-dependent to discuss in more detail here, but this area is a critical design consideration. As a rule of thumb, the n+ to p+ spacing can be made to be on the order of the epi thickness.

Strong latchup suppression measures can be taken at the "floor plan" stage of a layout. As was indicated above, the most severe latchup causing disturbances enter through the signal I/O pads. Therefore heavy guarding is done around these I/O regions. This guarding consists of placing well collars (already discussed) around most well structures, and placing collector "walls" (consisting of a stripe of well) around any structures that could potentially emit minority carriers into the substrate/epi region.

If the process does not use an epi layer, then stripes of substrate "plug" (p+ diffusion contacted to metal that is then connected to ground) are placed around the outside of well structures, causing majority carriers coming from the well to be shunted to ground.

If the conservative suppression measures used at the chip I/O were used throughout the entire chip, it would be impossible to attain circuit densities high enough for VLSI. Therefore, an interior region in which active p+ to n+ spacings are reduced and guarding is much relaxed is established in order to obtain higher circuit density.

In this situation, numerous pnpn latch paths exist that could easily be latched were a disturbance permitted to reach the latch path. Latching is prevented by guarding the border of this interior region so well that no disturbances can reach the interior through the substrate. This is illustrated schematically in Fig. 2.37. Devices in the exterior region are heavily guarded to attenuate the relatively severe disturbances that can enter through the signal I/O pads. The interior is shielded by a collector guard, as seen in Fig. 2.38. Therefore the only disturbances that enter are those on V_{DD} and ground, and those that enter in the form of radiation.

Susceptible exterior circuits are ESD "diodes" (which can inject minority carriers into the substrate or well if the signal overshoots V_{DD} or ground), output devices (especially those in a tri-state off state, wherein an output drain is floating), and transmission gates (which can also have floating drains directly attached to I/O pads).

The collector guard structure works as follows. Minority electrons emitted by an n+ emitter are intercepted by the n-well collector. The

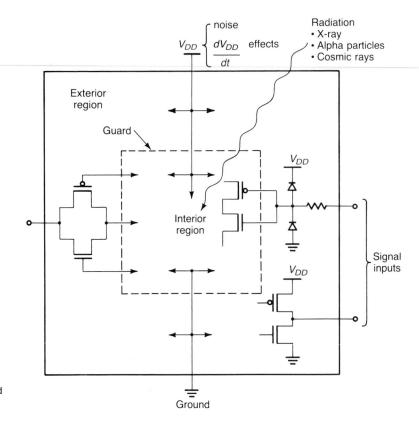

FIGURE 2.37 Schematic view of chip showing partitioning into interior and exterior regions.

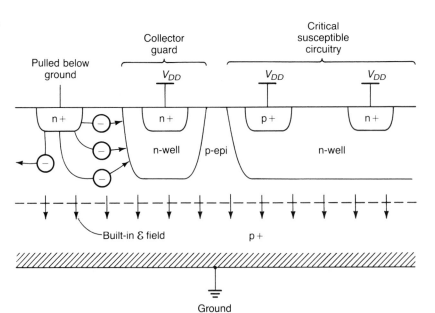

FIGURE 2.38 Collector guard structure.

built-in electric field, resulting from the p+ to p− gradient between substrate and epi, imposes a force on the minority electrons that tends to keep them in the epi region where they can be collected by the well guard. Attenuation of electrons that would otherwise reach critical susceptible circuitry is extremely high, about a million to one [Troutman 83].

Perfect guard structures around the interior regions may totally prevent disturbances from entering through the substrate. However, some disturbances can enter in other ways, including disturbances on V_{DD} and radiation effects.

Obviously, a V_{DD} power bus must enter an interior region in order to power the circuitry in the region. During turn on, as V_{DD} slews up to its full value from zero, capacitive displacement currents proportional to dV_{DD}/dt will flow in both the wells and the substrate. Usually, the effect of these displacement currents is stronger in well regions.

In this case, latchup is triggered when the displacement current in a well causes a voltage drop large enough (about 0.6 volts) to turn on one of the well vertical transistors, which subsequently would turn on a lateral transistor outside the well, resulting in regeneration and latchup. Latchup by this means is prevented by inserting into the well a large enough quantity of well "plugs" (ohmic contacts) that connect, in an n-well case, to V_{DD}. The resistance from any point to V_{DD} is never high enough to result in voltage drops that would turn on a base-emitter junction.

Radiation from external sources, such as from X-rays, alpha particles, or cosmic rays, can penetrate the chip and generate hole–electron pairs along the trajectory of the particle. In some situations this disturbance can be significant enough to cause latchup. An example of such exposure is a CMOS chip in a calculator that passes through X-ray inspection at an airport security check. In this case, prevention of latchup would be obtained by frequent enough well plugging and an appropriate substrate structure (either frequent enough substrate plugging, or the use of epi). Since the duration of the current spike from an alpha particle hit is extremely short, even if the magnitude of the disturbance is theoretically large enough to cause latchup, it is believed that the shortness of the pulse will prevent the loop from becoming unstable.

Second-order effects summary

ΔW	**Width Encroachment.** Smaller W lowers I_D, raises V_T and γ.
ΔL_{poly}	**Change in poly length.** Shorter L lowers V_T, γ, and raises I_D, σ.
X_J	**Junction depth** of source and drain diffusions. Larger X_J increases short channel effects.

L_{diff}	**Lateral Diffusion.** Increased L_{DIFF} shortens L, increases C_{GSO} and C_{DSO}.
C_{GSO}, C_{DSO}	**Gate overlap capacitances.**
μ_o	**Intrinsic mobility.** I_D scales with μ_o, μ_o decreases with increasing T.
f_v	**Vertical field mobility scaling factor.** Decreases μ and I_D for large V_{GS}.
f_h	**Horizontal field mobility scaling factor.** Lowers μ for large V_{DS}.
$\mathcal{V}_{\text{max}}$	**Maximum carrier velocity.** Causes I_D to saturate at lower currents. Foils square law dependence of $I_{D\text{sat}}$ on $V_{GS} - V_T$. Affects n-channel I_D more than p-channel I_D.
Λ	**Effective channel length change** due to channel length modulation. Effective L becomes shorter as V_{DS} increases past V_{DSAT}. Causes I_{DS} to increase with V_{DS} in saturation.
σ	**Coefficient of static feedback.** Accounts for the lowering of V_T as V_{DS} increases. As L decreases σ increases . Most important at low V_{GS}.
α	**Subthreshold figure of merit.** Measure of the voltage required to decrease I_D by factor of 10. As T increases, α increases. It can be very large for short channel devices.
γs	**Surface body effect parameter.** The body effect parameter can depend on V_{SB} if threshold adjust implants are used.
T	**Temperature.** As temperature increases, μ_o and V_T decrease, and α increases.
R	**Resistance.** Source resistance increases the effective V_T as I_D increases.
V_M	**Junction breakdown voltage.** Increases as substrate doping decreases.
V_{PT}	**Punch through voltage.** Increases as substrate doping and $|V_{SB}|$ increase.

TABLE 2.2 Threshold voltage response to increasing the value of the parameter

Parameter	V_{TN}	$|V_{TP}|$
T_{OX}	up	up
N_A or N_D	up	up
L	up	up
X_J	down	down
W	down	down
T	down	down
$|V_{DS}|$	down	down
$|V_{SB}|$	up	up

2.4 The modeling of interconnect

The majority of the area on a typical integrated circuit is comprised, not of active devices, but of interconnect. There can be several meters of wire on a typical LSI chip and it is important to understand its properties. In this section, we look at both the static and dynamic properties of the conductors used in integrated circuits.

Resistance

The resistance of a uniform wire of width W, thickness T, and length L is given by

$$R = \rho \frac{L}{TW}, \tag{2.170}$$

where ρ is the resistivity of the material. Since the thickness of the wires on an integrated circuit is reasonably uniform over the chip, we can absorb T into the resistivity and define a new variable ρ_S, called the sheet resistance, which has dimensions of ohms. We have

$$R = \rho_S \frac{L}{W}. \tag{2.171}\star$$

Typical sheet resistances for aluminum, heavily doped polycrystalline silicon, heavily n+ silicon and heavily p+ silicon are 0.04, 40, 15, and 25 $\Omega/\square$, respectively. The notation $\Omega/\square$ is read as "ohms per square." These somewhat strange units are explained by observing from Eq. (2.171) that all square geometries have the same resistance between opposing edges. The sheet resistance concept is particularly well suited to integrated circuits because it easily handles resistance nonuniformities in the thickness dimension such as would occur for implanted or diffused conductors. We have

$$\rho_S = \int_0^T \rho(x)dx. \tag{2.172}$$

The results above are valid for rectangular wires. For complicated shapes, one must solve Laplace's equation to determine the resistance. This degree of accuracy is almost never needed on digital integrated circuits because an integrated circuit design must, after all, be robust to processing variations. One important special shape we will consider is the right angle bend. On right angle bends the resistance is less than that computed by using the center line as the length. It is more accurate to count the corner square as 1/2 square rather than one square. For an extensive compilation of shape correction factors, see Hall [Hall 68,

Horowitz 83]. It is important to calculate the resistance from the effective geometries rather than from the mask specifications. For right angle bends in which one of the legs is wider than the other, Bain [Bain 82] has developed an approximation for the resistance of the corner rectangle. We have

$$R_{\text{corner}} = (0.46 + 0.1a)\,\rho_S, \tag{2.173}$$

where a is the ratio of the wide to narrow widths.

The three most common interconnect materials used on MOS integrated circuits are aluminum, heavily doped polycrystalline silicon, and heavily doped single-crystal silicon. Poly and single-crystal silicon have sufficiently dissimilar characteristics to justify their separate classification. Other materials sometimes used in MOS integrated circuits include gold, tungsten, WSi_2, and $MoSi$. The desirability of a particular conductor depends on its reliability, stability, conductivity, and compatibility with (possibly high-temperature) processing.

The resistivity of aluminum, as with most metals, is dominated at room temperature by collisions of the conduction electrons with lattice phonons [Kittel 66]. Because the number of phonons increases with temperature, the resistivity of metal increases with temperature. The resistivity of pure aluminum at room temperature is approximately 3×10^{-6} $\Omega \cdot$cm. In practical integrated circuits, the aluminum is alloyed with 1–2% silicon to lower the tendency of the aluminum to spike into the silicon substrate. A few percent copper is also often added to increase its resistance to metal migration. The motivations for introducing these impurities are discussed in the next section.

The interconnect paths in the single-crystal silicon substrate are formed by diffusing or implanting P, As, or B into the surface of the wafer at very high concentrations. The specific resistivity of the path depends on the dopant and the doping profile. Very high levels of impurity atoms cause the semiconductor to become degenerately doped. The Fermi level is moved into either the conduction band (for n++) or valence band (for p++), and the density of states distribution is smeared out. This causes the semiconductor to act very much like a metal alloy. The resistivity at these high doping levels is limited by both phonon and ionized impurity scattering. The effect of ionized impurity scattering is to lower the room temperature mobility of electrons in silicon from its intrinsic value of about 1350 $cm^2/(V\cdot s)$ to about 100 $cm^2/(V\cdot s)$ for an ionized donor density of 10^{19} cm^{-3}. The mobility of heavily doped n+ silicon decreases with increasing temperature.

The use of silicides to help lower the resistance of interconnect on the silicon and polysilicon levels of a chip is becoming more widespread. Typical silicides have resistances on the order of 2 $\Omega/\square$ and are placed on top of the higher resistance interconnect. For short hops, the contact resistance can dominate, making the silicide less effective in these applications.

The resistivity of polycrystalline silicon varies with the doping concentration in a way that is very different from that of single-crystal material. The resistance starts very high for low doping and decreases only slowly until the dopants saturate the grain boundaries where they are being trapped. After this point the resistance drops abruptly. Thin films of polysilicon evidence a basaltic texture in which the grains grow with a columnar shape from the underlying material. The average size of the grains increases with processing temperature. Inside the grains the conductivity is assumed to be that of single-crystal silicon. The interface between grains consists of layers of disordered atoms acting as a transitional region between grains of different crystallographic orientations. Several models for the complex physics at these interfaces have been proposed [Lu 81, Seto 75, Colinge 81]. A major effect appears to be the trapping of carriers caused by defects resulting from disordered or incomplete atomic bonding. This trapping not only lowers the number of free carriers available for conduction, but also creates a potential barrier between grains, which further impedes the flow of current. For low values of doping concentration, the resistivity of polysilicon decreases with temperature. But at doping levels on the order of 10^{20} cm^{-3}, the sign of the temperature coefficient reverses and becomes the same sign as for single-crystal silicon. Because of the potential barrier, the current–voltage characteristics of poly is actually slightly nonlinear.

Electromigration and power dissipation in conductors

Two current-related forces act on lattice atoms in conductors [Black 69, Chua 81, Heurle 71]. The electric field exerts a force on activated positive ions that is in the direction opposite to the direction of the flow of electrons. Because of the shielding effect of the electrons, this effect is quite small. The other force on the lattice atoms is momentum exchange with the electron wind being driven by the electric field. These collisions exert a force in the same direction as the electron flow. The effect of this wind is to move vacancies upstream where they condense to form voids, while ions move downstream, forming crystals, whiskers, and hillocks. This causes two problems. First, the coalescing of voids can cause a wire to open circuit. This process is unstable because a collection of voids increases the current density in that region, which in turn causes material to be removed at an even greater rate. A second mode of failure is due to the collection of material near the positive terminal. This extra material can short wires together. The median time to failure of a wire has the form

$$\mathrm{MTF} = \frac{\mathrm{Area}}{\beta J^M} \exp(\phi/kT), \qquad (2.174)$$

where ϕ and β are constants of the materials and crystal structure. M is a constant between 1 and 4 and is usually taken as 2. J is the current density. Symmetric ac current waveforms cause negligible metal migration. For pulsed circuits the rms value of current is used. The kT term relates to the number of activated ions. Thus the electromigration effect is exacerbated at high temperatures and can be nearly neglected at liquid nitrogen temperatures. For typical integrated circuits using aluminum interconnect, the current densities must be kept below 1 mA/μm^2 for reliable operation. We translate between the 1 mA/μm^2 constraint to a specification in terms of mA/μm of wire width by looking down on the chip and absorbing the thickness of the metal into the specification. The 1 mA/μm^2 number contains a safety factor for the degradation of the metal height over steps. Contact cuts from metal to poly or diffusion are particularly bad in this regard. The specification of current allowed to flow through a contact is usually given in terms of mA/μm of contact edge because the thin metal at the peripheral edge acts as the limiting factor for metal migration. There seems to be a correlation between the resistance to metal migration of a conductor and its melting temperature. Qualitatively, this is because the more firmly fixed in the lattice the atoms are, the higher the melting point and the higher their resistance to migration effects.

A second major electromigration effect on integrated circuits is due to the fact that silicon dissolves into aluminum. The dissolved silicon is easily moved by the electron wind away from the negative contact causing the aluminum to entrench into the vanishing silicon contact. Eventually, this will cause the aluminum connection to a source region to spike through to the substrate. This effect will occur despite a presolution of silicon in the aluminum. Special contact barrier metals such as Pt are sometimes placed between the Al and Si to prevent spiking and also to lower the contact resistance.

The power dissipated in a conductor per unit length is $\rho_S I^2/W$. For the metal and polysilicon layers, this heat must be removed from the conductor through the SiO$_2$ insulating layer. SiO$_2$ is a poor thermal conductor, roughly 100 times worse than silicon. The temperature rise of the conductor is approximately

$$\Delta T = \frac{\rho_S I^2 H}{W^2 \kappa_{\text{thermal}-\text{SiO}_2}} \tag{2.175}$$

where H is the height of the wire above the substrate and κ_{thermal} is the thermal conductivity. For example, if $H = 0.6$ μm, $W = 2$ μm, $\rho_S = 40$ $\Omega/\square$, and $I = 25$ mA, we have $\Delta T = 2700°$ C. Polyimides, which are organic dielectrics, have thermal resistivities ten times higher than that of silicon dioxide.

Interconnect capacitance

The capacitance and sometimes the resistance of the signal interconnect network dominate the phenomena contributing to logic gate delays. Not only are parallel plate contributions to the capacitance important but the capacitance due to fringing electric fields also constitutes a major contribution to the capacitance of narrow wires. For a 2 μm process the fringing capacitance of a metal or poly wire can be as high as the parallel plate capacitance.

The capacitance between a conducting wire and a ground plane or another wire is determined by solving Laplace's equation. Rather than solve this electric field problem, we will examine two special cases that have well known solutions. The first special case is that of a parallel plate capacitor. The capacitance is

$$C_{\text{plate}} = \frac{\epsilon W L}{H}.$$ (2.176)

The capacitance depends linearly on the area WL and the dielectric constant ϵ, and inversely on the distance H between the two plates. C_{plate} closely approximates the true capacitance when $W, L \gg H$.

The second special case occurs for narrow wires; that is, for $L \gg H$ and $W \leq H$. A simple expression for the capacitance can be obtained if we approximate the wire as a long cylinder. In this case, we have

$$C_{\text{cylinder}} = \frac{2\pi \epsilon L}{\ln\left(1 + \frac{2H}{W}\left(1 + \sqrt{1 + \frac{W}{H}}\right)\right)}.$$ (2.177)

Note that the capacitance of the cylinder depends only logarithmically on W and H. We have interpreted W as the diameter of the cylinder. An examination of these two special cases illustrates an important implication for the scaling of interconnect wires. For a given insulator height H, narrowing the wires through better lithographic and etching techniques will linearly decrease the capacitance only until the width of the wire becomes of the same order as its height. After that, further improvements are achieved at only a logarithmic rate.

Typical interconnect wires have a capacitance that is some combination of the parallel plate capacitance one calculates using the parallel plate formula and the fringing or edge capacitance. We can use Eqs. (2.176) and (2.177) to approximate the total capacitance of a wire. We define

$$C_{\text{wire}} = C_{\text{plate}} + 2C_{\text{edge}},$$ (2.178)

where C_{plate} is given by Eq. (2.176). C_{edge} can be approximated as

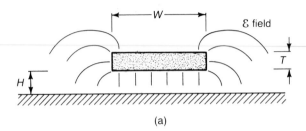

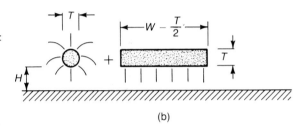

FIGURE 2.39   Fringing capacitance of narrow wires: (a) fringing electric field lines of a wire; (b) model of wire as a parallel plate capacitor plus a cylindrical conductor over a ground plane.

is shown in Fig. 2.39. We replace the sides of the wire with half-cylinders. The edge capacitance is then approximated by half the cylinder capacitance minus a small portion of the parallel plate solution we have replaced. We have

$$C_{\text{edge}} \approx \epsilon L \left\{ \frac{\pi}{\ln\left(1 + \frac{2H}{T}\left(1 + \sqrt{1 + \frac{T}{H}}\right)\right)} - \frac{T}{4H} \right\}. \qquad (2.179)$$

For $H = T$ and $\epsilon = 3.9\epsilon_0$, we have $C_{\text{edge}} \approx 0.044$ fF/μm. This number is fairly independent of T.

For wires that are as far apart as they are wide, the coupling capacitance between a wire and its two adjacent neighbors becomes equal to its capacitance to the ground plane when W/H is between 1 and 2 [Dang 81]. Lewis gives formulas for the coupling capacitances of VLSI wires [Lewis 84]. This interwire capacitance is very important in noise considerations.

Equation (2.179) is valid for polysilicon and metal lines. Interconnect that is diffused or implanted into the single-crystal silicon evidences an even stronger edge capacitance. This is because the field implant causes the depletion regions, which play the role of an insulator, to neck down at the sides of the wires. This depletion side-wall capacitance often dominates the interconnect capacitance connecting neighboring gates.

The depletion layer capacitance is nonlinear and has the form

$$C = \frac{C_J W L}{\left(1 + \frac{V}{\phi_J}\right)^{M_J}} + \frac{2 C_{JSW}(L+W)}{\left(1 + \frac{V}{\phi_{JSW}}\right)^{M_{JSW}}}, \qquad (\mathbf{2.180})$$

where V is the wire-to-substrate voltage, ϕ is the junction built-in potential, and M is a constant, roughly between 0.3 for a graded junction and 0.5 for an abrupt junction. The capacitance between crossing or stacked lines can be approximated by the techniques already presented, or more accurate results can be obtained by appealing to the microwave stripline and microstrip literature.

■■■■■■■■■■■■■■■ EXAMPLE 2.8

Counting parasitics of an nMOS superbuffer
The development of a circuit model of the superbuffer shown in Fig. 1.32 involves counting capacitance squares. In this example we will ignore resistance effects. The extracted schematic is illustrated in Fig. 2.40. Nodes are labeled both with their names and SPICE node numbers. Ground is always node 0.

Transistor parameters including gate, overlap, and source/drain capacitances will be part of the transistor models. We will look at these in a moment. Let us first examine the four fixed capacitances CIN, CK, CIO, and CLOAD. CIN has several components including both metal over field area and fringing capacitance C_{mf}, poly over field capacitance C_{polyf}, and metal over poly capacitance C_{mpoly}. A value of 0.044 fF/μm was assumed for the edge capacitance. All other values were taken from Appendix C, set to their maximum (worst-case) value. Note that the metal over the poly at the input contact was not counted because the two wires are shorted together. The metal over poly capacitance

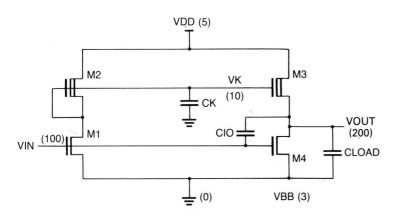

FIGURE 2.40   Model of a superbuffer. SPICE node numbers are shown in parentheses.

TABLE 2.3   Fixed capacitances

TYPE	CIN	CK	CIO	CLOAD
$C_{\mathrm{mf}}(\mathrm{area})(\mu\mathrm{m}^2/\mathrm{pF})$	200/0.0084	0/0	0/0	164/0.0069
$C_{\mathrm{mf}}(\mathrm{edge})(\mu\mathrm{m}/\mathrm{pF})$	100/0.0044	0/0	0/0	98/0.0043
$C_{\mathrm{mpoly}}(\mu\mathrm{m}^2/\mathrm{pF})$	16/0.0018	48/0.0053	8/0.0009	0/0
$C_{\mathrm{polyf}}(\mu\mathrm{m}^2/\mathrm{pF})$	48/0.0034	44/0.0031	0/0	0/0
$C_{\mathrm{mdiff}}(\mu\mathrm{m}^2/\mathrm{pF})$	0/0	0/0	0/0	56/0.0062
$C_{\mathrm{total}}(\mathrm{pF})$	0.0180	0.0084	0.0009	0.0174

occurs in the region over the pulldown transistors. Where the input poly crosses the output metal line there is a Miller capacitance CIO. The poly fringing capacitance was ignored. Table 2.3 lists the values of the fixed capacitances.

The transistor parameters extracted from the layout must be adjusted for ΔL_{poly} and ΔW before being entered into SPICE. The worst-case ΔL_{poly} is -0.125 μm and the worst-case ΔW is 0.9 μm (long narrow transistors). The diffusion capacitance is entered as areas (AD and AS) and perimeters (PD and PS) in SPICE. We count the length of the diffusion that abuts the transistor as diffusion edge (either PS or PD). This is not quite correct, but it is a reasonable approximation. We also do nothing special regarding the buried contact, despite the fact that it reaches more deeply into the substrate than do the source/drain junctions because of the P, rather than As, dopant. Again, this is a fair approximation. We are being a little optimistic on the buried contact and a little pessimistic on the transistor edge capacitance.

How we partition the diffusion capacitance between the drain of the pulldown and the source of the pullup is quite arbitrary, as long as it is all accounted for somewhere. Obviously, we need not calculate the ground and V_{DD} capacitances. The final transistor parameters are illustrated in Table 2.4. Table 2.5 shows a SPICE input file for the superbuffer.

TABLE 2.4   Transistor parameters

	M1	M2	M3	M4
Drawn W/L	13/2	6/6	15/6	23/2
SPICE W/L	11.2/2.25	4.2/6.25	13.2/6.25	21.2/2.25
AD	61	0	0	0
AS	0	0	110	0
PD	42	0	0	0
PS	0	0	62	0

TABLE 2.5   SPICE input file

```
SPICE DECK for Fig. 2.40
*
.MODEL NENHS NMOS LEVEL=3 RSH=0 TOX=330E-10 LD=0.19E-6 XJ=0.27E-6
+ VMAX=13E4 ETA=0.25 KAPPA=0.5 NSUB=5E14 UO=650 THETA=0.1
+ VTO=0.946 CGSO=2.43E-10 CGDO=2.43E-10 CJ=6.9E-5 CJSW=3.3E-10
+ PB=0.7 MJ=0.5 MJSW=0.3 NFS=1E10
*
.MODEL NDEPS NMOS LEVEL=3 RSH=0 TOX=330E-10 LD=0.19E-6 XJ=0.27E-6
+ VMAX=13E4 ETA=0.25 KAPPA=0.5 NSUB=50E14 UO=650 THETA=0.04
+ VTO=-2.078 CGSO=2.43E-10 CGDO=2.43E-10 CJ=6.9E-5 CJSW=3.3E-10
+ PB=0.7 MJ=0.5 MJSW=0.3 NFS=1E10
*
******************************************************************
VIN 100 0 PULSE(5 0 ONS 2NS)
VDD 5 0 5
VBB 3 0 -3
.OPTIONS DEFL=2.25E-6
.WIDTH IN=75 OUT=75
* node 10 is VK
CIN 100 0 0.018PF
CK 10 0 0.0084PF
CIO 10 200 0.0009PF
CLOAD 200 0 0.0174PF
CHACK 200 0 1PF
* CHACK is an additional output load to make things more realistic
M1 10 100 0 3 NENHS W=11.2U AD=61P PD=42U
M2 5 10 10 3 NDEPS W=4.2U L=6.25U
M3 5 10 200 3 NDEPS W=13.2U L=6.25U AS=110P PS=62U
M4 200 100 0 3 NENHS W=21.2U
.TRAN 0.3NS 15NS
.PLOT TRAN V(200) V(10) V(100) (0, 6)
.END
```

RC lines

RC ladder and tree structures occur very often in MOS integrated circuits. An *RC* ladder is illustrated in Fig. 2.41. The long poly or diffusion wire is an example, as is a string of pass transistors. These networks can be nonlinear, distributed, or both. In this section, we examine the dynamic behavior of linear *RC* ladder networks. By examining this specialized, but still important, case we will obtain both

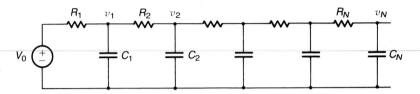

FIGURE 2.41 An RC ladder network (v_N is the output node).

an introduction to the literature as well as a good understanding of the more salient features of this class of circuits. Heat flow and diffusion can also be modeled by this type of network.

We begin by examining the voltage step response of a ladder of N resistors and N capacitors, as shown in Fig. 2.41. The input voltage rises from zero to V_0. We may write the voltage on the Nth node as

$$v_N(t) = V_0 - \sum_{j=1}^{N} C_j \frac{dv_j}{dt} \sum_{i=1}^{j} R_i, \qquad (2.181)$$

where $C_j \frac{dv_j}{dt}$ represents the current flowing to ground through the capacitor connected to the jth node, and $\sum_{i=1}^{j} R_i$ represents the voltage drop due to that current. The outer summation is the superposition of all the voltage drops. Rearranging Eq. (2.181) and integrating both sides from time $t = 0$ to time $t = T_D$, we have

$$\int_0^{T_D} (V_0 - v_N(t)) \, dt = \sum_{j=1}^{N} \sum_{i=1}^{j} R_i C_j v_j. \qquad (2.182)$$

T_D is the time it takes the voltage on the Nth node to reach v_N. We have assumed that the initial conditions on the node voltages were zero. It can be shown [Protonotarios 67] that

$$v_N \le v_j \le V_0 \quad \text{for all } j. \qquad (2.183)$$

That is, the voltages on the jth node are always less than or equal to v_{j-1}. Let us define a time constant T_E such that

$$T_E \equiv \sum_{j=1}^{N} C_j \sum_{i=1}^{j} R_i. \qquad (2.184)$$

From (2.183), we have

$$v_N T_E \le \sum_{j=1}^{N} \sum_{i=1}^{j} R_i C_j v_j \le V_0 T_E. \qquad (2.185)$$

The function $v_N(t)$ is a monotonically increasing function of time.

Therefore we also have

$$T_D(V_0 - v_N) \leq \int_0^{T_D} (V_0 - v_N(t))\, dt \leq T_D V_0. \qquad (2.186)$$

For a given output voltage v_N, T_D is the time it takes the output to achieve that voltage. Using Eqs. (2.182), (2.185), and (2.186), we can obtain bounds on that time. We have

$$\frac{v_N}{V_0} T_E \leq T_D \leq \frac{V_0}{V_0 - v_N} T_E. \qquad (2.187)$$

It is helpful to look at a special case. If all of the resistors have the value R_0 and all of the capacitors C_0, then

$$T_E = R_0 C_0 \frac{N(N+1)}{2}. \qquad (2.188)$$

We can see that the delay of a uniform RC ladder network increases quadratically with the number of elements. Physically, this means that the delay of an RC line increases quadratically with length. Another way of conceptualizing this effect is to realize that the delay increases as the total capacitance times the total resistance times an approximation factor, which depends on v_N/V_0 and the exact circuit configuration. Both the total capacitance and total resistance increase with line length. As $N \to \infty$, $T_E \to R_L C_L/2$, where R_L is the total line resistance and C_L is the total line capacitance. Rubinstein et al. have developed bounds that are tighter than those in Eq. (2.187). They have shown that, in addition to Eq. (2.187), it is also true [Rubinstein 83] that

$$T_D \geq T_E - T_R \left(1 + \ln \left(\frac{T_E}{T_R} \left(1 - \frac{v_N}{V_0} \right) \right) \right), \qquad (2.189)$$

$$T_D \leq \frac{T_E}{1 - \frac{v_N}{V_0}} - T_R \qquad (2.190)$$

and

$$T_D \leq T_E \left(1 - \ln \left(1 - \frac{v_N}{V_0} \right) \right) - T_R, \qquad (2.191)$$

where

$$T_R \equiv \frac{\sum_{j=1}^N C_j \left(\sum_{i=1}^j R_i \right)^2}{\sum_{j=1}^N R_j}. \qquad (2.192)$$

All of these bounds can be manipulated to give bounds on the voltage v_N for a given delay T_D.

■ EXAMPLE 2.9

SPICE model of a ROM

A ROM array has 256 words of 256 bits each. The word line is poly and the bit line is metal. The ground is distributed in n+ and is refreshed (tied to a metal ground line) every 15 words. Figure 2.42 illustrates the layout. We develop a model for the array to help predict the worst-case speed.

The RC delay along the poly word line will be significant. To avoid simulating 256 separate sections, we break the line into a smaller number of components. Eight is a reasonable number of sections, which can be verified by running a simulation with ten sections and seeing that the waveforms are not very different. If the resistance of one bit's worth of poly is R_{word}, each section will have a resistance of

$$32R_{\text{word}} + \frac{32}{15}R_{\text{refresh}},$$

where R_{refresh} is the resistance of the poly in the area of the ground refresh line. The capacitance has four components. First is the capacitance of the poly to fixed potentials; we call this C_{word}. Second is the coupling capacitance to the metal bit lines C_C. Third is the gate capacitance of the transistors. And fourth is the small capacitance due to the layout of the ground refresh circuitry C_{refresh}. Assuming that a word line can

FIGURE 2.42   ROM cell layout.

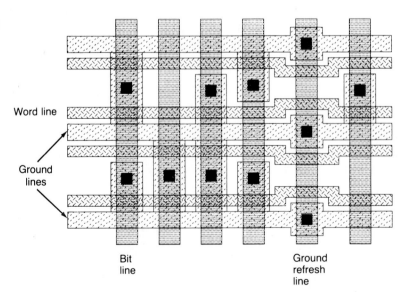

Word line

Ground lines

Bit line

Ground refresh line

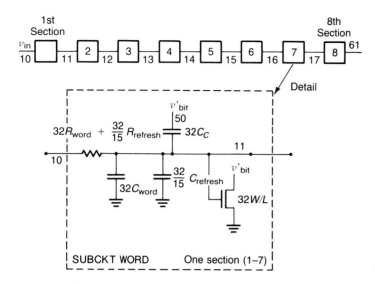

FIGURE 2.43   Word line model. The first seven are identical; the last is different.

be fully populated, the word line model is as shown in Fig. 2.43. Note that the bit line voltage v_{bit} is part of the model. We will need to find a way to generate v_{bit} later.

Because the problem is fairly complex, our objective is to generate a model to put into SPICE. Therefore we do not calculate the input capacitance of the transistor. We will let SPICE do that. One approximation we have made is that we assume that all the bit line voltages are the same. For the worst-case delay, all of the bit lines will be falling, but they will not fall at exactly the same time because of the poly RC delay. Nevertheless, assuming that they do fall at the same time is not a bad approximation for this problem since the C_C component of the delay is not that large. The model in Fig. 2.43 will be used in seven of the sections. The model of the eighth section will need to be more complicated.

The next part of the modeling effort involves the bit line and the generation of v_{bit}. The slowest bit line is the one between ground line straps at the far end of the array (bit line number 249 or so). R_S is the ground resistance of one cell. In counting squares to compute R_S, one unfortunately needs to include the effects of ΔW. In the worst-case all 15 transistors are on. We would like to find a way not to simulate all 15 transistors. By symmetry, equal current flows to the left and right of the eighth transistor. If we conceptually short together equipotential nodes, we can model the pulldown network as shown in Fig. 2.44. Seven of the transistors and seven of the resistors are wired in parallel. R_C

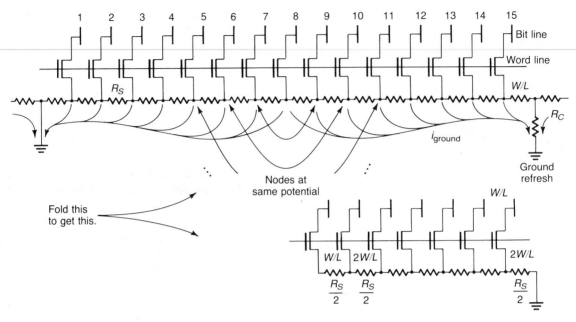

FIGURE 2.44 Using symmetry to reduce the number of elements in the ground refresh model. Nodes at the same potential can be shorted together.

is the resistance of one contact cut. Since it is so small (see Table 2.6) we will ignore it. Figure 2.45 shows the model for the cell that actually pulls down on the bit line. Voltage-controlled voltage sources are used to generate v'_{bit} and v'_{word} without loading down the bit and word lines. In all sections, v'_{bit} is used.

TABLE 2.6 Simulation model parameters

Parameter	Value	Units	Comments
R_C	4	Ω	not important
R_{word}	142	Ω	$40 \times 8/2.25$
R_S	258	Ω	5×50
R_{refresh}	213	Ω	5.3×40
W	4.2	μm	$6 - 2 \times 0.9$
L	2.25	μm	$\Delta L = -0.125 \ \mu$m
C_{bit}	11×10^{-4}	pF	$C_{\text{mdiff}} = 11 \times 10^{-5} \ \text{pF}/\mu\text{m}^2$
C_{word}	2.8×10^{-4}	pF	$C_{\text{polyf}} = 7 \times 10^{-5} \ \text{pF}/\mu\text{m}^2$
C_C	8.8×10^{-4}	pF	$C_{\text{mpoly}} = 11 \times 10^{-5} \ \text{pF}/\mu\text{m}^2$
C_{refresh}	3.12×10^{-3}	pF	$32 \ \mu\text{m}^2 C_{\text{polyf}} + 8 \ \mu\text{m}^2 C_{\text{mpoly}}$
AD	4920	μm^2	$52 \times 42 + 76 \times 36$
PD	3176	μm	$52 \times 26 + 76 \times 24$

As with the transistors, we do not model the drain capacitance on the bit line, but give the areas and perimeters to SPICE. Assume that in the worst case 70% of the bit locations are populated. The worst way these 180 locations can be arranged is if all 128 contact cuts to source diffusions exist, with 76 of these contacts having only one transistor connected, and the other 52 contacts having two transistors each.

We now calculate actual values for these parameters using the 2 μm nMOS example process from the appendixes. We are interested

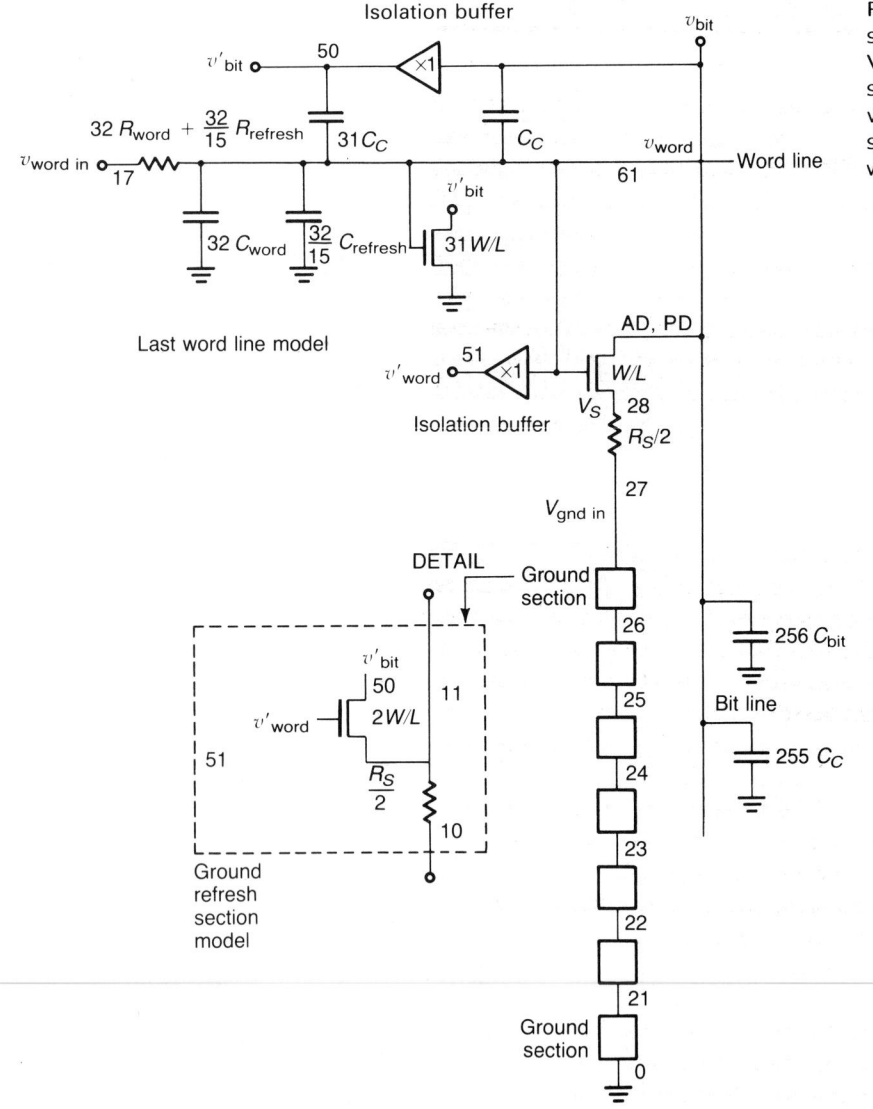

FIGURE 2.45   The eighth section of the ROM model. Voltage-controlled voltage sources are used to generate voltages needed in other sections of the model without capacitive loading.

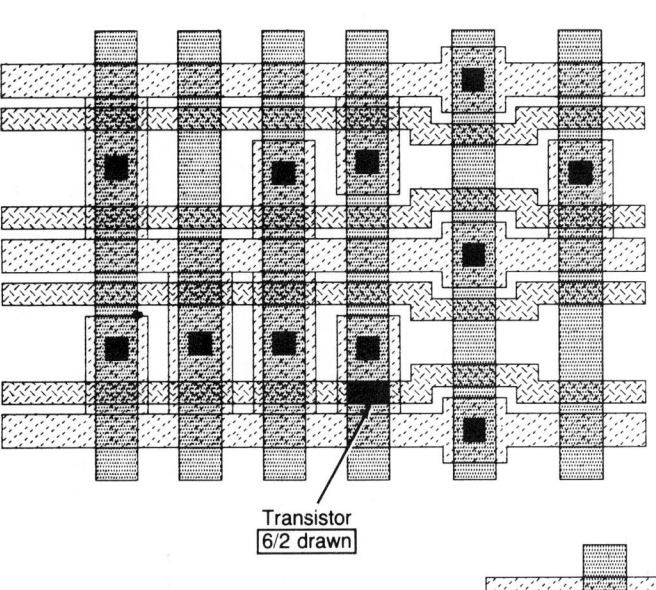

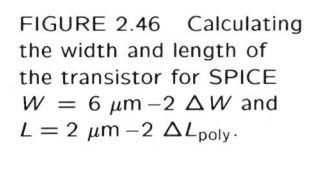

FIGURE 2.46 Calculating the width and length of the transistor for SPICE $W = 6\ \mu\text{m} - 2\ \Delta W$ and $L = 2\ \mu\text{m} - 2\ \Delta L_{\text{poly}}$.

Transistor
6/2 drawn

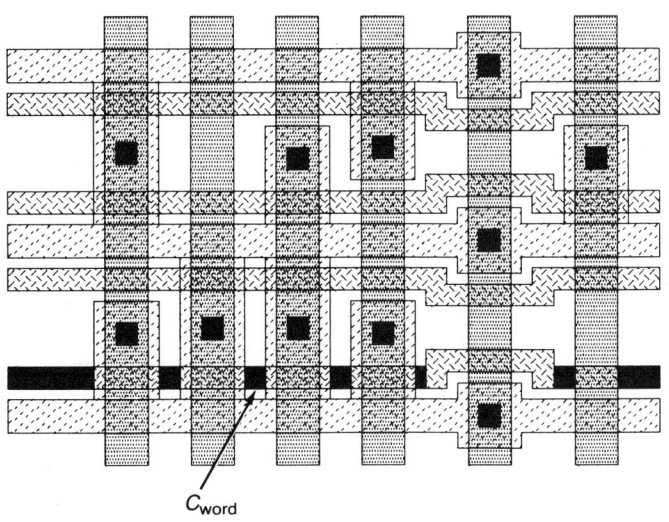

FIGURE 2.47 Word line capacitance, $C_{\text{word}} = 4\ \mu\text{m}^2 \times C_{\text{poly}f}$.

C_{word}

in the slow case, so we use the maximum values for all the parasitic capacitances and the maximum poly resistance.[22]

Solving for the various parameters in the model, we obtain the values given in Table 2.6.

The counting of squares in some cases in which there are strange geometries can be difficult. In these cases, accuracy exceeding the parameter's importance is unproductive. Figures 2.46 through 2.53 illustrate how the various parameters are extracted from the layout.

---

[22] [Read this footnote after reading Chapter 4.] While the worst-case ΔL_{poly} for the transistor is negative for the bit line model, the worst-case poly RC actually occurs for ΔL_{poly} positive if fringing capacitance is taken into account. The simulation of this process corner would require the doctoring of the slow process "file."

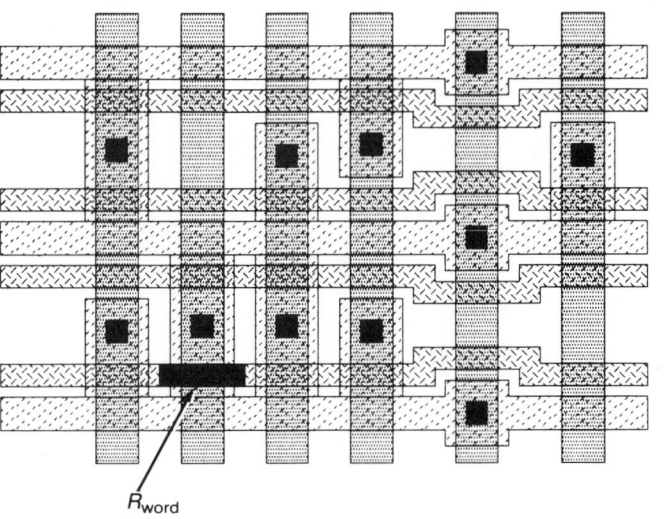

FIGURE 2.48 Word line resistance, $R_{word} = (8 \ \mu m^2 \times R_{poly})/(2 \ \mu m - 2 \ \Delta L_{poly})$.

R_{word}

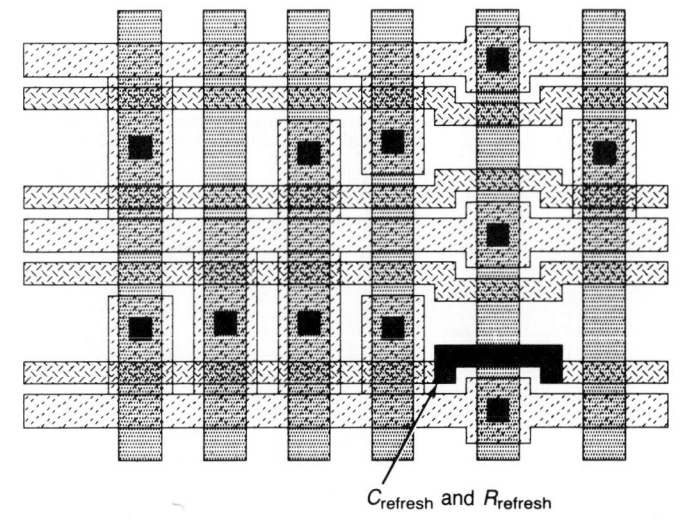

FIGURE 2.49 Refresh capacitance and resistance: $C_{refresh} = 32 \ \mu m^2 \, C_{poly \ f} + 8 \ \mu m^2 \, C_{mpoly}$; $R_{refresh} = R_{poly}(16 \ \mu m - 4 \ \mu m)/(2 \ \mu m - 2 \ \Delta L_{poly})$. In the calculation of the resistance we count each corner as half a square.

$C_{refresh}$ and $R_{refresh}$

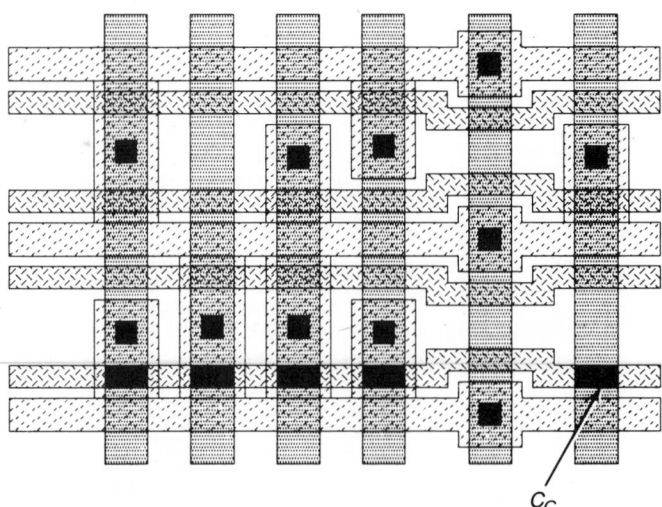

C_C

FIGURE 2.50 Coupling capacitance is metal over poly: $C_C = 8 \ \mu m^2 \, C_{mpoly}$.

147

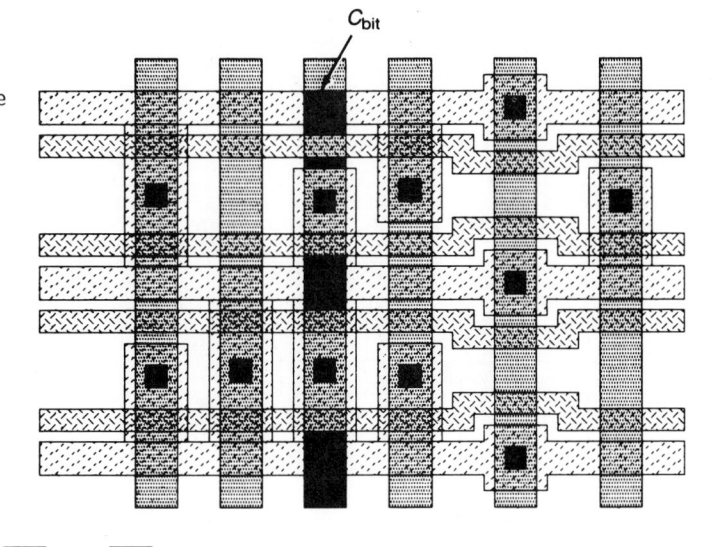

FIGURE 2.51 Metal bit line capacitance is mostly metal over n+ with some metal over field, which we ignore: $C_{bit} = 20 \ \mu m^2 \ C_{mdiff}/2$.

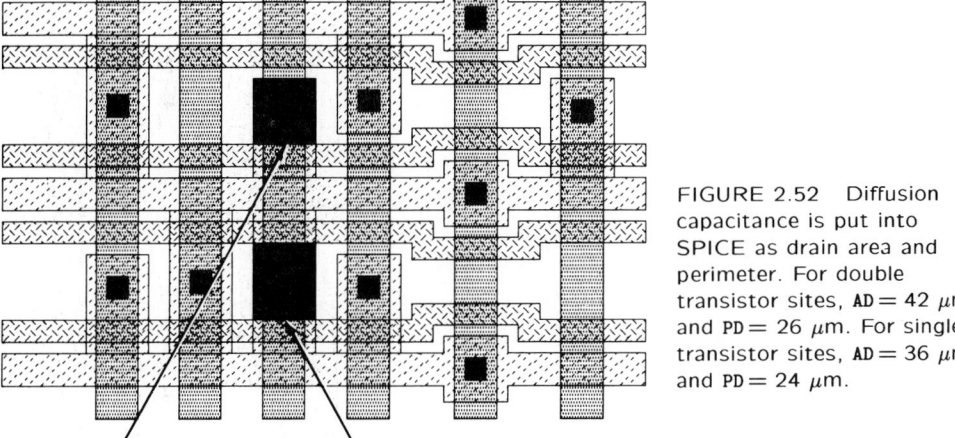

n+ for one transistor   n+ for two transistors

FIGURE 2.52 Diffusion capacitance is put into SPICE as drain area and perimeter. For double transistor sites, $\text{AD} = 42 \ \mu m^2$ and $\text{PD} = 26 \ \mu m$. For single transistor sites, $\text{AD} = 36 \ \mu m^2$ and $\text{PD} = 24 \ \mu m$.

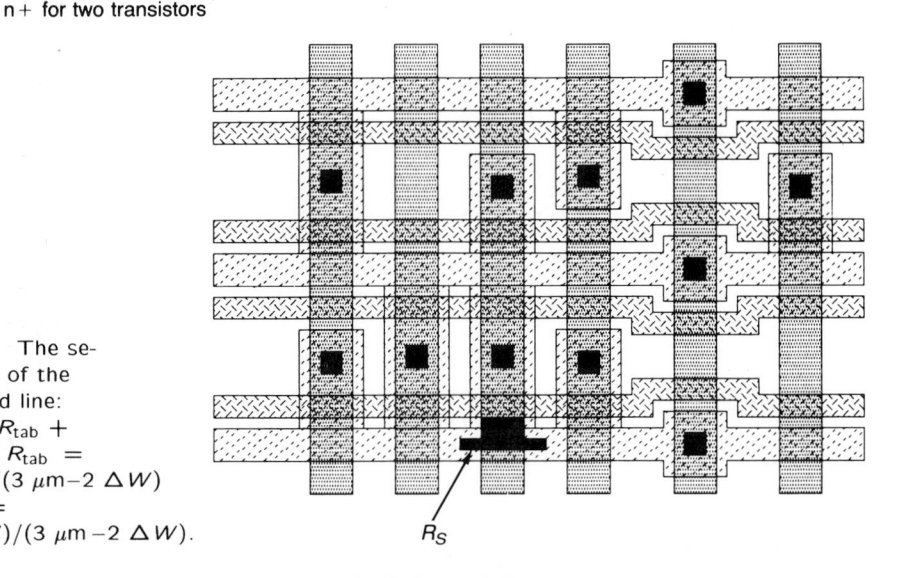

FIGURE 2.53 The series resistance of the diffused ground line: $R_S \approx R_{diff}(2R_{tab} + R_{middle})$ where $R_{tab} = (1 \ \mu m + \Delta W)/(3 \ \mu m - 2 \ \Delta W)$ and $R_{middle} = (6 \ \mu m - 2 \ \Delta W)/(3 \ \mu m - 2 \ \Delta W)$.

148

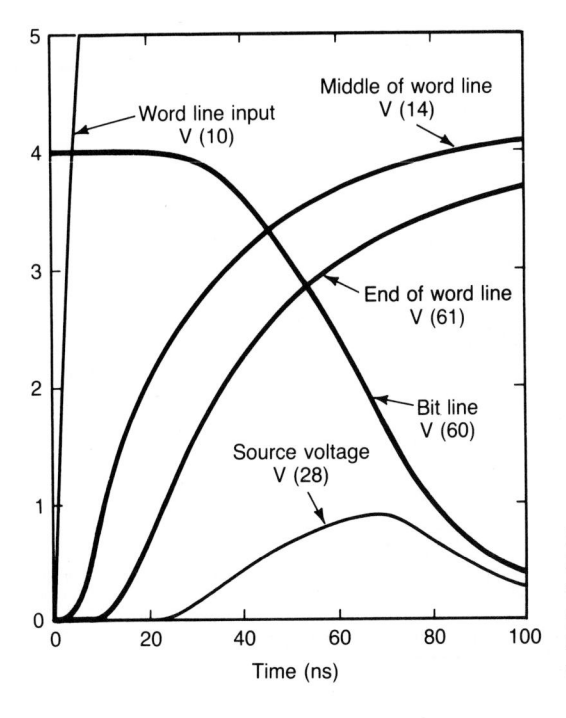

FIGURE 2.54   Simulated ROM waveforms. Observe the source voltage of the pulldown device; its rise has a serious impact on the speed.

To compute C_{bit}, we look only at the metal over n+ in the area of the grounds. There is a little bit of metal over field but not enough to worry about. There is 10 μm^2 of metal over n+ per word. In C_{refresh}, we counted the poly over field, which assumes both sides are programmed.

Table 2.7 shows an input SPICE deck and Fig. 2.54 illustrates the output of a SPICE simulation.

TABLE 2.7   ROM simulation SPICE deck

```
ROM simulation model
*
* node names worth remembering
* VBIT 60
* VBIT' 50
* VWORD 61
* VWORD' 51
*
VBB 2 0 -3V
*
* step input
VIN 10 0 PWL(ONS 0 6NS 5 500NS 5)
*
*
* word line section model                        (Continued)
```

TABLE 2.7   *(Continued)*

```
* VIN VOUT VBIT' VBB
.SUBCKT WORD 10 11 50 2
RWORD 10 11 5006.2
CWORD 11 0 0.009PF
* CRFSH = 32/15 C-REFRESH
CRFSH 11 0 0.0067PF
* 32 cells worth of CC
CC 50 11 0.0282PF
* no need for AD, PD since VBIT' is driven from VCCS
MWORD 50 11 0 2 NENHS W=134.4U L=2.25U
.ENDS WORD
*
* rfsh line section model
* VIN VOUT VWORD' VBIT' VBB
.SUBCKT RFSH 10 11 51 50 2
* RRFSH = RS/2
RRFSH 10 11 129.17
MRFSH 50 51 11 2 NENHS W=8.4U L=2.25U
.ENDS RFSH
*
* the whole thing
*
* 7 word line sections
X1 10 11 50 2 WORD
X2 11 12 50 2 WORD
X3 12 13 50 2 WORD
X4 13 14 50 2 WORD
X5 14 15 50 2 WORD
X6 15 16 50 2 WORD
X7 16 17 50 2 WORD
* the last word line section is different
RWORD 17 61 5006.2
CWORD 61 0 0.009PF
* CRFSH = 32/15 C-REFRESH
CRFSH 61 0 0.0067PF
* 31 cells worth of CC
CCS 50 61 0.0273PF
CC 60 61 0.0009PF
MWORD 50 61 0 2 NENHS W=130.2U L=2.25U
*
* 7 refresh sections
X11 0 21 51 50 2 RFSH
X12 21 22 51 50 2 RFSH
```

```
X13 22 23 51 50 2 RFSH
X14 23 24 51 50 2 RFSH
X15 24 25 51 50 2 RFSH
X16 25 26 51 50 2 RFSH
X17 26 27 51 50 2 RFSH
* the last refresh section is different (W/L vs 2W/L)
RRFSH 27 28 129.17
MRFSH 60 61 28 2 NENHS W=4.2U L=2.25U AD=4920.0P PD=3176.0U
CBIT 60 0 0.2816PF
CCBIT 60 0 0.2244PF
*
* unity gain isolation buffers
EBIT 50 0 60 0 1.0
EWORD 51 0 61 0 1.0
*
* precharge the bit line
.IC V(60)=4
*
.TRAN 1NS 100NS
.PRINT TRAN V(10) V(14) V(28) V(60) V(61)
*.PLOT TRAN V(10) V(14) V(28) V(60) V(61)
.OPTION LIMPTS=1001
*
*****************************************************************
* worst-case slow parameters
*
.MODEL NENHS NMOS LEVEL=3 RSH=0 TOX=330E-10 LD=0.19E-6 XJ=0.27E-6
+ VMAX=13E4 ETA=0.25 KAPPA=0.5 NSUB=5E14 UO=650 THETA=0.1
+ VTO=0.946 CGSO=2.43E-10 CGDO=2.43E-10 CJ=6.9E-5 CJSW=3.3E-10
+ PB=0.7 MJ=0.5 MJSW=0.3 NFS=1E10
*
.MODEL NZEROS NMOS LEVEL=3 RSH=0 TOX=330E-10 LD=0.19E-6 XJ=0.27E-6
+ VMAX=13E4 ETA=0.25 KAPPA=0.5 NSUB=4.0E14 UO=680 THETA=0.1
+ VTO=0.526 CGSO=2.43E-10 CGDO=2.43E-10 CJ=6.9E-5 CJSW=3.3E-10
+ PB=0.7 MJ=0.5 MJSW=0.3 NFS=1E10
*
.MODEL NDEPS NMOS LEVEL=3 RSH=0 TOX=330E-10 LD=0.19E-6 XJ=0.27E-6
+ VMAX=13E4 ETA=0.25 KAPPA=0.5 NSUB=50E14 UO=650 THETA=0.04
+ VTO=-2.078 CGSO=2.43E-10 CGDO=2.43E-10 CJ=6.9E-5 CJSW=3.3E-10
+ PB=0.7 MJ=0.5 MJSW=0.3 NFS=1E10
*
.END
```

Note the word line delay and the ground line noise.

2.5   The removal of heat

The heat generated in a VLSI circuit is insidious and difficult to remove. If one designs an nMOS VLSI circuit without special attention to the chip's power budget, the average gate could easily dissipate on the order of 1 mW. Multiply this by the number of gates on a VLSI chip and the power is enormous. As pointed out by Keyes [Keyes 75], sunlight, which contains only about 0.1 W/cm^2, can warm objects appreciably; a 60 watt light bulb, with a surface area of 120 cm^2 and a power dissipation of only 0.5 W/cm^2, is too hot to touch. Imagine, then, the problem of dissipating several watts of power from a chip a few millimeters on a side.

The cost of dissipating power on a chip is primarily attributable to the price of the high-quality power supply when the chip's power dissipation is below one watt. Above 1 watt, the cost of cooling becomes a dominant consideration. The exotic packages, fans, and refrigeration equipment needed when the chip's power dissipation exceeds 1 watt quickly drives up the costs associated with power dissipation. About 8 watts per chip seems to be a practical limit for today's packages. Packages that can dissipate up to 25 W/cm^2 are in the planning stages. Compare this number to the 100 W/cm^2 seen on the heat shield of a space vehicle re-entering the earth's atmosphere. Nevertheless, powers approaching the kilowatt range have been successfully removed from a silicon chip by cutting grooves in the back of the silicon and forcing liquid through the excellent heat exchanger that results [Tucherman 82].

To remove heat from an object, there must be a temperature differential between that object and the heat sink. Still air can transfer about 0.001 W/(cm^2° C). Thus for a package with a surface area of 10 cm^2 we expect to see a temperature differential of about 100° C/W. By using forced air, the power that can be dissipated rises to between 0.01 and 0.03 W/(cm^2° C). For these low values of power dissipation, the package and the silicon are at very nearly the same temperature.

If the power dissipated through air cooling is insufficient, liquid cooling can be used. For low temperature differentials between the heat source and the liquid, the heat is transferred by convection. For example, Freon, which boils at 48° C can transfer 20 W/cm^2 with a temperature differential of 20° C. For large temperature differentials, nucleate boiling occurs. Eventually, the boiling becomes so fierce that the vapor formed at the liquid/heat source interface becomes an insulating barrier against further efficient heat transfer.

The maximum temperature of the chip is constrained by the twin concerns of reliability and the maximum affordable transistor parameter (chiefly mobility and voltage threshold) degradation. For reliability reasons, the maximum temperature of a chip in plastic is

about 100° C, and for a chip in a ceramic package it is 120° C. The actual temperature of a packaged chip can be measured by performing pulsed I–V measurements on a calibrated p–n diode. Measurements must be pulsed to prevent heating of the chip. The voltage drop across a forward-biased diode is measured at a given current. By putting the device in an oven, the voltage versus temperature of the diode can be calibrated. Then, when the chip is in operation, the voltage measurement can again be made and the junction temperature read from the calibration chart.

We define Θ_{ja} as the thermal resistance between the junction and the ambient. For most commercial packages, this is nearly the thermal resistance of the package. The thermal resistance Θ_{ja} is the difference in temperature (in ° C) between a chip and the ambient still air when 1 watt is being dissipated in the chip. The thermal resistance of popular packages is between about 30 and 70° C/W. A forty-pin plastic DIP has a Θ_{ja} of about 62° C/W and a 64 pin ceramic chip carrier has a thermal resistance of about 36° C/W [Murtuza 82]. These thermal resistances are influenced by the package's environment. Thus the layout and positioning of the printed circuit board, for instance, influence the effective thermal resistance of the package. Significant heat can be removed through the leads. In forced air (18 m^3/min), Θ_{ja} lowers by about 10 to 15° C/W. The thermal resistance between a fair-sized die and the package is on the order of 3° C/W and is usually negligible for air-cooled chips. On the other hand, poor packaging can result in a much higher thermal resistance, which would be a problem in any context.

Chip packaging has recently been receiving a lot of research and development attention. One can anticipate rapid advances over the next few years. The thresholdlike behavior of power costs make defining power budgets relatively straightforward—designing within these constraints is, of course, another story.

2.6   Perspective

Device electronics is one of the foundations of VLSI circuit design. There is an intimate relationship between the physics of MOS devices and their terminal characteristics. The MOS transistor is a voltage-controlled nonlinear resistor. A myriad of phenomena affects the various parts of the voltage–current characteristics, but most of these effects can be understood in terms of charge distributions and carrier velocities. One key parameter is the saturation current, which limits the speed that capacitors can be charged by a transistor. Another key parameter is

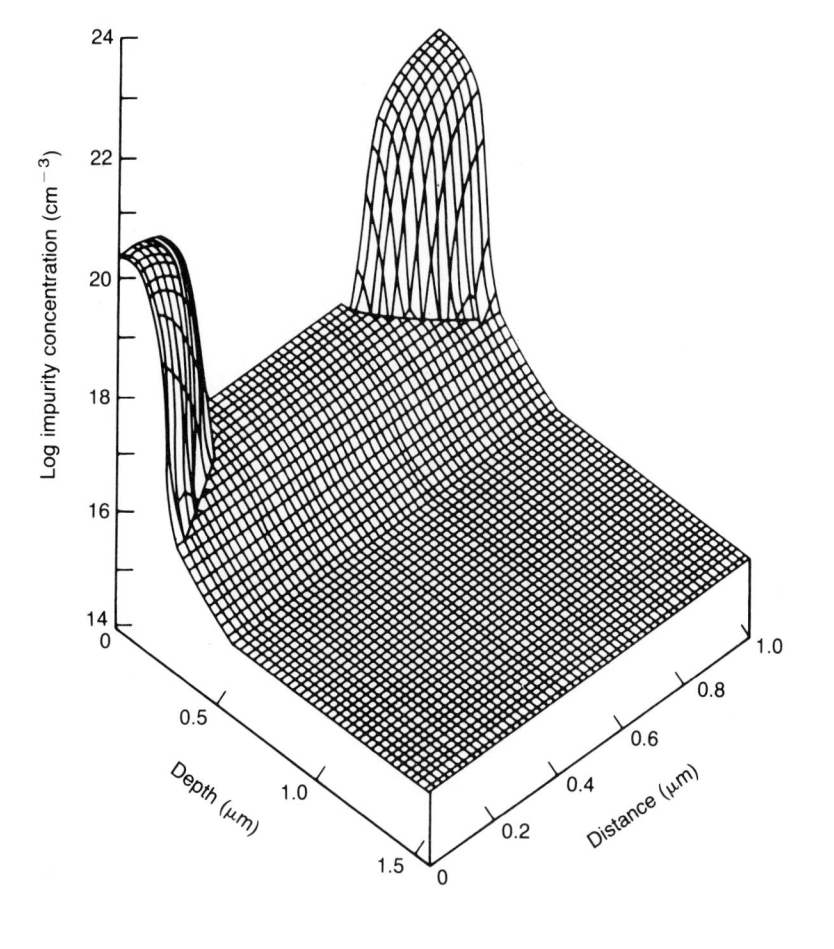

FIGURE 2.55 Surface plot of the simulated net doping concentration of an enhancement-mode transistor in a 1 μm technology. (W. Fichtner, "Device Scaling & Low Temperature Processing," VLSI SCIENCE & TECHNOLOGY (1948), page 133.) This figure was originally presented at the Spring 1984 Meeting of The Electrochemical Society, Inc., held in Cincinnati, Ohio.

the threshold voltage. We have seen that the threshold voltage is not a constant, but varies with the applied voltages and transistor dimensions. Figure 2.55 illustrates the simulated doping profile of a 1 μm nMOS transistor. In Fig. 2.56, we see a plot of the electrostatic potential for this device in saturation. Figures 2.57 and 2.58 illustrate the electron distribution. Note how the number of carriers decreases as one travels along the channel from the source to the drain. Figure 2.59 illustrates the lateral electric field that propels the carriers. Note how it peaks near the drain end of the channel.

Most of the MOS transistor's capacitance comes from the gate oxide, but we have seen two other important effects. The varactor effect causes the input capacitance of a MOS device to increase abruptly at threshold. When a MOS transistor is in saturation, the drain-to-gate capacitance is extremely low. This is important in the design of depletion loads.

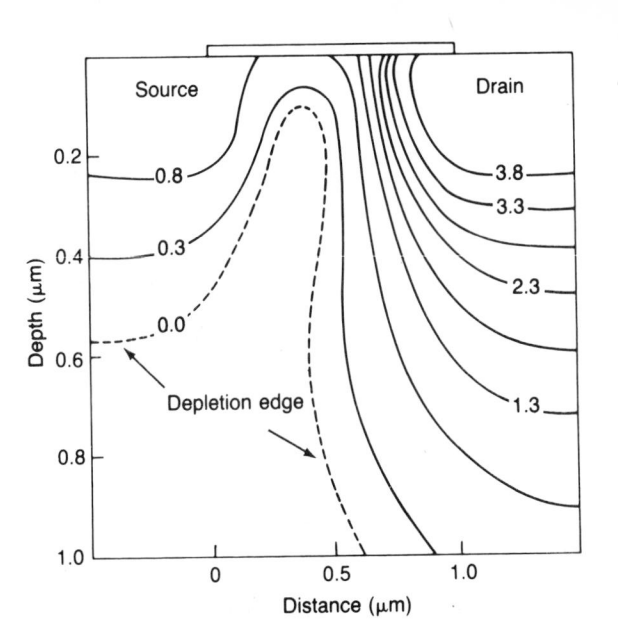

FIGURE 2.56 Plot of the electrostatic potential in the device of Fig. 2.55 in saturation with $V_{GS} = 1$ V, $V_{DS} = 3$ V, and $V_{BS} = 0$ V. $T_{OX} = 260$ Å and $V_{TX} = 0.6$ V. (W. Fichtner, "Device Scaling & Low Temperature Processing," VLSI SCIENCE & TECHNOLOGY (1948), page 133.) This figure was originally presented at the Spring 1984 Meeting of The Electrochemical Society, Inc., held in Cincinnati, Ohio.

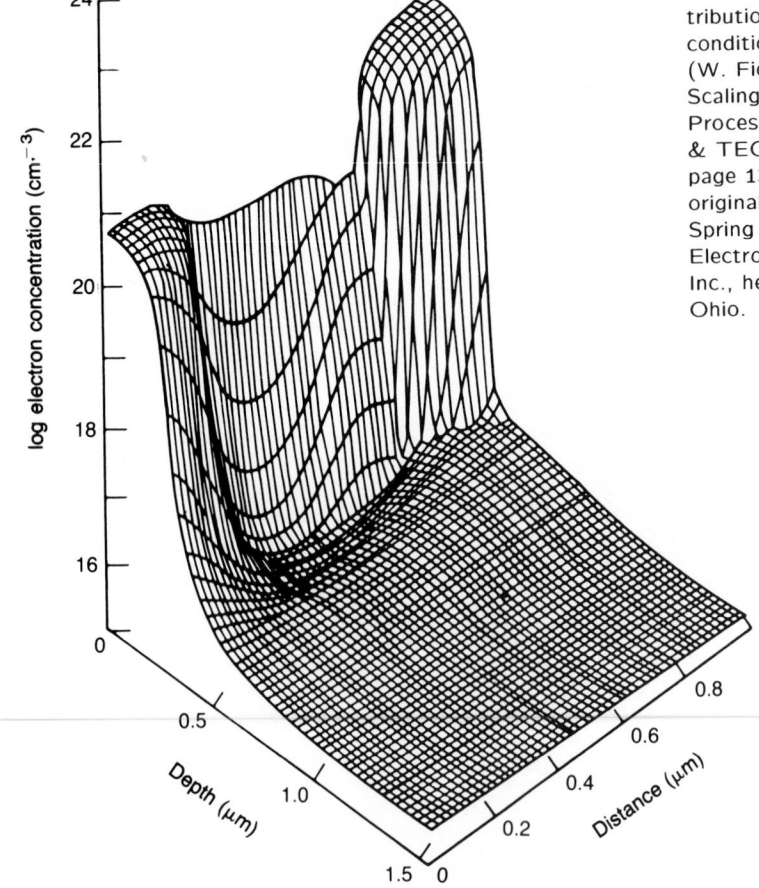

FIGURE 2.57 Electron distribution for the same bias conditions as for Fig. 2.56. (W. Fichtner, "Device Scaling & Low Temperature Processing," VLSI SCIENCE & TECHNOLOGY (1948), page 133.) This figure was originally presented at the Spring 1984 Meeting of The Electrochemical Society, Inc., held in Cincinnati, Ohio.

155

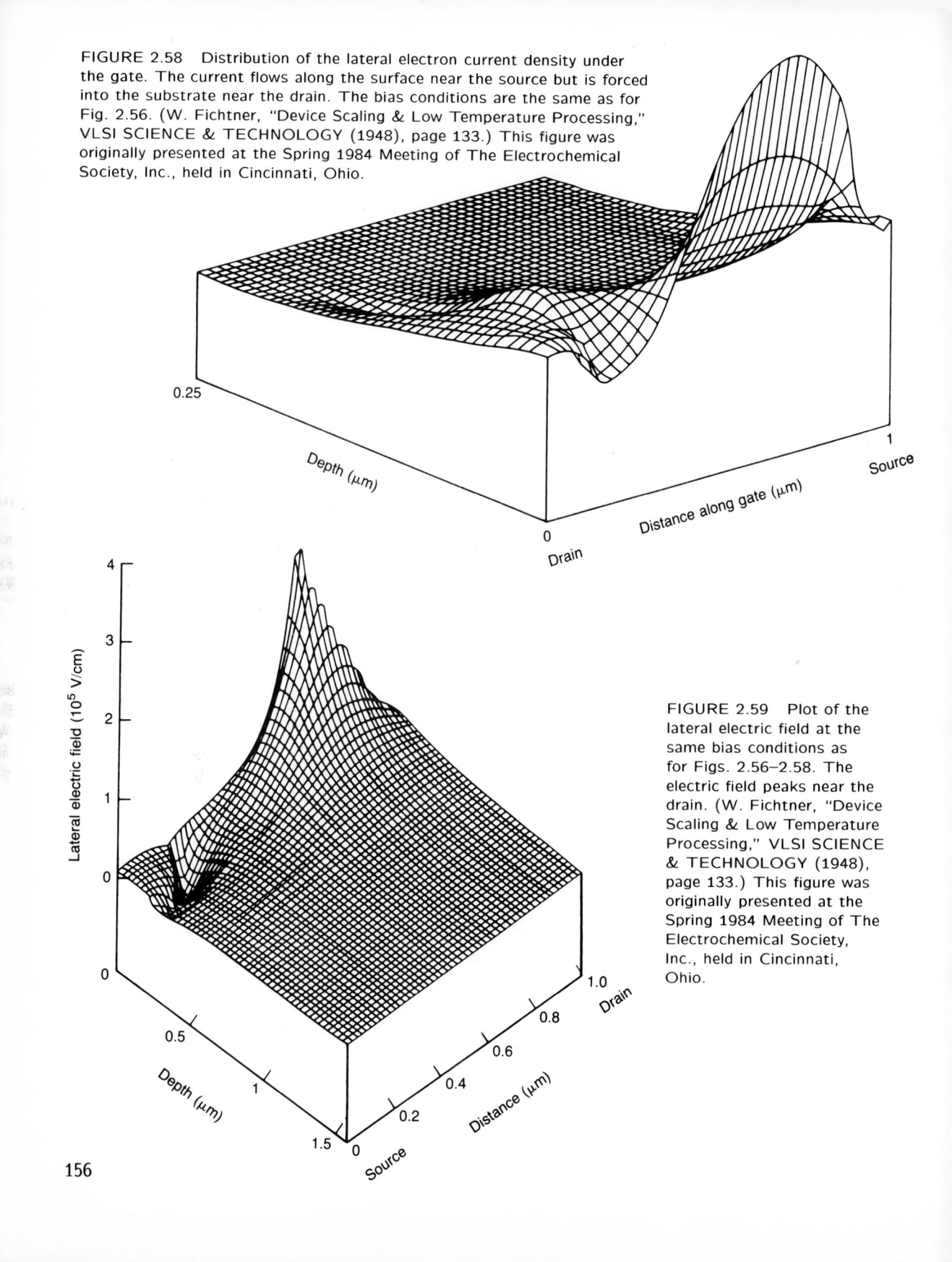

FIGURE 2.58   Distribution of the lateral electron current density under the gate. The current flows along the surface near the source but is forced into the substrate near the drain. The bias conditions are the same as for Fig. 2.56. (W. Fichtner, "Device Scaling & Low Temperature Processing," VLSI SCIENCE & TECHNOLOGY (1948), page 133.) This figure was originally presented at the Spring 1984 Meeting of The Electrochemical Society, Inc., held in Cincinnati, Ohio.

FIGURE 2.59   Plot of the lateral electric field at the same bias conditions as for Figs. 2.56–2.58. The electric field peaks near the drain. (W. Fichtner, "Device Scaling & Low Temperature Processing," VLSI SCIENCE & TECHNOLOGY (1948), page 133.) This figure was originally presented at the Spring 1984 Meeting of The Electrochemical Society, Inc., held in Cincinnati, Ohio.

Interconnect is far from ideal. Conductors have resistance and capacitance. The fringing capacitance can be of the same magnitude as the parallel plate capacitance. Long RC lines obey the diffusion equation and have quadratic delay versus length characteristics.

Heat dissipation is a problem, especially in nMOS technology. Chips are limited to power budgets of 1–5 watts for this reason.

The physics of semiconductor devices comprises one set of fundamental constraints on the performance of VLSI circuits. Other constraints arise from the limitations of the manufacturing process (Chapter 3) and the reliable transformation between the electrical domain and the digital abstraction (Chapter 4).

Problems

2.1 A process has $T_{OX} = 250$ Å and a substrate resistivity of 50 Ω·cm. Calculate ϕ_{Fp}, $2\Phi_{Fp}$, X_{Pmax}, V_{FB}, V_{TE}, γ, and δ. Assume $N_f = 1.5 \times 10^{11}$ cm^{-2} and $Q_{it} = \rho(z) = 0$. Do your calculations for $V_{BB} = 0$ and -2.5 V.

These calculations are valid for a "natural" transistor. We will now investigate some properties of the parasitic transistors formed by poly and metal lines over the substrate. Assume that the metal is aluminum and the poly is degenerately doped n-type. The gate oxides are the field and CVD oxides. Calculate V_{FB}, γ, and V_{TX}, assuming $V_{BB} = 0$. $T_{CVD} = 5000$ Å and $T_{field} = 7000$ Å.

If your calculations are correct, you will find that these parasitic transistors are likely to be always turned on. To get around this problem the thresholds of these devices are raised by placing an implant under the field oxide. This raises the effective substrate doping seen by these devices. What is the value of N_A needed to raise the field thresholds above 15 V? What are γ and the avalanche breakdown voltage V_M for this doping level? $\triangle$

2.2 The three-transistor memory cell in Fig. 2.60 can be made to work if one takes account of the body effect. Explain how one would read and write this memory cell. $\triangle$

2.3 In the Section 2.2, simplifications were made for the bulk charge term in Q_B. It is possible to carry out the integration with the complete equation for Q_B. Using the complete expression for Q_B, show that

FIGURE 2.60

$$I_{DS} = SK'\left(\left(V_G - V_{FB} - 2\Phi_{Fp} - V_S\right)^2 - \left(V_G - V_{FB} - 2\Phi_{Fp} - V_D\right)^2\right.$$

$$\left. - \frac{4\gamma}{3}\left(\left(V_D + 2\Phi_{Fp}\right)^{1.5} - \left(V_S + 2\Phi_{Fp}\right)^{1.5}\right)\right).$$

To find the point of saturation, V_{Dsat}, we set $Q_N = 0$. Also show that

$$V_{Dsat} = V_G - V_{FB} - 2\Phi_{Fp} + \frac{1}{2}\gamma^2\left(1 - \sqrt{1 + 4(V_G - V_{FB})/\gamma^2}\right).$$

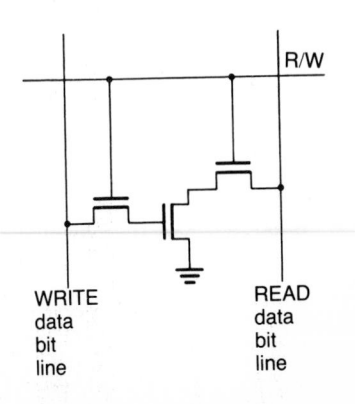

R/W

WRITE
data
bit
line

READ
data
bit
line

2.4 By our theory, a channel is formed when $V_G > V_{TX}$. But if V_S and V_D are greater than zero, then no current flows as long as $V_G < V_S + V_{TX}$. If there is a channel, then why can current not flow?

2.5 Define $g_m \equiv \partial i_D / \partial v_{GS}$ and $g_{mb} \equiv \partial i_D / \partial v_{BS}$. Show that $g_{mb} = b g_m$.

2.6 The equations governing thermal conductance are exactly the same as those governing capacitance. For thin wires ($W/H \leq 10$), the thermal resistance is actually less than that given in Eq. (2.175) because of thermal spreading. In Section 2.4 we looked at the increase in the capacitance of a wire due to the spreading of the electric field lines. Use these results to develop a more accurate expression for ΔT.

2.7 Figure 2.6 illustrates the C–V characteristics of a MOS capacitor. This capacitance can be broken into three components as illustrated in Fig. 2.29. Using the definitions of C_N and C_B given in Eqs. (2.155) and (2.156), plot these three capacitances as a function of V_G.

2.8 Compare one aspect of the SPICE level 2 and 3 models by setting up a reasonable SPICE model deck (perhaps by editing the models given in the appendixes) with $\gamma = 0.6$. Put together a model of a 1 meter wide by 3 meter long transistor (which we are confident will exhibit no short channel effects) and look at the current with $V_{GS} = V_{DS} = 5$ V. Then break the model into three series 1 meter by 1 meter transistors with the gates tied together. Examine the current under the same bias conditions. Do these experiments for both level 2 and level 3 SPICE models. Explain the (somewhat strange) results you obtain. △

2.9 Why is the capacitance of a MOS capacitor not equal to C_{OX} at $V_G = V_{FB}$? (*Hint:* This is not covered in the text; the reasons have to do with the depletion region approximation and a quantity called the Debye length.) △△

2.10 By manipulating the ROM model developed in Example 2.9, discover the following percentage effects of various changes in the design. (After each change go back to the original model.)
  a. Setting $R_{\text{word}} = 0$
  b. Setting $C_C = 0$
  c. Placing the ground straps every seven bits
  d. Driving the ROM from the middle △

2.11 Room temperature corresponds to about 25 mV and the ionization potential for the common donor and acceptor atoms in silicon is on the order of 40 mV. Nevertheless, at room temperature almost all donors and acceptors are ionized. Why? △△

2.12 How many sections are required to model the transistor illustrated in Fig. 2.61 using a model of the form illustrated in Fig. 2.62? An accuracy of 5% is desired. Use the nMOS 2 μm enhancement-mode transistor model from the appendixes.

2.13 For an advanced MOS process, we try to make γ better. What gets worse?

2.14 Charlie Analog claims that he can control MOS transistor thresholds much better when $T_{OX} = 100$ Å than when $T_{OX} = 1000$ Å. Is this reasonable? Why?

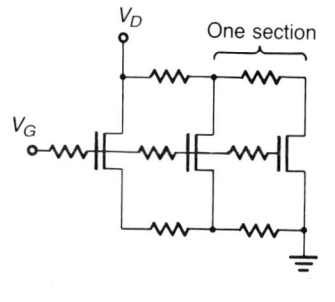

300 μm

V_D

3 μm
2 μm

V_G

A/A   Poly

FIGURE 2.61

FIGURE 2.62

V_D One section

V_G

2.15   Janet DeVice claims that values of Θ used in our example $2\,\mu$m processes are too low. Use the equations in this chapter to estimate Θ for these processes. Was she right? When $V_{GS} = V_{TE}$, what is the expression for the electric field in the oxide?

2.16   Why does $\mathcal{E}$, which represents a force on an electron, cause a velocity $(\mathcal{V} = \mu\mathcal{E})$ rather than an acceleration? $\triangle$

2.17   Assume we take the $2\,\mu$m CMOS process and do nothing else but halve all of the dimensions, including the depletion widths, but still run it at 5 volts. Assuming that the devices do not break down or punch-through, how would the current of a minimum length, nonminimum width, scaled device compare, keeping the shapes the same, in the various regimes? For instance, for the n-channel device with $V_{GS} = 5$ V, how do the currents compare at $V_{DS} = 0.1$ V, at $V_{DS} = 5$ V? Compare a 20/2 in the $2\,\mu$m technology to a 10/1 in the scaled $1\,\mu$m technology. What about at $V_{GS} = 0.5$ V? Explain your answers! What are the important effects?

References

[Akers 82] L. A. Akers and J. J. Sanchez, "Threshold Voltage Models of Short, Narrow, and Small Geometry MOSFETs: A Review," *Solid-State Electronics* **25**: 621–641, 1982.

[Baccarani 84] G. Baccarani, M. Wordeman, and R. H. Dennard, "Generalized Scaling Theory and Its Application to a $\frac{1}{4}$ Micrometer MOSFET Design," *IEEE Trans. Electron Devices* **ED-31**: 452–462, 1984.

[Bain 82] L. I. Bain, private communication, 1982.

[Balk 74] P. Balk, "Layered Dielectrics in the MOS Technology," *Inst. Phys. Conf. Series 1973* **19**: 51–82, 1974.

[Barnes 79] J. J. Barnes, K. Shimohigashi, and R. W. Dutton, "Short-Channel MOSFET's in the Punchthrough Current Mode," *IEEE J. Solid-State Circuits* **SC-14**: 368–374, 1979.

[Baum 70] G. Baum and H. Beneking, "Drift Velocity Saturation in MOS Transistors," *IEEE Trans. Electron Devices* **ED-17**: 481–482, 1970.

[Bell 70] *Physical Design of Electronic Systems* I, "Design Technology, Pt. II, Thermal Design," Members of Technical Staff, Bell Telephone Laboratories, Prentice-Hall, Englewood Cliffs, N.J., 1970.

[Black 69] J. R. Black, "Electromigration Failure Modes in Aluminum Metalization for Semiconductor Devices," *Proc. IEEE* **57**: 1587–1594, 1969.

[Botchek 83] C. M. Botchek, *VLSI Basic MOS Engineering* **1**, Pacific Technology Group, Saratoga Calif., 1983.

[Brews 80] J. R. Brews, W. Fichtner, E. H. Nicollian, and S. M. Sze, "Generalized Guide for MOSFET Miniaturization," *IEEE Trans. Electron Devices Lett.* **EDL-1**: 2–4, 1980.

[Chua 81] S. J. Chua, "Current and Resistivity Dependence of Electromigration from a Statistical Analysis of Metalization Failure Data," *Solid State Electronics* **24**: 173–178, 1981.

[Colinge 81] J. P. Colinge, E. Demoulin, F. Delannay, and M. Lobet, "Grain Size and Resistivity of LPCVD Polycrystalline Silicon Films," *J.*

Electrochem. Soc.: Solid-State Science and Technology **128**: 2009–2014, 1981.

[Dang 79] L. M. Dang, "A Simple Current Model for Short-Channel IGFET and Its Application to Circuit Simulation," *IEEE J. Solid-State Circuits* **SC-14**: 358–367, 1979.

[Dang 81] R. L. M. Dang and N. Shigyo, "Coupling Capacitances for Two-Dimensional Wires," *IEEE Trans. Electron Devices Lett.* **EDL-2**: 196–197, 1981.

[Davis 81] J. R. Davis, *Instabilities in MOS Devices*, Gordon and Breach, New York, 1981.

[Dennard 74] R. H. Dennard, F. H. Gaensslen, H. N. Yu, V. L. Rideout, E. Bassons, and A. R. LeBlanc, "Design of Ion-Implanted MOSFET's with Very Small Dimensions," *IEEE J. Solid-State Circuits* **SC-9**: 256–267, 1974.

[Elmasry 82] M. I. Elmasry, "Capacitance Calculations in MOSFET VLSI," *IEEE Trans. Electron Devices Lett.* **EDL-3**: 6–7, Jan. 1982.

[Estreich 80] D. B. Estreich, "The Physics and Modeling of Latch-up and CMOS Integrated Circuits," Stanford Electronics Labs, Stanford, Calif., Tech. Rep. G201-9, November 1980.

[Estreich 82] D. B. Estreich and R. W. Dutton, "Modeling Latch-up in CMOS Integrated Circuits," *IEEE Trans. Computer-Aided Design* **CAD-1**: 157–162, 1982.

[Fang 70] F. F. Fang and A. B. Fowler, "Hot Electron Effects and Saturation Velocities in Silicon Inversion Layers," *J. Appl. Phys.* **44**: 1825-1831, 1970.

[Gibbons 75] J. F. Gibbons, W. S. Johnson, and S. W. Mylroie, *Projected Range Statistics: Semiconductors and Related Materials*, (2nd ed.), Halstead Press, Stroudsbury, Pa., 1975.

[Gibbons 80] J. F. Gibbons and K. F. Lee, "One-Gate-Wide CMOS Inverter on Laser-Recrystallized Polysilicon," *IEEE Trans. Electron Devices Lett.* **EDL-1**: 117–118, 1980.

[Glaser 77] A. B. Glaser and G. E. Subak-Sharpe, *Integrated Circuit Engineering*, Addison-Wesley, Reading, Mass., 1977.

[Goetzberger 66] A. Goetzberger, "Ideal MOS Curves for Silicon," *Bell Syst. Tech. J.* **45**: 1097–1123, 1966.

[Gray 69] P. E. Gray and C. L. Searle, *Electronic Principles: Physics, Models, and Circuits*, Wiley, New York, 1969.

[Greeneich 83] E. W. Greeneich, "An Analytical Model for the Gate Capacitance of Small-Geometry MOS Structures," *IEEE Trans. Electron Devices* **ED-30**: 1838–1839, 1983.

[Gregory 73] B. L. Gregory and B. D. Shafer, "Latch-up in CMOS Integrated Circuits," *IEEE Trans. on Nuclear Science* **NS-20**: 293–299, 1973.

[Grove 67] A. S. Grove, *Physics and Technology of Semiconductor Devices*, Wiley, New York, 1967.

[Hall 68] P. M. Hall, "Resistance Calculations for Thin Film Patterns," *Thin Solid Films, An International Journal on their Science and Technology* **1**: 277–295, Elsevier, Amsterdam, 1967/68.

[Hamilton 75] D. J. Hamilton and W. G. Howard, *Basic Integrated Circuit Engineering*, McGraw-Hill, 1975.

[Heil 35] O. Heil, British Patent 439,457 (1935).

[Heurle 71] F. M. D'Heurle, "Electromigration and Failure in Electronics: An Introduction," *Proc. IEEE* **59**: 1409–1418, 1971.

[Hoeneisen 72] B. Hoeneisen and C. A. Mead, "Fundamental Limitations in Microelectronics—I. MOS Technology," *Solid-State Electron.* **15**: 819–829, 1972.

[Horowitz 83] M. Horowitz and R. W. Dutton, "Resistance Extraction from Mask Layout Data," *IEEE Trans. Computer-Aided Design* **CAD-2**: 145–150, 1983.

[Irvin 62] J. C. Irvin, "Resistivity of Bulk Silicon and of Diffused Layers in Silicon," *Bell Sys. Tech. J.* **41**: 388, 1962.

[Jacoboni 77] C. Jacoboni, C. Canali, G. Ottaviani, and A. A. Quaranta, "A Review of Some Charge Transport Properties in Silicon," *Solid State Electronics* **20**: 77–89, 1977.

[Keyes 75] R. W. Keyes, "Physical Limits in Digital Electronics," *Proc. IEEE* **63**: 740–767, 1975.

[Kittel 76] C. Kittel, *Introduction to Solid State Physics*, p. 171, Wiley, New York, 1976.

[Kriegler 74] R. J. Kriegler, "Ion Instabilities in MOS Structures," *Proc. 12th IEEE Reliability Phys. Sym.*, pp. 250–259, New York, 1974.

[Kroell 76] K. E. Kroell and G. H. Ackermann, "Threshold Voltage of Narrow Channel Field Effect Transistors," *Solid-State Electron.* **19**: 77–81, 1976.

[Lewis 84] E. T. Lewis, "An Analysis of Interconnect Line Capacitance and Coupling for VLSI Circuits," *Solid State Electronics*, to be published.

[Lilienfeld 30] J. E. Lilienfeld, U. S. Patent 1,745,175 (1930).

[Lu 81] N. C.-C. Lu, L. Gerzberg, C.-Y. Lu, and J. D. Meindl, "Modeling and Optimization of Monolithic Polycrystalline Silicon Resistors," *IEEE Trans. Electron Devices* **ED-28**: 818–830, 1981.

[Masuda 79] H. Masuda, M. Nakai, and M. Kubo, "Characteristics and Limitations of Scaled Down MOSFET's Due to Two Dimensional Field Effect," *IEEE Trans. Electron Devices* **ED-26**: 980–986, 1979.

[Merkel 77] G. Merkel, "CAD Models of MOSFETs," *Process and Device Modeling for Integrated Circuit Design*, Wiele, Engl, and Jespers (eds.), Noordhoff, Leyden, the Netherlands, 1977.

[Meyer 71] J. E. Meyer, "MOS Models and Circuit Simulation," *RCA Rev.* **32**: 42–63, March 1971.

[Muller 77] R. S. Muller and T. I. Kamins, *Device Electronics for Integrated Circuits*, Wiley, New York, 1977.

[Murtuza 82] M. Murtuza, "Computer-generated Models Abridge Thermal Analysis of Packaged VLSI," *Electronics* **55**: 145–148, Feb. 10, 1982.

[Nicollian 82] E. H. Nicollian and J. R. Brews, *MOS Physics and Technology*, Wiley, New York, 1982.

[Ochoa 79] A. Ochoa, W. Dawes, and D. Estreich, "Latchup Control in CMOS Integrated Circuits," *IEEE Trans. Nucl. Sci.* **NS-26**: 5065–5068, 1979.

[Ogura 82] S. Ogura, C. F. Codella, N. Rovedo, J. F. Shepard, and J. Riseman, "A Half Micron MOSFET Using Double Implanted LDD," *IEEE International Electron Device Meeting*, pp. 718–721, San Francisco, 1982.

[Paulos 82] J. J. Paulos, D. A. Antoniadis, and Y. P. Tsividis, "Measurement of Intrinsic Capacitances of MOS Transistors," *IEEE J. Solid-State Circuits*, pp. 238–239, 1982.

[Paulos 83] J. J. Paulos and D. A. Antoniadis, "Limitations of Quasi-Static Capacitance Models for the MOS Transistor," *IEEE Trans. Electron Devices Lett.* **EDL-4**: 221–224, 1983.

[Protonotarios 67] E. N. Protonotarios and O. Wing, "Theory of Nonuniform *RC* Lines, Part II: Analytic Properties in the Time Domain," *IEEE Trans. Circuit Theory* **CT-14**: 13–20, 1967.

[Ratnakumar 81] K. N. Ratnakumar, J. D. Meindl, and D. L. Scharfetter, "New IGFET Short-Channel Threshold Voltage Model," *IEEE International Electron Device Meeting*, pp. 204–206, Washington, D.C., 1981.

[Rubinstein 83] J. Rubinstein, P. Penfield, Jr., and M. A. Horowitz, "Signal Delay in *RC* Tree Networks," *IEEE Trans. Computer-Aided Design* **CAD-2**: pp. 202–211, 1983.

[Rung 82] R. D. Rung, H. Momose, and Y. Nagakubo, "Deep Trench Isolated CMOS Devices," *IEEE International Electron Device Meeting*, pp. 237–240, San Francisco, 1982.

[Sabnis 79] A. G. Sabnis and J. T. Clemens, "Characterization of Electron Mobility in the Inverted ⟨100⟩ Si Surface," *IEEE International Electron Device Meeting*, pp. 18–21, Washington, D.C., 1979.

[Sah 64] C. T. Sah, "Characteristics of the Metal-Oxide-Semiconductor Transistors," *IEEE Trans. Electron Devices* **ED-11**: 324–345, 1964.

[Schroeder 80] J. E. Schroeder and A. Ochoa, Jr., and P. V. Dressendorfer, "Latch-up Elimination in Bulk CMOS LSI Circuits," *IEEE Trans. on Nuclear Science* **NS-27**: 1735–1738, 1980.

[Schwarz 83] S. A. Schwarz and S. E. Russek, "Semi-Empirical Equations for Electron Velocity in Silicon: Part II—MOS Inversion Layer," *IEEE Trans. Electron Devices* **ED-30**: 1634–1639, 1983.

[Serhan 83] G. I. Serhan and S.-Y. Yu, "A Simple Charge-Based Model for MOS Transistor Capacitances: A New Production Tool," *IEEE Trans. Computer-Aided Design* **CAD-2**: 48–51, January 1983.

[Seto 75] J. Y. W. Seto, "The Electrical Properties of Polycrystalline Silicon Films," *J. Applied Phys.* **46**: 5247–5254, 1975.

[Shickman 68] H. Shickman and D. Hodges, "Modeling and Simulation of Insulated-Gate Field-Effect Transistor Switching Circuits," *IEEE J. Solid-State Circuits* **SC-3**: 285–289, 1968.

[Shockley 48] W. Shockley and G. L. Pearson, "Modulation of Conductance of Thin Films of Semiconductors by Surface Charges," *Phys. Rev.* **74**: 232, 1948.

[Singh 76] B. R. Singh and K. Singh, "Instability Phenomena in Thin Insulating Films on Silicon," *Microelectronics and Reliability* **15**: 385–398, 1976.

[Smith 80] P. Smith, M. Inoue, and J. Frey, "Electron Velocity in Si and GaAs at Very High Electric Fields," *Appl. Phys. Lett.* **37**: 797–798, 1980.

[Snow 68] E. H. Snow and B. E. Deal, "Polarization Effects in Insulating Films on Silicon—A Review," *Trans. Met. Soc. AIME* **242**: 512–522, 1968.

[Sodini 82] C. G. Sodini, T. W. Ekstedt, and J. L. Moll, "Charge Accumulation and Mobility in Thin Dielectric MOS Transistors," *Solid-State Electronics* **25**: 833–841, 1982.

[Swanson 72] R. M. Swanson and J. D. Meindl, "Ion-Implanted Complementary MOS Transistors in Low-Voltage Circuits," *IEEE J. Solid-State Circuits* **SC-7**: 146–153, 1972.

[Sze 81] S. M. Sze, *Physics of Semiconductor Devices*, Wiley, New York, 1981.

[Troutman 79] R. R. Troutman, "VLSI Limitations from Drain-Induced Barrier Lowering," *IEEE J. Solid-State Circuits* **SC-14**: 383–390, 1979. (Also punch-through)

[Troutman 83] R. R. Troutman, "Epitaxial Layer Enhancement of n-Well Guard Rings for CMOS Circuits," *IEEE Trans. Electron Devices Lett.* **EDL-4**: 438–440, 1983.

[Troutman 84] R. R. Troutman and H. P. Zappe, "Layout and Bias Considerations for Preventing Transiently Triggered Latchup in CMOS," *IEEE Trans. Electron Devices* **ED-31**: 315–321, 1984.

[Tsividis 80] Y. Tsividis, "Relation Between Incremental Intrinsic Capacitances and Transconductances in MOS Transistors," *IEEE Trans. Electron Devices* **ED-27**: 946–948, 1980.

[Tucherman 82] D. B. Tucherman and R. F. W. Pease, "High-Performance Heat Sinking for VLSI," *IEEE Trans. Electron Devices Lett.* **EDL-2**: 126–129, l981.

[Vladimirescu 80] A. Vladimirescu and S. Liu, "The Simulation of MOS Integrated Circuits using SPICE2," Memo. UCB/ERL M80/7, University of California, Berkeley, October, 1980.

[Walczyk 83] F. Walczyk and J. Rubenstein, "A Merged CMOS/Bipolar VLSI Process," *IEEE International Electron Device Meeting*, pp. 59–62, Washington, D.C., 1983.

[Wallmark 74] J. T. Wallmark and L. G. Carlstedt, *Field-Effect Transistors in Integrated Circuits*, Halsted Press, New York, 1974.

[Ward 78] D. E. Ward and R. W. Dutton, "A Charge-Oriented Model for MOS Transistor Capacitances," *IEEE J. Solid-State Circuits* **SC-13**: 703–707, 1978.

[Ward 82] D. E. Ward and K. Doganis, "Optimized Extraction of MOS Model Parameters," *IEEE Trans. Computer-Aided Design* **CAD-1**: 163–168, 1982.

[Woods 74] M. H. Woods, "Instabilities in Double Dielectric Structures," *Proc. 12th IEEE Reliability Phys. Sym.*, pp. 259–266, New York, 1974.

[Wordeman 79] M. R. Wordeman, "Characterization of Depletion-mode MOSFETs," *IEEE International Electron Device Meeting*, pp. 26–29, Washington, D.C., 1979.

[Yang 78] E. S. Yang, *Fundamentals of Semiconductor Devices*, McGraw-Hill, New York, 1978.

[Yang 82] P. Yang and P. K. Chatterjee, "SPICE Modeling for Small Geometry MOSFET Circuits," *IEEE Trans. Computer-Aided Design* **CAD-1**: 169–182, 1982.

[Yau 74] L. D. Yau, "A Simple Theory to Predict the Threshold Voltage of Short-channel IGFET's," *Solid State Electronics* **17**: 1059–1063, 1974.

[Yuan 82] C. P. Yuan and T. N. Trick, "A Simple Formula for the Estimation of the Capacitance of Two-Dimensional Interconnects in VLSI Circuits," *IEEE Trans. Electron Devices Lett.* **EDL-3**: 391–393, 1982.

PROCESSING

A VL$I manufacturing facility can cost anywhere from the tens of millions to hundreds of millions of dollars to construct. With this facility, and an increasingly sophisticated assortment of processing techniques and equipment, one produces parts that sell for about five dollars. In this chapter, we examine some issues related to the manufacturing and integrity of VLSI chips. The keystone of the technology is mass production. The economic forces in the manufacturing process have historically been in the directions of process optimization, larger wafers, minimum die area, and minimum number of mask levels that will produce a product with the required specifications. Extreme focus on the last two objectives will cause the design time to increase, however, and this too has associated costs. It is often difficult to judge how to best minimize the total product cost.

Fabricating a VLSI circuit requires well over one hundred separate processing steps. These steps are tightly coupled in the sense that changing any one aspect of the process is likely to cause changes in other seemingly unrelated parts of the process. For instance, oxide growth times affect impurity distribution profiles, while the rate of oxide growth, in turn, depends on the impurity concentration. This sort of phenomenon makes it difficult to decouple the effects of changing any one parameter. Because of the economics of VLSI, optimization of the product cost, rather than independence among process steps, is the goal of most process engineers.[1]

---

[1] Flexibility is lost by global optimization, so there is some recent research in designing process "modules" that can be put together in different ways for different projects.

In this chapter, we will gain an appreciation for the connections between processing technology and VLSI circuit performance. Not only is the existence of processing technology a necessary precondition for the construction of VLSI circuits but uncertainties and tolerances in the manufacturing process produce parametric circuit variations. Manufacturing processes cannot unerringly produce identical objects.

As with any artist, the better a circuit designer's appreciation for the medium, the better the results. This chapter discusses some of the more prominent textures of MOS fabrication technology.

3.1 Semiconductor device fabrication techniques

In this section, we examine a number of the processing techniques available to the process engineer for the production of VLSI chips [Sze 83, Colclaser 80, Glaser 77]. In the next section, we will examine how these techniques are composed to build MOS integrated circuits.

Crystal growth

The growth of a single-crystal silicon boule is the first step in the substrate preparation process. The Czochralski [Rea 81] method is typically used to pull a silicon crystal from a crucible filled with molten silicon and the appropriate dopants. The crucible is made of graphite. The process takes place in an inert atmosphere, such as argon, while RF or resistive heating provides the energy. The radial resistivity variations are about ±10%. Resistivities from 30 kΩ·cm to 0.0001 Ω·cm can be obtained by this technique, in combination with float zone refining for the higher resistivity values.

Next, the crystal is ground to a standard cylindrical size. X-ray diffraction determines the crystallographic orientation,[2] and flats are ground into the cylinder to record this information. The flats used to denote a p-type $\langle 100 \rangle$ crystal are shown in Fig. 3.1. The boule is then sawed into wafers. In subsequent steps, scratches are removed by etching, the backside of the wafer is sandblasted, and the frontside is polished to a roughness less than ±6 μm. The wafers are then cleaned. The final wafers are on the order of 300 μm thick.

FIGURE 3.1 The major and minor flats that denote a p-type $\langle 100 \rangle$ wafer. The polished side is up.

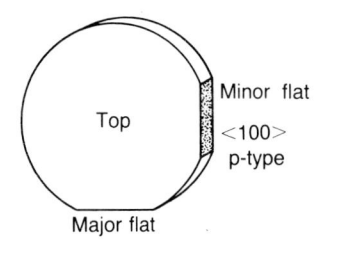

Top

Minor flat

<100>
p-type

Major flat

---

[2] Silicon has the same crystal structure as diamond. As in the case of diamond, one may cleave silicon along many different planes. Not all of these planes are equal. For instance, the density of atoms at the surface is different for different crystal orientations. Two different crystallographic orientations are denoted by "$\langle 111 \rangle$" and "$\langle 100 \rangle$."

At this point, an epitaxial layer[3] can be grown if desired. In this process a gas such as silane (SiH_4) is passed over the wafers, where, under the influence of RF heat, it is decomposed. A single-crystal silicon film between 0.5 and 100 μm thick can be grown on top of the existing single-crystal wafer by this technique. The growth rate of the film is related to the temperature. In one of the more mature processes, $SiCl_4$ is decomposed at $1200°$ C for a growth rate on the order of 1 μm/min. SiH_2Cl_2 can be used around $1080°$ C with about one third the growth rate. SiH_4 can be used around $1000°$ C. Going to lower temperatures ensures that dopants move as little as possible. One can grow an epitaxial layer that changes from a degenerately doped n-type substrate to 50 Ω·cm material in about 2 μm.

One of the big advantages of epitaxy is that the epi layer can have a doping quite different from the substrate. A typical n-type dopant in the epitaxial layer is arsenic and the typical p-type dopant is, of course, boron. Buried layers can be diffused into the substrate before the epitaxial growth. Arsenic and antimony are typical buried layer dopants. This technique is very common in bipolar circuits; it is being used somewhat in CMOS circuits to build, for instance, retrograde wells. In retrograde wells the doping increases with distance from the surface. This has advantages for latch-up resistance. With heavily doped buried layers, there may be crystal damage. Because epitaxy is done at high temperature, the vapor pressure of the dopants in the substrate can be an issue. Wafers upstream give off dopants that are absorbed by wafers downstream. In addition, the dopants in the gas stream get depleted. Oxides are sometimes grown on the back of wafers to limit out-gassing. In order to account for these effects, the gas flow and temperature of the wafers must be continually adjusted during growth. Epitaxial growth is a key step in the fabrication of twin-well and other latch-up resistant CMOS technologies.

Photolithography and mask making

Photolithography is the process linking together the many techniques that will selectively remove or add material to the wafers. The mask contains the information we want to transfer to the wafer. For geometries larger than about 1 μm, optical (actually UV) techniques can be used. The most inexpensive masks are made of UV-transparent glass with a 2–4 μm-thick coat of photographic emulsion. These masks are typically used in contact printing and wear out quickly. In contact printing the mask is brought in intimate contact with the wafer. The advantage of contact printing is that the optics are fairly straightforward.

---

[3] An epitaxial layer is a thin layer of material grown on top of a substrate. The word comes from the root *epi,* meaning "on."

Hard masks use a 1000–2000 Å-thick coating of either chromium or iron oxide. These masks are typically used in projection alignment systems.[4] Iron oxide has the advantage that it is semitransparent to visible light while being opaque to UV light. This makes the alignment of masks much easier. Masks for X-rays usually have their patterns etched in gold on a beryllium or silicon substrate. The push to go from UV to X-rays comes because the shorter wavelength of X-ray radiation allows for better resolution.

Pattern generation for the mask can be done optically or by an electron beam machine. The optical technique uses a precision mechanical instrument that flashes rectangles through an adjustable iris and onto a photographic plate. Once a single mask level is exposed for one chip, it is transferred many times onto a second mask via a step-and-repeat technique. A mask might contain hundreds of copies of a chip. This was the standard technique of the 1970s. The final mask will be used to simultaneously print all of the chips on a given wafer. Because there are many processing steps that occur simultaneously for all chips on a given wafer, there has been an evolution toward larger wafers to increase productivity.

The electron beam technique is similar except that an electron beam is used to directly write the mask. The electron beam is driven by computer and, unlike optical or X-ray techniques, must write each pixel sequentially. The step-and-repeat technique can then be used, or the electron beam machine can write all of the locations on the mask. The masks (called reticles) out of the electron beam machine are much more expensive and of higher quality than their optically generated predecessors. Figure 3.2 illustrates a mask. When a reticle is used directly for printing, a projection aligner is used. The mask lasts much longer since it never touches the wafer. The optics of a projection aligner are quite sophisticated and must become more so as the radius of the wafers increase. Exposure takes place in a strip of UV light that moves across the wafer. Because of this, an early generation of projection aligners had a preferred direction. That is, geometries in z could be defined more finely than those in y. This feature has been discontinued.

Because there is no constraint that every chip location on the electron beam written mask contain the same information, the electron beam technique can be used to write multiproject [Conway 80, Jansen 81] masks—that is, masks with many different chip designs. This has been key to the development of silicon foundries in which designs from many different customers are processed with the same mask set and on the same wafers. In another technique, the mask out of the electron beam machine (usually with geometries 5 or 10× their final size on the wafer)

---

[4] In a projection alignment system, the mask image is projected onto the wafer, which it never touches.

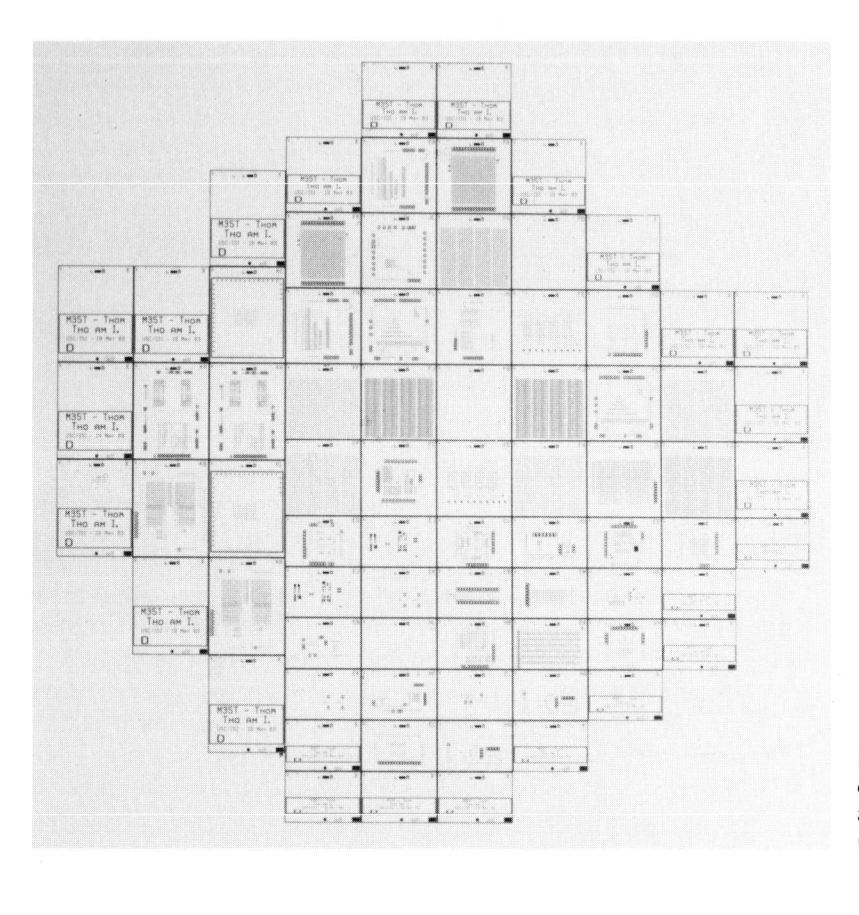

FIGURE 3.2   Photograph of the active area mask for a 4 μm nMOS technology. (Courtesy of MOSIS.)

can be stepped and repeated on the wafer rather than on a second mask. This type of aligner is called a wafer stepper. It allows for alignment at each die location and, because of the smaller exposure field, can have better optics.[5] It should also be noted that the electron beam machine can write directly on the wafer without any masks at all. This technique is slow, but appropriate in some research contexts.

Patterns are transferred from the mask to the wafer by first coating the wafer with photoresist and then exposing that photoresist through the mask. Photoresist is an organic polymer that is sensitive to light or electron beams. It comes in two types: positive and negative. In negative photoresist, the polymer crosslinks where the UV light impinges on it and forms long organic molecules. In positive photoresist, the material is broken into smaller molecules where the light strikes. In either case, the photoresist is developed by washing it in a weak organic solvent that

---

[5] Another reason for a small exposure field is that a wafer may grow or shrink by several microns during processing.

is capable only of dissolving the shorter molecules. Thus with positive photoresist the material is removed where it is exposed and vice versa for negative resist. Because of reflections, lines in positive photoresist tend to shrink, while those in negative photoresist expand. For this reason positive photoresist is preferred for defining fine geometries.

Some special photoresists are sensitive to X-rays or electron beams rather than UV. Some resists are inorganic. Key qualities of a resist include its adhesion, resolution, sensitivity, and resistance to solvents. One cannot always find everything one wants in a single resist. Therefore multilevel resist systems are sometimes used. In these systems, only the top layer is exposed to the masking optics. Sublayers are defined by chemical etching, using the top layer of photoresist as the mask. Since the bottom layers of the resist conform to the wafer geometries and fill up the valleys, the top layer of resist can be on a very flat surface. For these reasons, multilevel resists have about 50% better resolution than single-layer resist systems. Organic photoresists can not be used over about 200° C.

Photoresist is applied to the wafer on a spinner, a machine that applies a drop of photoresist and, by spinning the wafer under controlled velocity and acceleration, spreads the photoresist into a nearly uniform layer. Exposure takes place in a mask aligner. This machine, which is usually the bottleneck in any fabrication line, allows alignment marks on the wafer to be aligned with alignment marks on the mask. This provides layer-to-layer registration. In a manual machine, an operator views the mask and wafer through a stereo microscope. Alignment marks, usually concentric squares, are used as alignment keys. In a more modern machine, such as the wafer stepper, computer-readable alignment marks are provided on the masks and the alignment is done automatically. Since wafers are exposed one at a time, mask alignment is a relatively expensive technique compared to, say, epitaxial growth where 20 wafers can be processed at once.

After the exposure and development of the resist, a post-bake operation which increases the resist's adhesion and hardness is performed. The resist can then be used to protect the surface of the wafer from, for instance, the effects on an etch. After the resist has served its purpose it is stripped in a solvent capable of attacking the long polymer chains.

The selective removal of material

Of the many techniques for material removal, etching is certainly the most well established. Chemical or "wet" etching has the advantage of excellent selectivity. Selectivity refers to the propensity for the etchant to etch the material one wants to remove rather than the material one does not want removed. For example, buffered HF will attack SiO_2 but not silicon. Dilute HF was used to strip off the SiO_2 over-glass and aluminum

in order to take the photomicrograph of the T11 shown in Fig. 1.2(b). *HF is very dangerous and should be used only with extreme caution. HF does not burn but rather seeps down to the bone, killing tissue and nerves* [Straub 83]. *Obviously, one cannot store HF in glass bottles.*

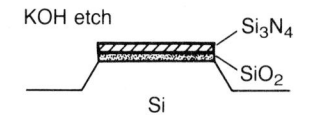

FIGURE 3.3   Preferential etching of silicon using KOH.

Many etches will preferentially etch certain crystallographic planes in silicon. This allows some interesting silicon structures to be built. This technique is, for instance, the basis of VMOS. In one commonly used CMOS process, KOH is used to etch the field areas as shown in Fig. 3.3. KOH etches ⟨100⟩ material quickly and ⟨111⟩ material almost not at all. When the oxide is grown, the surface becomes almost planar, as shown in Fig. 3.4. The bird beak encroachment is still severe, however. Techniques to reduce this encroachment have been investigated [Oldham 82, Chin 82].

Plasma, sputtering, and reactive ion etching are variations on the same theme in which a gas is excited by RF or dc means and the excited ions blast away at the silicon surface. In sputter etching the gas is inert and removes material mechanically. In plasma etching the gas is chemically active and removes material more or less isotropically—like wet etching. Reactive ion etching (RIE) is sputtering with chemically active ions. The advantage of RIE is that the electric fields can cause the ions to impinge the surface vertically. This causes anisotropic etching with the steep vertical walls needed for very fine linewidths. The main drawback to these techniques is that the selectivity is less than ideal.

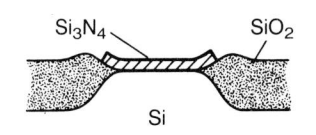

FIGURE 3.4   Bird-beak formation in the field oxide. Notice how planar the top surface is due to the preferential etch.

In some cases photoresist can be used directly to mask the removal of material, while in other cases a more indirect approach is necessary. For instance, photoresist can be used to mask the etch of SiO_2, which can in turn be used to mask the etching of silicon. A key place this indirect technique is used is in the definition of the gate region in a silicon gate MOS process. After the poly is etched, the poly is used as a mask first to etch the gate oxide and later to mask the source and drain implant. These last two steps are self-aligned to the poly. Self-aligned processing steps are key to achieving close manufacturing tolerances.

The addition of material

Many techniques are available for the addition of material to the silicon wafer. The wafer may be put in an oven and oxidized. For most oven and clean steps, wafers are processed by the "boat" load. A boat contains 10 to 25 wafers. As the oxidation process occurs, silicon becomes SiO_2. Since the ratio of volumes is approximately 1 : 2, we find that the SiO_2 is above the original silicon surface by about the same amount as it is below. The SiO_2 is amorphous. Silicon dioxide can be grown at any temperature between 850° C and 1000° C. At 900° C one can grow 400 Å of oxide in about 5.5 min, while it takes 190 min to grow 4000 Å of oxide at the same temperature. Note that the growth rate

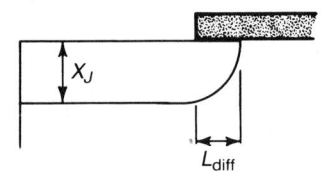

FIGURE 3.5   Lateral diffusion at a corner. Typically, $L_{diff} = 0.7 X_J$.

decreases with increasing thickness. This is because the oxygen must diffuse through the silicon dioxide to reach the silicon surface. Very thick oxides take a long time to grow because of the quadratic time behavior of the diffusion equation. Growing 4000 Å of oxide at 1000° C takes about 48 min. Because oxide growth is a high-temperature process, it alters the concentration of impurities in the vicinity of the silicon/silicon dioxide interface. A key question is whether the dopant prefers to reside in the silicon or the silicon dioxide. For instance, as the oxide grows down into the silicon surface, boron is depleted from the surface, while the concentration of phosphorus increases in the silicon just below the interface. The impurities in the oxide do not seem to be important. Oxidation of the silicon surface can be masked by Si_3N_4.

The silicon surface can be doped by diffusing in impurities from the surface. The exact impurity profile depends on the type of source but the impurity distribution tends to drop off as $\exp(-x/\sqrt{Dt})$ in the bulk, where x is the distance into the wafer, D is the diffusion coefficient, and t is the diffusion time. The diffusion coefficient is a strong function of temperature. The diffusion coefficients for boron and phosphorus are about equal, while that of arsenic is about one order of magnitude smaller. This means that all else being equal, phosphorus will diffuse about π times as deep as arsenic.[6] Diffusion of these impurities can be masked by SiO_2 or Si_3N_4. The maximum doping level is determined by the solid solubility limit and this can effect the minimum resistance and doping profile. Given a corner, as shown in Fig. 3.5, one will typically see about 70% as much lateral diffusion as vertical.

Another technique for adding material is ion implantation. Ion implantation can be masked by almost any material, including photoresist. This low-temperature technique was discussed extensively in Section 2.3. An annealing step is necessary afterward to repair damage to the lattice and to electrically activate the impurities. Annealing temperatures range from 900° C to 1000° C. High temperatures seem to be needed in order to give the ions enough energy to find their place in the lattice and become electrically active. Since we do not want these atoms to diffuse, a variety of rapid annealing techniques have been invented so that the lattice temperature is raised for only 10 to 20 seconds. Energy is added to the silicon via lasers, electron beams, or radiant heat. There is an important second-order effect that is especially severe with boron. Implanted boron is very efficient at traveling along the silicon crystallographic planes (channeling). This leads to a deep minimum junction depth. This problem and some techniques to achieve improved results are discussed by Liu and Oldham [Liu 83]. The fact that ion implantation can be masked by photoresist and the fact that it is a mostly low temperature process combine to make ion implantation one of the more important techniques for VLSI fabrication. Its major limitation

---

[6] $\pi \approx \sqrt{10}$, for large values of π.

is the depth to which implantation can be done (less than a couple of microns). For this reason we expect that long diffusions will continue to be used for deep steps, such as well fabrication in CMOS. Of course, ion implantation can be used to insert the impurities to be diffused.

Thin film deposition is another technique for the addition of material. In one method, called evaporation, the material to be deposited is heated until it evaporates. It then falls like calm snow and condenses on the cool wafer surface. Step coverage with this technique is very poor. This can be used to advantage to selectively deposit some materials, such as aluminum, in a technique called lift off. With lift off, the places where one does not want material are masked by photoresist. After the material is deposited (everywhere), the photoresist is dissolved, and together with the material that landed on top of it, is washed away leaving material only where there was no photoresist. This technique is illustrated in Fig. 3.6.

Sputtering is another common technique for adding material. In fact, most aluminum is deposited by sputtering. The target material to be added is blasted with an argon beam. The material lands everywhere, including on the wafer, which is biased to attract the ions of the target material and keep secondary electrons away so that they do not cause heating.

The technique of plating involves the immersion of the material to be added and the wafer in an electrolyte solution. Current is forced to flow in the direction that causes ions to be attracted to the wafer. Plating is a critical technique in the formation of very thick metal layers, such as those used in beam lead devices. As their name implies, beam lead devices have wires sticking out from the chip that are fabricated as part of the chip processing. In general, the problem with plated metal is that it is very rough because of dendrite growth.

The epitaxial growth technique already discussed is a special case of chemical vapor deposition (CVD). Good step coverage is attained by reacting the CVD gas in the same chamber as the wafer. Silicon, silicon dioxide, silicon nitride, and many other materials can be deposited in this way. Typical CVD temperatures range from 380 to 600° C. A special case of CVD is LPCVD, an acronym of "low pressure" CVD. Some LPCVD techniques can provide excellent conformal depositions. (The opposite of the anisotropic depositions needed for lift off techniques, conformal contours are called for when good step coverage is an issue.) To obtain the lowest CVD temperatures, optical energy is sometimes added to assist the chemical reaction.

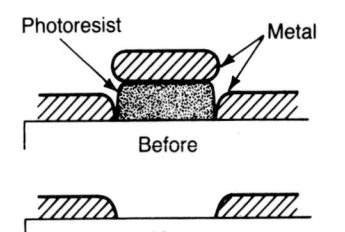

FIGURE 3.6 Metal lift-off. The metal that lands on a photoresist is washed away when the resist is removed. The metal deposition must have poor step coverage for this technique to work.

Finished wafers and packaged chips

A finished wafer is shown in Fig. 3.7. After the wafer is processed it must go through the steps of assembly and packaging. These steps include

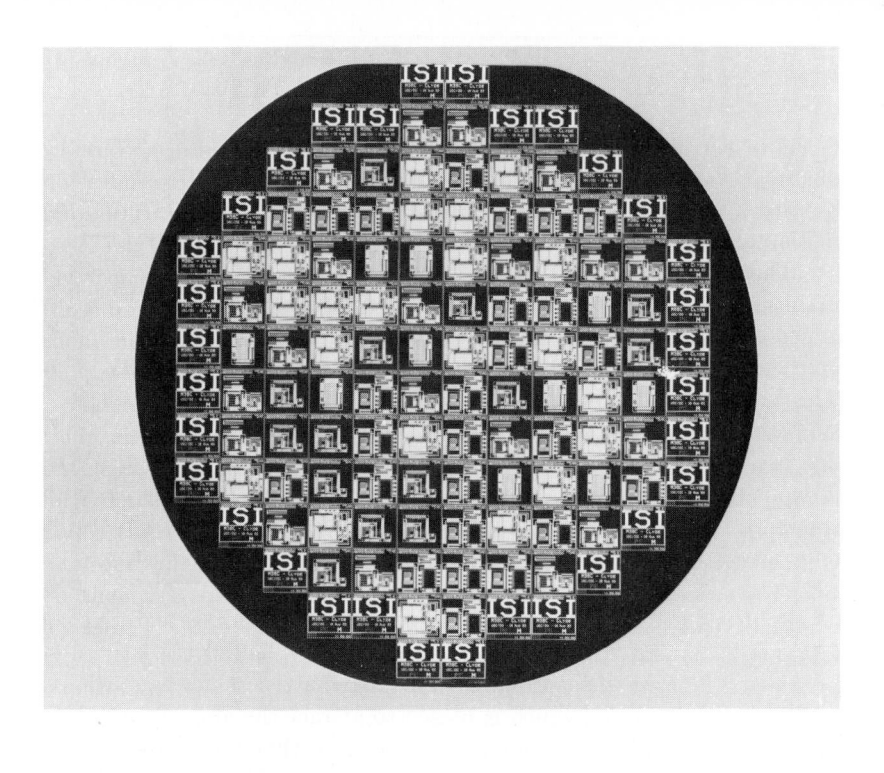

FIGURE 3.7 A finished
multiproject wafer. The
technology is 5 μm p-well
CMOS with two layers
of poly and one of metal.
(Courtesy of MOSIS.)

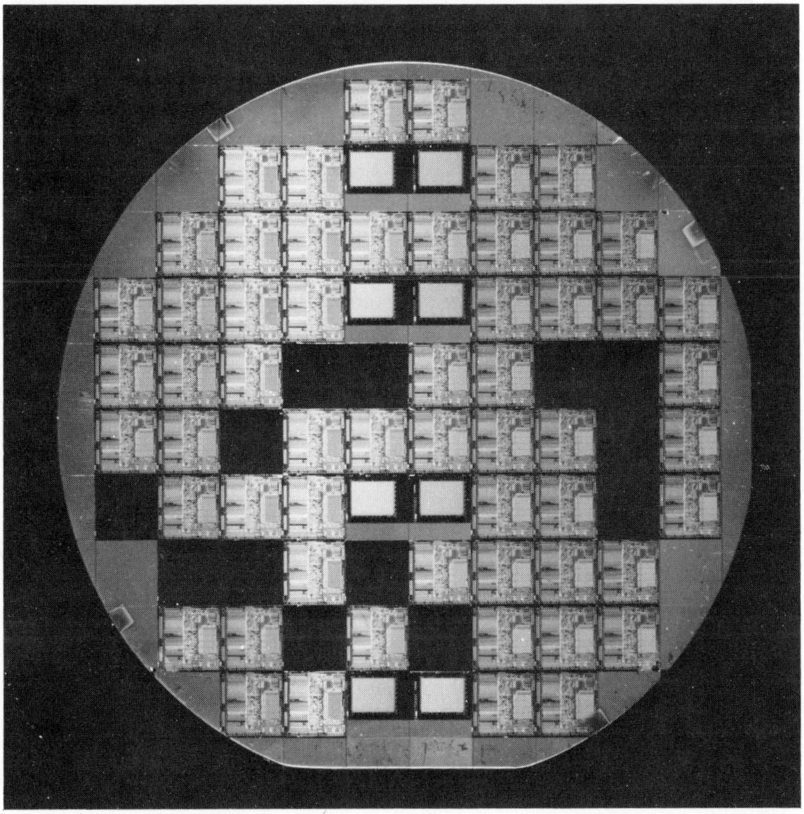

FIGURE 3.8 A collection
of unbonded chips.
(Copyright ©Digital
Equipment Corporation
1985. All rights reserved.
Reprinted with permission.)

174

FIGURE 3.9 A bonded chip.
(Copyright ©Digital Equipment
Corporation 1985. All rights
reserved. Reprinted with permission.)

wafer probing, die separation, die attach, wire bonding, and package sealing.

Wafers are tested on an automatic test system that applies test patterns to the chip through probes. The time required to apply a sufficiently large number of test stimuli to ascertain that the device under test is working correctly can have a significant negative impact on the cost of the final product. When bad chips are found, they are marked with a drop of ink.

After the wafers have been tested and inked, they are either cut apart with a diamond saw or scribed with a diamond stylus. After being scribed, the wafers are placed on a tape and broken apart. This technique is similar to the technique used to cut glass. A small collection of chips is shown in Fig. 3.8. The chips can be mounted either by scrubbing[7] them into a package with a high-temperature solder (usually a gold eutectic) or, more commonly, with epoxy.

The most common method for connecting the chips to the outside world is wire bonding. There are two common forms. Thermocompression bonding uses gold wire at about 240° C. This technique is most successful when attaching gold to gold. It can, however, be used with aluminum bonding pads, providing the temperature is kept low. The welds formed are of high quality. The bonding is a two-step process wherein the wire is first bonded to the chip bonding pad and then to the package. Ultrasonic bonding is a second technique. In this case, aluminum wires 20–30 μm in diameter are used. This is the more common technique for MOS VLSI. Whereas most semiconductor processing techniques are done by the wafer or boat load, bonding is an expensive exception. Shown in Fig. 3.9 is a bonded chip before the package is covered and sealed.

---

[7] Placing the chip on a melted solder preform and ultrasonically vibrating the chip to break the solder skin.

Another bonding technique, used extensively at IBM, is solder bump technology. Dies are attached to the package by solder bumps that have been plated onto the bonding pads. The pads are typically located near the center of the chip to accommodate thermal expansion effects.

Packages can be metal, ceramic, or plastic. Metal packages have the best thermal characteristics, and plastic the worst. Plastic is also the cheapest. Ceramic chip carriers that hold and interconnect several die are in use at several companies.

3.2   Two basic MOS processes

The techniques discussed in the last section can be combined in various creative ways to build MOS integrated circuits. We take up the process of building a MOS circuit at the point where we have wafers of single-crystal silicon. A typical wafer is a disk approximately 100 mm or more in diameter and 300 μm thick. It has a mirror polish on the top surface. Because silicon processing is extremely susceptible to contamination, the entire fabrication process takes place in a nearly dust-free environment. Many of the processing steps are interspersed with cleaning procedures to remove contaminants from the wafers. We will first discuss a basic nMOS process and then go back and discuss the modifications necessary to build an n-well CMOS process.

A basic nMOS process

The first step in fabricating an nMOS integrated circuit is typically to place the p-type wafer in an oxidation furnace where a thin buffer layer (300 Å) of high quality SiO_2 is grown. In a local oxidization process, the next step is to deposit a thin layer of Si_3N_4. Without the SiO_2, the Si_3N_4 would cause surface state problems. The silicon dioxide layer is called either a stress relief oxide or a buffer layer. The wafer is now ready for the first lithographic step. Photoresist is spun on and exposed using the first mask, as illustrated in Fig. 3.10. This is the active area mask (also called the n+ or diffusion mask). It defines what will eventually be sources, drains, channels, and diffused cross unders. The photoresist is developed and used to mask the Si_3N_4 from a plasma etch. The wafers are brought to the ion implanter where they receive the boron-high field implant, as illustrated in Fig. 3.11. The photoresist and nitride mask the implant from doping the channels, sources, and drains. The photoresist is then stripped.

The wafers are placed back in the oxidation furnace and the thick field oxide is grown. The nitride does not permit ambient oxygen to reach the silicon surface. To first order, the thick field oxide grows only

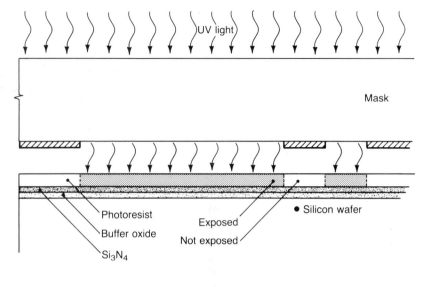

FIGURE 3.10   Exposing the active area mask with UV light.

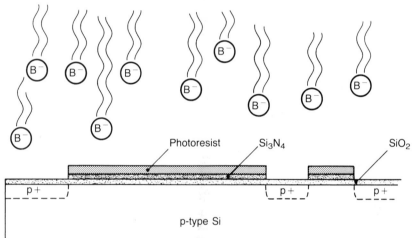

FIGURE 3.11   The boron field implant is masked by the photoresist defined by the active area mask.

where the nitride is not, as illustrated in Fig. 3.12. There are, however, some very important second-order effects. The high field implant diffuses part way into the masked region. This is partly responsible for the ΔW effect on transistors, which was discussed in Section 2.3. The other effect is that the oxide encroaches into the nitride region from the sides. Because one molecule of SiO_2 takes up approximately twice the volume of one molecule of Si, the bird beak structure shown in Fig. 3.12 is formed. The smooth taper of the bird beak provides for good step coverage in later material dispositions. On the other hand, the magnitude of the encroachment is a severe problem on processes

FIGURE 3.12 Growth of the field oxide is masked by the nitride patch. The buffer oxide under the nitride helps lower the stress on the silicon surface. The nitride is removed after this step.

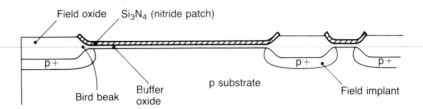

below about 4 μm. Note how the high-temperature step of growing the silicon dioxide causes the redistribution of the boron ions. Obviously, if one is trying to define very fine structures, the movement of impurities defined in previous steps is undesirable. This is the reason behind the trend toward lower-temperature processing.

The nitride, the buffer oxide, and a little of the field oxide is then removed. A new very thin oxide is then grown. This will be the gate oxide. Because the oxide eats into the silicon, the silicon/silicon dioxide interface under the gate will be virgin material. In fact, most processes grow and strip several gate oxides to assure purity. Thermally grown oxides such as the field oxide and gate oxide are of high quality. They have higher electric field breakdown thresholds and are typically more resistant to etches than oxides deposited by chemical vapor deposition. To be reliable, a gate oxide should not operate with a field of more than 2 MV/cm.

The next few steps are for adjusting the thresholds of the different transistor types. Thick photoresist is spun on the wafer and exposed through the ion implant mask. After being developed, the thick photoresist acts as a mask to the ions. The implant can go through the thin gate oxide, which protects the surface of the silicon from contamination. Implants occur at the intersection of the active area and the implant masks. Implants can also be used to adjust the doping profile in the regions that will eventually be channels. In advanced processes, implants are used to control the doping profile of the source and drain diffusions near the transistors. This is to minimize short channel and hot electron effects.

Photoresist is again spun onto the wafer and the buried contact mask is used to define where polysilicon and n+ will be able to contact. The photoresist is developed and the buried contacts are etched in the usual manner. The buried contact etch needs only to remove the small amount of oxide in the gate region. Holes in the oxide are etched only at the intersection of the active area and buried contact masks.

We should point out again that there are many possible variations to the "typical" process flow we are describing. For instance, in the next step, we deposit, by chemical vapor deposition, heavily phosphorus-doped polycrystalline silicon. Undoped poly could have been deposited instead, and then doped. Alternatively, the undoped poly could be

MASK
Poly
Depletion implant
Enhancement implant
Buried contact
Active area

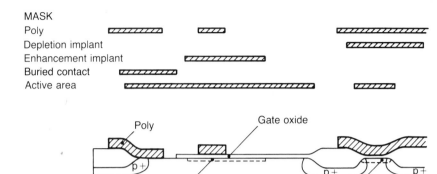

FIGURE 3.13   An nMOS wafer after the growth of the gate oxide, the threshold-adjust implants, the buried contact etch, the deposition and doping of the poly, and the poly etch. The masks for these steps are shown schematically above the wafer.

etched and then doped simultaneously with the sources and drains. Phosphorus is used in the polysilicon because it yields a lower minimum resistivity. The poly is defined with the poly mask and etched, as illustrated in Fig. 3.13. The poly is then used as a mask for the thin gate oxide. The gate oxide is removed everywhere except where it is protected by the polysilicon, as shown in Fig. 3.14. Thus the gate is self-aligned to the poly.

In the next step, the source and drain regions are defined with a heavy implant dose of arsenic. Arsenic is used because it has a lower diffusion coefficient than phosphorus and hence, for the fabrication of fine geometries, tends to stay where it is put. This allows the generation of shallow junctions and limits lateral diffusion.

The wafer is then slightly reoxidized. This does several things. It anneals out the damage caused by the ion implantation and it causes the various impurity atoms to diffuse. The phosphorus from the poly diffuses into the silicon, which is underneath buried contacts, and causes deep junctions in these areas. A thin, strong oxide is formed on the polysilicon and on the source and drain regions. The sources and drains are the region of heavy arsenic implant. The arsenic implant occurs at the intersection of the active area mask and the complement of the poly mask. By the same physics that caused the bird beaks in the nitride mask steps, the gate oxide becomes slightly thicker in the regions next

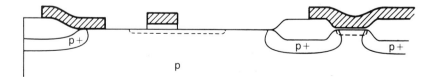

FIGURE 3.14   Etching the gate oxide using the poly as a mask. This is what makes silicon-gate nMOS "self-aligned."

FIGURE 3.15 Diffusing the sources and drains. The poly will oxidize slightly and the buried contact will be driven into the substrate. The n+ is often implanted. Note that the poly will get an extra dose of As.

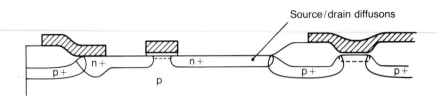

to the sources and drains. Since this region is over the n+ that laterally diffuses into the transistors, this process helps to lower the gate overlap capacitances. (Again, in advanced processes this effect is not very strong because the oxidation time is kept short.) At this point we have the transistors illustrated in Fig. 3.15. The remaining steps have to do with interconnect.

More oxide is then deposited by chemical vapor deposition. A phospho-silicate glass is often used at this stage, instead of pure SiO_2, to passivate the chip. The P_2O_5-loaded SiO_2 tends to keep sodium ions away from the silicon surface and thereby stabilizes the device thresholds against flat-band voltage shifts due to Na contamination. It does this by "gettering" or binding up the sodium ions. The wafer is heated and the CVD glass is reflowed in order to form a smooth surface for the remaining process steps. The P in the glass helps the reflow process. For shallow junction processes, this reflow step is avoided because of its detrimental effect on the junction depth. This, however,

FIGURE 3.16 A technique for forming self-aligned contacts from metal to poly: (a) nitride on top of poly on top of oxide; (b) patterned nitride; (c) half-etched poly; (d) thermally oxidized poly; (e) nitride removed, and metal contacts made where desired.

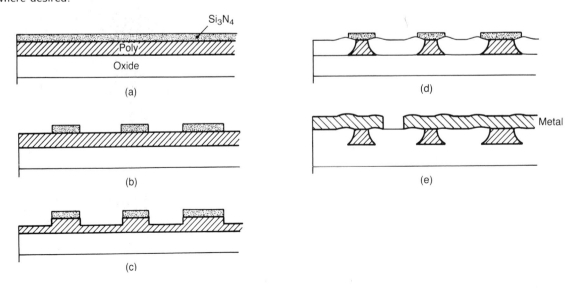

causes step coverage problems. The contact hole mask is next. (The reflow is sometimes done after the contacts are etched.) Contact holes are etched down to the n+ and poly. Note that these contacts do not have to go through the thermally grown thick field oxide.

There is a trend toward the use of self-aligned contacts because contacts are becoming more and more of a problem for advanced technologies, in terms both of resistance and area. Figures 3.16 and 3.17 illustrate some of the elements of a self-aligned contact technology. In Fig. 3.16, we see a technique for making self-aligned contacts to poly [Rideout 79, 80, Sakamoto 80, Muramoto 78]. The key to this technique is the selective oxidation of poly in regions where one does not want contacts. The removal of the nitride layer exposes the poly in regions where one would like to make contact. In Fig. 3.17, we see a related technique used to contact poly, where the contact can occur over gate material. This technique is successful because some CVD oxides can be made to etch an order of magnitude faster than thermally grown

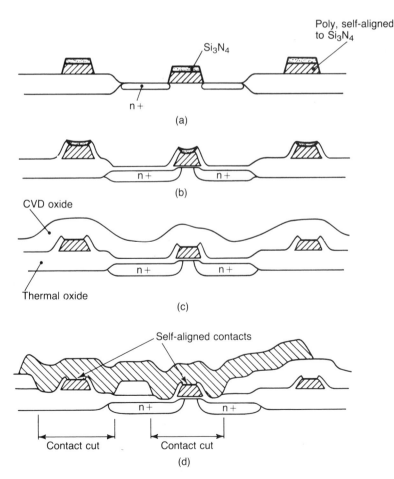

FIGURE 3.17   A second technique for forming self-aligned contacts. The key to this technique is that thermally grown oxides may be much more resistant to etches than CVD oxides. (a) Nitride is used to mask the poly etch that in the conventional manner is used to define the diffusion and gate oxide. (b) The wafer is thermally oxidized. (c) CVD oxide is deposited. (d) Oversize contacts are etched in the CVD oxide. The etch rate slows when the thermal oxide is reached. The metal is deposited and patterned.

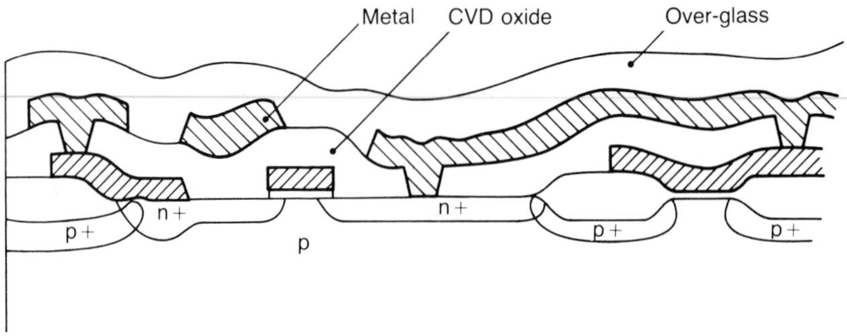

FIGURE 3.18 A finished nMOS wafer with conventional contacts. Though this is the planar process, the top is not very planar.

oxide. Thus one can open a large contact window over the poly and be assured that one will not etch down to the substrate or diffused silicon regions. Both of these techniques have the problem that they require high-temperature processing.

Aluminum is then sputtered onto the wafer and patterned with the metal mask. This aluminum layer forms the third tier of interconnect. A final layer of "over-glass" is then added for protection. This over-glass is typically Si_3N_4 over SiO_2. Figure 3.18 illustrates a cross section at this point in the process. Large holes are etched in the over-glass to provide access to the bonding pads.

The following is a summary of the basic nMOS process steps.

- Grow buffer oxide and deposit Si_3N_4. Spin on photoresist.
- Expose *active area mask* to define eventual channel and n+ regions. Develop photoresist and etch Si_3N_4.
- Do field implant. Then strip photoresist.
- Grow field oxide using the Si_3N_4 as a mask. Strip remaining Si_3N_4.
- Grow gate oxide. Pattern and etch buried contacts using *buried contact mask*.
- Use *threshold adjust masks* to define different transistor types. The threshold adjust is performed by ion implantation and masked by photoresist. The implant goes through the gate oxide but not the field oxide, which is much thicker.
- Deposit the poly and pattern with the *poly mask*. Use poly as a mask to etch the gate oxide.
- Implant arsenic to form n+ regions. Source/drain regions are self-aligned to the poly. Reoxidize the wafer to activate the implants and drive the buried contact junction.
- Deposit CVD oxide and pattern metal contact cuts with the *contact mask*. Etch contacts.
- Deposit metal. Pattern with *metal mask*. Etch metal.

- Deposit over-glass. Pattern with *over-glass mask*. Etch holes for bonding wires.
- Probe wafers and ink bad die.
- Saw and dice wafers.
- Package, bond, seal, and retest.

────────────── EXAMPLE 3.1

A cross section drawn through an in-line buried contact
Figure 3.19 illustrates top and cross-sectional views of an in-line buried contact on a depletion mode pullup. Because the length of the transistor is poorly controlled, it must be drawn quite long if its conductance is at all critical. This technique is popular in bleeder configurations. Note the deep diffusion of the phosphorus under the buried contact.   ■

An n-well CMOS process

The basic difference between the nMOS processing sequence we just described and an n-well CMOS processing sequence is the first step. In CMOS processing, the first step is to define and diffuse the wells

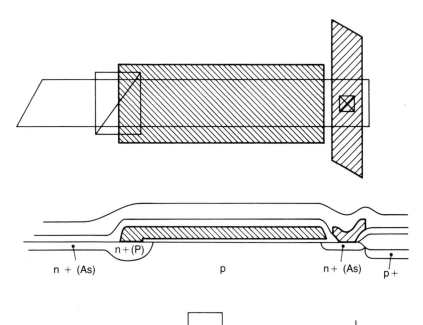

FIGURE 3.19   An "in-line" buried contact viewed from the top and side. The length of the transistor is poorly controlled.

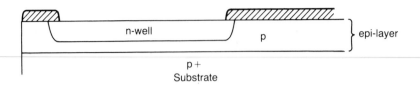

FIGURE 3.20   An n-well diffusion in a CMOS process with epi.

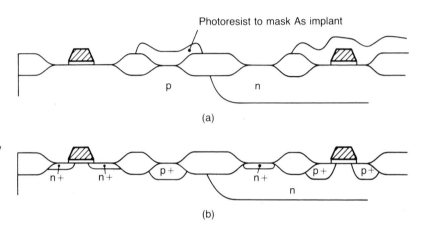

FIGURE 3.21   (a) Before and (b) after illustration of the source/drain implant. Photoresist is shown masking the n+ implant. Note that the poly is slightly counter-doped with B in the region of p-type transistors. This sometimes increases the poly's resistance, but it remains n-type.

illustrated in Fig. 3.20. Silicon dioxide or silicon nitride can be used as a diffusion mask. Typically, buried contacts are not used since the poly will be either all n-type or all p-type and thus only one sex of contact will be ohmic.[8] Field implants are also less common in CMOS processes larger than about 2 μm. After the threshold adjust implants, the poly is deposited, doped, and patterned. The n+ mask is used to define the active areas that will be implanted with arsenic. Figure 3.21 illustrates a CMOS chip before and after the n+ and p+ diffusions. Photoresist can be used as a mask. The p+ mask is then used for the p+ source and drain definition. The p+ mask can be optically generated as the complement of the n+ mask, or positive and negative photoresist can be used with the same mask. In some cases, masks that are not the complements of each other will be used. This is required if one desires the ability to build Schottky barrier diodes.[9] Note that the n+ poly will be counterdoped

---

[8] If n+ poly is used, then buried contacts in the n-well (providing one keeps them from shorting to the well) will form p–n diodes that actually face in the right direction to be useful. An inverter built with this technique will have a forward-biased diode between the p- and n-channel devices. The minority carrier injection from the p–n diodes, however, tends to cause latch-up.

[9] A Schottky barrier diode is a diode formed at a semiconductor/metal interface. Murphy's law of processing says that one will obtain Schottky barrier diodes when one wants to form ohmic (linear resistive) contacts, and vice-versa.

in the p+ area. The poly remains n+ because the original n+ dose was very high, but the poly can have a slightly higher resistance in these areas. Figure 1.14 illustrates a cross section of a typical CMOS layout. Note that the n+ sources and drains have a smaller junction depth than the p+ sources and drains. This is a result of the differences in dopant diffusivity and the propensity for the boron to channel.

Multiple layers of metal are sometimes used. The challenge is that if silicon dioxide is used between layers of aluminum, it can not be reflowed in the usual manner without melting the first layer of aluminum. Special dielectrics, deposition techniques, and refractory metals have each been used by manufacturers to deal with this problem. The dielectric polyimide is sometimes used. It is an organic material that is compatible with low-temperature processing and has the nice property that it produces a very flat top surface for the next processing step. Refractory metals that have been used for MOS include Ta, W, Mo, $TiSi_2$, $TaSi_2$, and PtSi.

The following is a summary of the basic CMOS process steps.

- Grow oxide. Spin on photoresist and expose with *n-well mask*.
- Develop photoresist and etch oxide. Strip photoresist, then diffuse n-well except where masked by oxide. Strip oxide.
- Grow buffer oxide and deposit Si_3N_4. Spin on photoresist.
- Expose *active area mask* to define eventual channel and diffusion regions. Develop photoresist and etch Si_3N_4.
- Grow field oxide using the Si_3N_4 as a mask. Strip remaining Si_3N_4.
- Grow gate oxide.
- Use *p+ and n+ masks* to adjust transistor thresholds.
- Deposit the poly and pattern with the *poly mask*. Use poly as a mask to etch the gate oxide.
- Use the *n+ mask* to define the source/drain regions to receive an arsenic implant. Also defines n+ plugs to n-well.
- Use the *p+ mask* to define the source/drain regions to receive a boron implant. Also defines p+ plugs to substrate.
- Deposit CVD oxide and pattern metal contact cuts with the *contact mask*. Etch contacts.
- Deposit metal. Pattern with *metal mask*. Etch metal.
- Deposit over-glass. Pattern with *over-glass mask*. Etch holes for bonding wires.
- Probe wafers and ink bad die.
- Saw and dice wafers.
- Package, bond, seal, and retest.

3.3   Design rules

Design rules are an enforced methodology for the composition of mask geometries. Design rules come about through the interaction of electrical and reliability constraints with the capabilities of the fabrication technology. A typical design rule might be "poly-to-poly spacing must be greater than or equal to 2 μm." The complexity of a set of design rules can vary anywhere from the two-page "lambda" design rules of Mead and Conway [Mead 80, plates 2,3] to sets with well over 50 pages of documentation. It is a matter of the number of special cases one is willing to explain, the degree of process characterization, and philosophy. Figure 3.22 illustrates a few simple design rules.

Each design rule is based on the interaction of several phenomena. At a minimum, a proper set of design rules must take into account the characteristics of the photolithography process, etching capabilities, registration tolerances, and electrical constraints. The registration tolerance is related to the expected misalignment variance between mask levels. If one expects conformance to design rules to be checked by a computer program, this must also be taken into account in the definition of the design rules.

By virtue of the central role of photolithography in process technology, photolithographic constraints play an important role in the determination of design rules. A key metric is the minimum geometry

FIGURE 3.22   Some typical design rules.

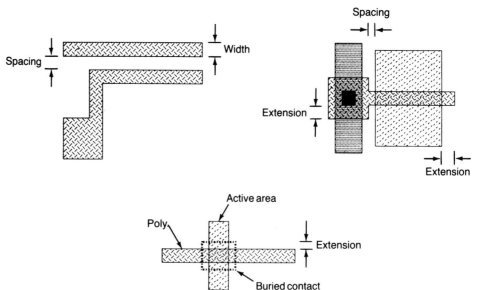

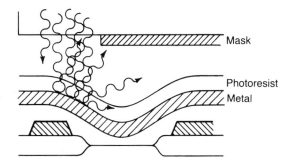

FIGURE 3.23   Reflections off metal side walls can cause "masked" photoresist to be unintentionally exposed.

that can be resolved in photoresist or, more practically, the tolerancing associated with small photoresist geometries. In theory, this particular metric should apply uniformly to rules covering all layers. In 1982, state-of-the-art dimensional tolerances on a planar surface had a 3σ tolerance of less than 0.5 μm, where σ is the standard deviation. There is one complication. The planarity of the surface can have a major effect on the quality of the image. As the semiconductor process progresses from level to level, the topmost surface can take on a nonuniform topography associated with the underlying layers. To the extent that this happens, the line spacing and width rules may need to be relaxed on the upper layers because optical diffraction limits the depth of field of the image to about 5 μm. This is one of the driving forces behind multilayer resist systems.

Another photolithographic effect associated with topography involves reflections of the UV light used for exposure. Since the metal goes over many hills and valleys, the lithography is clearly better in regions of the chip where there are no layers other than the metal. In an area where there are many steps, one has the possibilities of very strange rules called reflection rules. Recall the fabrication sequence wherein the metal is deposited on top of the mountainous chip surface and then coated with photoresist. One now has the frightening possibility that the light that impinges on the wafer from the transparent areas of the mask can be reflected sideways off the valley walls; this would expose supposedly masked regions of the chip, as illustrated in Fig. 3.23. Rules that try to account for this phenomenon by changing the metal spacing rules as a function of the number and kind of steps are called reflection rules. With this sort of rule, one might find that two metal lines could be 3 or 5 μm apart, but not 4 μm.

It is extremely difficult to develop CAD software to check for these sorts of rules. One may coat the aluminum with an antireflection coating to avoid this problem. Figure 3.24(a) illustrates metal lines, which cover horizontal poly lines, defined without an antireflection coating. Figure 3.24(b) illustrates these lines when an antireflection coating is used. One may also have artifact rules on any or all of the mask layers. An artifact rule states that all geometries on a single mask layer must be larger than some minimum dimension. This is simply to make masks easier to inspect for obvious flaws or bugs. (There can also be a secondary issue of small geometries breaking off and landing some place fatal.)

Once the pattern is defined and developed in photoresist, the underlying layer, or layers, must be etched. This process adds additional

FIGURE 3.24   Electron micrographs of metal lines defined (a) without anti-reflection coating and (b) with anti-reflection coating. A 10 μm marker appears on the micrograph. (Courtesy of Raytheon Co.).

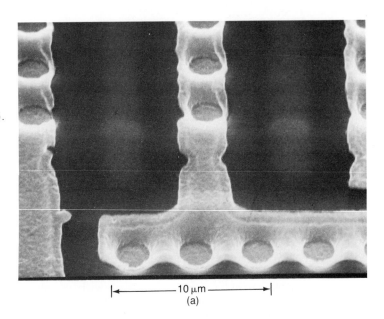

|← ———— 10 μm ———— →|
(a)

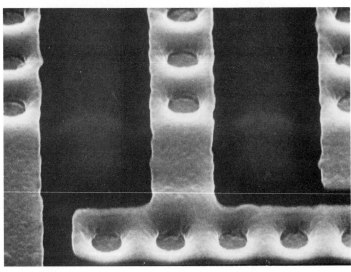

|← ———— 10 μm ———— →|
(b)

uncertainty to the line-width definition. At the most basic level, we want to prevent shorts between adjacent wires and open circuits on the wire itself. Traditional wet chemical processes are preferential, but isotropic. Because they remove material laterally as well as vertically, tight dimensional control can be difficult, especially for thick films. Plasma and RIE dry etching techniques have the characteristic of preferential vertical etching, leading to improved dimensional control capability. In all cases, dimensional control is easier for thinner films and varies somewhat with the material.

One of the complications of all etching processes is that the etch rate is not uniform across the wafer, but depends somewhat on the context in which a geometry sits. For instance, a thin line with nothing in its vicinity may etch faster than a thin line in a densely laid out region because the etchant chemistry is changed by the etching process. Because contact cut dimensions are critical to the layout density and difficult to control, many process engineers calibrate their processes to run best with the minimum-size contact. Thus contacts much larger than this tend to bloom and be uncontrolled. Therefore contacts might be required to be minimum size in at least one dimension, and possibly two. Some contacts will be over-etched so that all contacts will be guaranteed to be sufficiently etched.

Except for the very first masking layer, all subsequent layers must be aligned to a previous masking or process step. Until about 1980, this alignment was done manually by an operator matching concentric boxes called alignment keys. More recently, automatic alignment systems with better and more reproducible results have come into use. The wafer stepper technology also plays a key role here because of its advantage of being able to do a die-by-die alignment of each stepped pattern.

The choice of the alignment sequence has important consequences in the development of design rules. Consider, for instance, masking and process steps A, B, and C. Suppose B is aligned to A with a tolerance of ± 1.0 μm, and C is also aligned to A with the same tolerance. Then the worst-case registration of B to C is ± 2 μm. If we chose to align C to B, then the C-to-A registration would be similarly degraded. In practice, the tolerances are statistical and are not added algebraically. Usually a square root relationship is used; that is,

$$R = R_0 \sqrt{n}, \qquad (3.1)$$

where R is the registration, R_0 is the single level alignment between layers, and n is the number of levels of indirection. Figure 3.25 illustrates alignment trees for typical nMOS and CMOS processes. The distance between nodes is n. This data is re-expressed in Table 3.1. A major decision is whether the contacts are aligned to poly or active area.

Material properties and interactions generate additional constraints. We will now look at some of these.

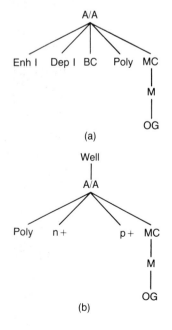

FIGURE 3.25 Alignment trees for (a) nMOS and (b) CMOS. The abbreviations are A/A = active area, Enh I = implant, BC = buried contact, MC = metal contact cut, M = metal, and OG = over-glass.

TABLE 3.1 MOS alignment tables

nMOS ALIGNMENT TABLE								
n	A/A	ENH	DEP	BC	POLY	MC	M	OG
Active/Area	0	1	1	1	1	1	2	3
ENH implant		0	2	2	2	2	3	4
DEP implant			0	2	2	2	3	4
Buried Contact				0	2	2	3	4
POLY					0	2	3	4
Metal Contact						0	1	2
Metal							0	1
Over-glass								0

CMOS ALIGNMENT TABLE								
n	W	A/A	POLY	n+	p+	MC	M	OG
Well	0	1	2	2	2	2	3	4
Active/Area		0	1	1	1	1	2	3
POLY			0	2	2	2	3	4
n+				0	2	2	3	4
p+					0	2	3	4
Metal Contact						0	1	2
Metal							0	1
Over-glass								0

Various levels have associated electrical constraints that may limit minimum dimensions. Metal has the simplest constraints. The only requirements are that adjacent lines must not come in physical contact (shorts) and that a minimum-width line must be wide enough to maintain its continuity (opens).

A more complex set of constraints is imposed on diffused lines. Due to the selective field oxidation, considerable shrinking will take place in the defined lines. If ΔW is not accounted for, a minimum drawn line could completely disappear. Minimum spacing of diffused lines is also more complicated since shorting can occur due to junction punch through even if the lines are not physically touching.

Design rules can relate to either a single layer or several layers. The design rule constraints on the active area mask consist only of line width and spacing rules since this is the first layer. Implant mask rules concern the spacing to, or extension beyond, transistors. One must make sure that, even with misalignment and uncertain photoresist exposure, implanted transistors are implanted and nonimplanted transistors are not implanted.

Buried contact rules are similar to implant rules in that one wants buried contacts in the regions defined, and nowhere else. They are complex because one has the possibility of poly, active area, and buried contact geometries—none of which overlap when laid out, and all moving or bloating in such a way that the three intersect on the final chip. Also, two buried contacts require more spacing than normal diffusion due to the larger lateral diffusion of P over As.

Polysilicon has the usual minimum width and spacing rules. In addition to these, there are rules for spacing to unrelated active area and extension beyond the gate area. For the buried contact, one has a minimum area of coincidence of poly with the active area mask. To guarantee a minimum contact area, one can have rules for the extensions of active area or poly, or both, in various directions beyond the area of coincidence. There is usually a minimum poly-to-active area spacing rule for the cases when one is running wires as opposed to building transistors. These rules exist so the capacitance of the poly wires will be low. It also affects yield because, if the poly is misaligned with respect to the active area so that there is thin oxide under part of the poly, the amount of gate area on the chip will be very large.

Contact cuts connect the metal layer to either the poly or active area layers through the CVD oxide. It is important that the contact cut land in the right place, so there are rules for the extension of active area and poly beyond the contact cut. For a metal-to-active area contact, the poly must be a fair distance away or it can get shorted to the metal.

We also do not want metal contacts on, or near, gates. This is because the aluminum can spike into the poly, along the grain boundaries, and from there spike through the oxide. Aluminum is a getter for oxygen. This is an advantage for forming aluminum-to-silicon contacts because the aluminum will spike through the inevitable thin oxide in the contact area. On the other hand, this characteristic can cause gate shorts. There is the secondary issue that the aluminum can change the work function of the gate material. Barrier metals may remove these limitations. Butting contacts, which consist of metal over contact over slightly overlapping poly and active area, are a matter of some debate. They are claimed to be a reliability hazard because the metal can short to the substrate in the region near the poly/n+ interface. This is illustrated in Fig. 3.26. Contacts to large metal areas can also cause problems because the metal can dissolve some silicon, and a lot of metal can dissolve more. Moving the contacts to thin tabs off the side of the main metal area is one solution, and providing "sacrifice" pieces of polysilicon is another. These techniques are illustrated in Fig. 3.27.

Above the metal is the passivation or over-glass layer. Pad window size and spacing relate to the bonding equipment. One also typically has prescriptions for how far nonrelated layers must be from the bonding pad to avoid possible injury.

In CMOS circuits, latch-up considerations generate two types of rules. One involves the spacing between the n+ in the p-type substrate and the p+ in the n-well, and the other involves how often to plug the

FIGURE 3.26   A butting contact. Spiking of the Al into the substrate near the n+/poly interface is a long-term reliability hazard.

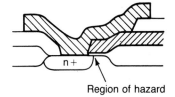

Region of hazard

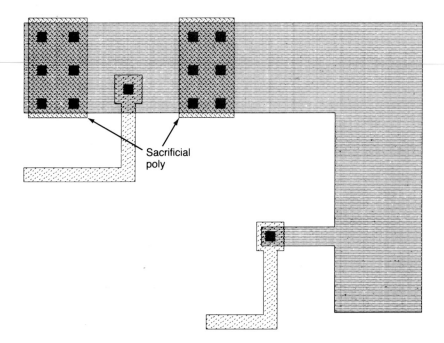

FIGURE 3.27   Techniques for contacting an active-area wire to a large metal bus. The desolution of Si into Al can cause reliability problems.

well. A well plug attaches the n-well, through an n+ contact to metal, to V_{DD}. Rules on how often to plug the well take two forms. The first form dictates that the well should be plugged every so many microns; while the second rule, designed to prevent large voltage drops in long thin wells, dictates that one should not have more than so many squares of well area without a plug.

In addition to the circuitry, chips must typically contain scribe marks and alignment marks, and may contain test structures, logos, initials, and a copyright mark.

Design rules are derived for a new process by a combination of calculation and experience. When a process is young, the design rules can be fairly dynamic as the process engineer optimizes the intrinsic process yield. After a few designs are done in a process, the design rules must stabilize.

■■■■■■■■■■■■■ EXAMPLE 3.2

Design rule calculations

As an example of a design rule derivation we calculate the rules associated with metal-to-poly contacts. The process has the following characteristics.

LEVEL	LINE WIDTH TOLERANCE	ALIGNED TO
Poly	$\pm 0.25\ \mu$m/edge	Active area
Metal	$\pm 0.40\ \mu$m/edge	Contacts
Contact	$^{+0.6}_{-0.0}\ \mu$m/edge	Active area

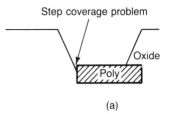

(a)

(b)

FIGURE 3.28 Problems caused by the misalignment and overetching of the contact cut to poly: (a) a step coverage problem; and (b) an etch pocket.

The contact line width tolerance is for a 2 μm $\times$ 2 μm drawn contact. Note the bias built into contacts. The alignment tolerance $R_0 = \pm 0.75\ \mu$m.

In order to achieve a legal contact, we assume that the contact edge must not overlap the poly. If it did, we might have a step coverage problem, as shown in Fig. 3.28(a) or, worse yet, we might create an etch pocket, as seen in Fig. 3.28(b). Since processing by-products tend to collect in the etch pocket (also called a garbage hole), it poses a long-term reliability hazard. In order to achieve a specified maximum value of contact resistance, we assume that the minimum contact area between the poly and metal is 3 μm^2.

We first consider the contact-to-poly registration R_{pc}. Since both are aligned to the active area, n equals 2. Therefore, we have

$$R_{pc} = \sqrt{2} R_0 = \pm 1.06\ \mu\text{m}.$$

In the worst case, the poly is small and the contact is large. We are looking for the minimum poly extension beyond contact distance. Adding up the worst-case bloats, shrinks, and translations, we have

$$1.06 + 0.25 + 0.6 = 1.91.$$

The minimum extension is 1.91 μm which we round to 2 μm. The worst-case contact-to-poly geometries are illustrated in Fig. 3.29.

Next we examine the metal. Since the metal is aligned to the contacts, n equals 1 in this case. If we assume that the worst translation happens in both y and z, then each side of the overlap area must be $\sqrt{3}$ μm long, which is equivalent to saying that the edge must be 0.73 μm past the center, as illustrated in Fig. 3.30. Adding up the errors, we have

$$0.73 + 0.75 + 0.40 = 1.88.$$

The metal must extend 1.88 μm, past the center of the contact. We round this to 2 μm. Thus the metal surround on contacts must be 1 μm. Note that if we wanted to guarantee the entire 4 μm^2 of contact area,

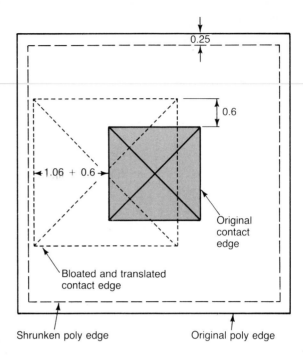

FIGURE 3.29   The solid lines illustrate the ideal geometries. The dashed lines are the worst case combination of bloats, shrinks, and misalignments.

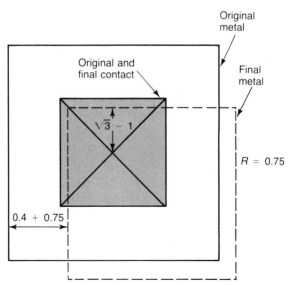

FIGURE 3.30   Worst-case situation for metal-over-contact cut. Contact area is minimized.

we would require a metal surround of

$$0.75 + 0.40 = 1.15 \ \mu\text{m}.$$

Now imagine that R_0 was decreased from ± 0.75 to $\pm 0.5 \ \mu$m. This would not change the poly surround problem, but look again at the metal surround. The value $1.88 \ \mu$m becomes $1.63 \ \mu$m, which still rounds to $2 \ \mu$m. But examine a common case in which the metal is used as a

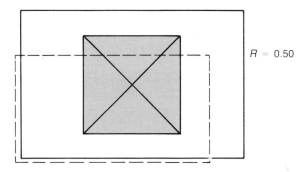

$R = 0.50$

FIGURE 3.31   Worst-case situation for metal wire running past a contact cut.

wire that continues past the contact more than 1.0 μm in two opposing dimensions. In this case, the worst-case misalignment is perpendicular to the wire. In order to have 3 μm^2 of contact area, the metal would need to cover 1.5 μm $\times$ 2 μm of contact area, as shown in Fig. 3.31. Thus the minimum metal extension perpendicular to the wire direction would need to be 0.4 μm, rounded up to 0.5 μm, in this case. This would be an enormous layout advantage in a technology that allowed a 3 μm metal width on wires.

This second example illustrates two points. First, the design can be made potentially more compact by allowing a finer granularity to the design rules. If all design rules must be multiples of 1 μm, then we gain nothing by a 0.4 μm rule since it must be rounded up to 1 μm. Second, the complexity of the design rules is somewhat a matter of philosophy. For instance, in this last case, we have potentially two metal surround rules. The first is that metal must extend 1 μm past the contact in all directions; the second is that the metal can extend 0.5 μm past the contact in two opposing directions, provided that it extends 1.0 μm past the contact in the other two directions. Do we want to include this second rule in the design rules? What would one do if, as can happen, this special case required a 1.5 μm extension in the long dimension?   ■

████████████████████ EXAMPLE 3.3

Design rule trade-offs
There are many trade-offs in composing design rules. Some parameters are pushed because they are important, others are not. Given CMOS technology, for example, we might examine the distance between the active area inside the well and the active area outside the well. On one hand, this distance determines latchup sensitivity, while on the other, it determines layout density. In our example process, the minimum distance was made just large enough so that one metal-to-poly contact would fit in the area between these active area regions—on the edge of the well. Since this contact is almost always needed, it did not pay to push the active areas any closer together.   ■

3.4 Yield

Yield is the most important metric characterizing the success of a given manufacturing process. Chip area is the principal parameter, under the designer's control, that affects yield (assuming one obeys the design rules!). The cost of manufacturing a chip grows faster than its area.

Several properties and figures of merit can be associated with the manufacture of an integrated circuit. In this section, we will consider only four. An integrated circuit has an area A_0, a probability P_0 that a given chip will work, a yield of Y_0 good devices per wafer, and a cost C, attributable to that yield loss. In general, we have

$$y \propto \frac{P}{A} \qquad (3.2)$$

and

$$C \propto y^{-1}. \qquad (3.3)$$

To obtain a simple yield model for integrated circuits, we assume that the probability that a chip is good depends only on its being free of point defects. We further postulate that the defect distribution may be described by Poisson statistics. Under these assumptions, the probability P that a chip is good has the form

$$P = e^{-A/A_{00}}, \qquad (3.4)$$

where A_{00} is a constant that depends on the quality of the process and is inversely related to the defect density. However, A_{00} is a random variable—which complicates the analysis quite a bit. A_{00} increases, perhaps quadratically, with the line width and decreases with the number of critical processing steps and the defect density. When the variance of A_{00} is large, P takes the form [Price 70] expressed by

$$P \approx \frac{1}{1 + A/A_{00}}. \qquad (3.5)$$

Equation (3.5) predicts a much greater yield than Eq. (3.4), and seems to be more consistent with experience.

Let us examine the case where a chip of size A_0 has a probability P_0 of being defect-free. This corresponds to a yield of Y_0 and a cost of C_0. For the cost, we have

$$C \propto \frac{A}{P} \propto A \left(1 + \frac{A}{A_{00}}\right) \qquad (3.6)$$

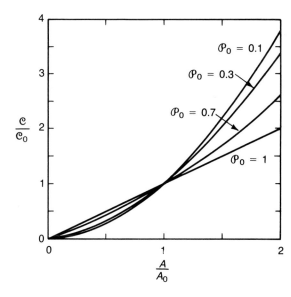

FIGURE 3.32 Normalized cost versus area, assuming the Price model of yield.

or

$$\frac{C}{C_0} = \frac{A}{A_0}\left(\mathcal{P}_0 + \frac{A}{A_0}(1 - \mathcal{P}_0)\right).$$ (3.7)

Normalized cost versus normalized area is plotted in Fig. 3.32 for several values of $\mathcal{P}_0$. We note from Eq. (3.7) that even if all chips were perfect, there would still be an increased cost associated with increased device area. When $\mathcal{P}_0$ is small, the cost of increased area increases even more quickly. The sensitivity of the cost to the area is

$$\frac{dC/C_0}{dA/A_0}\bigg|_{A=A_0} = 2 - \mathcal{P}_0.$$ (3.8)

Thus the cost increases twice as fast when $\mathcal{P}_0$ is small as when it is large. (Note that if Eq. (3.4) was in force, then Eq. (3.8) would evaluate to $1 - \ln \mathcal{P}_0$, an even more pessimistic function.)

The simple picture of chip yield given here is complicated in the real world by several other effects, including the spatial correlation of defects (such as scratches), the disproportionately higher rate of failure in I/O circuitry, packaging faults, parameter mismatching faults, and so forth. A positive correlation of defects helps the yield in the sense that for the same number of defects, one would much rather have all the defects on one die and none on the others. The theory is also not valid is for very low yields.

An important consideration affecting the choice of a design methodology is that defects that exist where there is no circuitry rarely

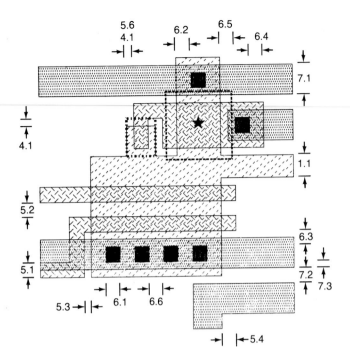

FIGURE 3.33 A layout illustrating numerous nMOS design rules in Appendix C.

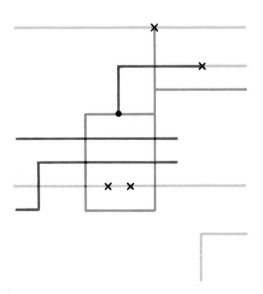

FIGURE 3.34 A stick diagram for the layout illustrated in Fig. 3.33. Stick diagrams should be done in color.

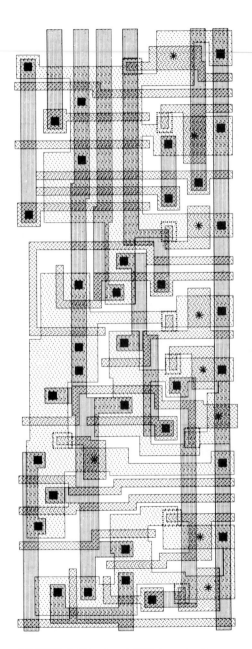

FIGURE 3.35 An nMOS layout of a piece of random logic. It is in the area of random logic that the difference between the novice- and expert-layout artist is most apparent.

cause device failures. For this reason, polycell [Kang 81] and gate-array type design methodologies, which leave large areas of the chip void of transistors, typically have much higher yields than a custom chip of the same die size. The use of fault tolerance to improve yield causes another complication [Smith 81]. When fault tolerance is used correctly, yield and area can increase together.

3.5   Layout mechanics

Layout is a skill easy to learn and difficult to master. The governing laws of the process are the design rules. Figure 3.33 illustrates an nMOS layout drawn according to the design rules in Appendix C. The application of the various rules are labeled by the rule number. In this layout, the depletion and enhancement mode threshold adjust masks are suppressed. Nevertheless, the design rules for these layers must, of course, be followed.

Layout planning is very important, as is the ability to investigate alternative designs quickly. Often one uses a shorthand notation for the layout to aid in manipulating the artwork. This shorthand, called sticks notation, is a mental place-holder for the the fully instantiated artwork. Stick layouts have no fixed rules. The key is to make them accurately reflect one's best guess as to the final layout. For instance, draw large contact cuts and use color in the layout. The common colors are metal: blue; poly: red; active area: green; contact cuts: black; depletion implant: yellow; well: brown; and p+: yellow. Figure 3.34 illustrates a stick figure for the layout in Fig. 3.33. Figures 3.35 and 3.36 illustrate two expert layouts, one nMOS and one CMOS, done by Kathy Tunis.

3.6   Perspective

Mass production is the central theme of device fabrication, and photolithography is the central processing step. Other steps concern the addition and removal of material from the chip. Electrical constraints, coupled with the chip manufacturing mechanics, give rise to geometrical constraints on the layout. Out of these constraints are developed design rules. Manufacturablity depends on chip size as well as adherence to the design rules. Because not all chips are free of defects, the manufacturing cost of a chip grows faster than its size.

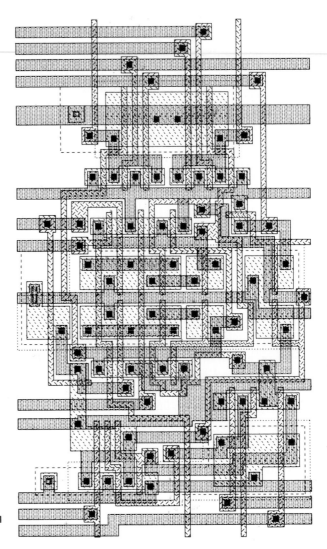

FIGURE 3.36 A CMOS layout of the same logical function as in Fig. 3.35.

Problems

3.1 Describe how to fabricate field implants in a CMOS process, without defining additional masks.

3.2 Describe how to fabricate n+ guard rings under the field oxide at the edge of all n-wells. Use no additional masks, though some existing masks may be modified.

3.3 Assume, in Example 3.2, that contact cuts are aligned to poly rather than to n+, and rederive the poly-to-metal contact rules.

3.4 In triple modular redundancy, three identical circuits are fabricated and a majority voting circuit determines the output. Assume the voting circuitry takes negligible area. Assume that A_{00} is constant over the area of the total circuit and that Poisson statistics apply. Derive an expression for the yield as a function of P_0, where P_0 is the probability of a single circuit functioning correctly. Plot the yield versus P_0 both with and without triple modular redundancy. For what range of P_0 does the use of triple modular redundancy pay off in terms of cost?

3.5 A chip packaged in a $40°\,C/W$ ceramic package is known to dissipate 1 watt in a $23°\,C$ ambient. Assuming that all variation in device current with temperature is due to mobility reduction, predict the power dissipation and junction temperature when the chip is operating in a $70°\,C$ ambient.

3.6 The MOS processes in the appendixes require a 5 μm pitch on diffused lines when they are run in the straightforward way. How can diffused lines be run with a 4 μm pitch without violating any of the design rules? $\triangle$

3.7 Do layouts of some of the circuits shown in Chapter 1. Draw cross sections through interesting parts.

3.8 Derive the minimum-drawn spacing design rule between a silicon gate transistor and a metal-to-diffusion contact. Assume the process has the same characteristics described in Example 3.2.

3.9 Lay out a pair of cross-coupled NOR gates in a technology in which V_{DD} exceeds the thick field parasitic transistor thresholds. The input and output voltages swing from rail to rail. $\triangle$

3.10 Consider the issue of buried contacts in CMOS technologies. If we use n-type poly, buried contacts to p+ material will result in a p–n diode that can be useful, providing the poly does not short to the n-type well. What process steps are required to keep these shorts from occurring? Another approach is to use poly that is p+ over the n-well and n+ over the p-type substrate. What is the problem with this and why does using a silicide on the poly help? $\triangle$

3.11 Extract the circuit schematics of the layouts in Figs. 3.35 and 3.36. $\triangle$

3.12 Invent a self-aligned contact technology that does not require high-temperature processing. $\triangle\triangle$

References

[Antoniadis 77] D. A. Antoniadis, S. E. Hansen, R. W. Dutton, and A. G. Gonzalez, "SUPREM 1—A program for IC process modeling and simulation," Stanford Electronics Laboratories, Stanford, Calif. Tech. Rep. 5019-1, 1977.

[Antoniadis 79] D. A. Antoniadis and R. W. Dutton, "Models for Computer Simulation of Complete IC Fabrication Process," *IEEE J. Solid-State Circuits* **SC-14**: 412–422, 1979.

[Bayless 83] M. Bayless, W. Devanney, W. Waller, and D. Laurent, "An Automated Methodology for Generating Self-Consistent Layout Rules for VLSI Designs," *IEEE International Electron Device Meeting*, 250–254, Washington, D.C., 1983.

[Blodgett 83] A. J. Blodgett, Jr., "Microelectronic Packaging," *Scientific American* **249**: 86–96, 1983.

[Burger 67] R. M. Burger and R. P. Donovan (eds.), *Fundamentals of Silicon Integrated Device Technology*, Prentice-Hall, Englewood Cliffs, N.J., 1967.

[Chin 82] K. Y. Chin, R. Fang, J. Lin, J. L. Moll, C. Lage, S. Angelos, and R. Tillman, "The SWAMI—a Defect Free and Near-Zero Bird's-Beak Local Oxidation Process and Its Application in VLSI Technology," *IEEE International Electron Device Meeting*: 224–227, San Francisco, 1982.

[Chin 83] D. Chin, S-Y Oh, S-M Hu, R. W. Dutton, and J. L. Moll, "Two-Dimensional Oxidation," *IEEE Trans. Electron Devices*, **ED-30**: 744–749, 1983.

[Colclaser 80] R. A. Colclaser, *Microelectronics Processing and Device Design*, Wiley, New York, 1980.

[Conway 80] L. Conway, A. Bell, and M. E. Newell, "MPC79: The Large-Scale Demonstration of a New Way to Create Systems in Silicon," *LAMBDA* **1**: (2), 10–19, 1980.

[Dennard] R. H. Dennard, F. H. Gaensslen, H. N. Yu, V. L. Rideout, E. Bassons, and A. R. LeBlanc, "Design of Ion-Implanted MOSFET's with Very Small Dimensions," *IEEE J. Solid-State Circuits* **SC-9**: 256–267, 1974. (Scaling theory)

[Ghandhi 69] S. K. Ghandhi, *The Theory and Practice of Microelectronics*, Wiley, New York, 1969.

[Gibbons 80] J. F. Gibbons and K. F. Lee, "One-Gate-Wide CMOS Inverter on Laser-Recrystallized Polysilicon," *IEEE Trans. Electron Devices Lett.* **EDL-1**: 117–118, 1980.

[Gupta 83] D. C. Gupta, (ed.), *Silicon Processing*, Amer. Soc. Testing and Materials (ASTM), Philadelphia, Pa., 1983.

[Hu 84] S. M. Hu, "On Yield Projection for VLSI and Beyond, I. Analysis of Yield Formulas," *IEEE Electron Devices Newsletter* **69**: 4–7, 1984.

[Hunter 81] W. R. Hunter, T. C. Holloway, R. K. Chatterjee, and A. F. Tasch, Jr., "A New Edge-defined Approach for Submicrometer MOSFET Fabrication," *IEEE Trans. Electron Devices Lett.* **EDL-2**: 4–6, 1981.

[IEEE 81] *Proc. IEEE* **69**: February, 1981.

[Jansen 81] W. D. Jansen and D. G. Fairbairn, "The Silicon Foundry: Concepts and Reality," *LAMBDA* **2**: (1), 16–26, 1981.

[Kang 81] S. M. Kang, "A design of CMOS polycells for LSI circuits," *IEEE Trans. Circuits and Systems* **CAS-28**: 838–843, 1981.

[Lee 80] Y. K. Lee and J. D. Craig, "Polyimide Coatings for Microelectronic Applications," *Division of Organic Coatings and Plastics Chemistry, ACS Preprints* **43**: 451–458, 1980.

[Liu 83] T. M. Liu and W. G. Oldham, "Channeling Effect of Low-Energy Boron Implant in (100) Silicon," *IEEE Trans. Electron Devices Lett.* **EDL-4**: 59–62, March, 1983.

[Liu 84] S. S. Liu, G. E. Atwood, E. Y. So, B. J. Wu, R. W. Leftwich, and K. R. Hasserjian, "1.5 μ Scaled CMOS Microcomputer Technology," *IEEE International Solid-State Circuits Conf.*: 156–157, San Francisco, 1984.

[Lyon 81] R. F. Lyon, "Simplified Design Rules for VLSI Layouts," *LAMBDA* **2**: (1), 54–59, 1981.

[Mikkelson 81] J. M. Mikkelson, L. A. Hall, A. K. Malhotra, S. D. Seccombe, and M. S. Wilson, "A NMOS VLSI Process for Fabrication of a 32-Bit CPU Chip," *IEEE J. Solid-State Circuits* **SC-16**: 542–547, October, 1981.

[Muramoto 78] S. Muramoto, T. Hosoya, and S. Matsuo, "A New Self-Aligning Contact Process for MOS LSI," *IEEE International Electron Device Meeting*, 185–188, Washington, D.C., 1978.

[Oldham 82] W. G. Oldham, "Isolation Technology for Scaled MOS VLSI," *IEEE International Electron Device Meeting*, 216–219, San Francisco, 1982.

[Parrillo 80] L. C. Parrillo, R. S. Payne, R. E. Davis, G. W. Reutlinger, and R. L. Field, "Twin-Tub CMOS—A technology for VLSI Circuits," *IEEE International Electron Device Meeting*, 752-755, Washington, D.C., 1980.

[Pashley 77] R. Pashley, W. Owen, K. Kokkonen, R. Jecmen, A. Ebel, C. Ahlquist, and P. Schoen, "A High-Performance 4K Static RAM Fabricated in an Advanced MOS Technology," *IEEE International Solid-State Circuits Conf.*, 22–23, New York, 1977.

[Peltzer 83] D. L. Peltzer, "Wafer-Scale Integration: The Limits of VLSI?" *VLSI Design* **4**: (5), 43–47, 1983.

[Price 70] J. E. Price, "A New Look at Yield of Integrated Circuits," *Proc. IEEE*: 1290–1291, 1970.

[Rea 81] S. N. Rea, "Czochralski Silicon Pull Rate Limits," *J. Cryst. Growth* **54**: 267–274, 1981.

[Rideout 79] V. L. Rideout and V. J. Silvestri, "MOSFET's with Polysilicon Gates Self-Aligned to the Field Isolation and to the Source and Drain Regions," *IEEE Trans. Electron Devices*, **ED-26**: 1047–1052, 1979. (Self-aligned contacts)

[Rideout 80] V. L. Rideout, J. J. Walker, A. Cramer, "A One-Device Memory Cell Using a Single Layer of Polysilicon and a Self-Registering Metal-to-Polysilicon Contact," *IBM J. Res. Develop.* **24**: 339–347, 1980.

[Rung 81] R. D. Rung, "Determining IC Layout Rules for Cost Minimization," *IEEE J. Solid-State Circuits* **SC-16**: 35–43, February, 1981.

[Saito 82] K. Saito and E. Arai, "Experimental Analysis and New Modeling of MOS LSI Yield Associated with the Number of Elements," *IEEE J. Solid-State Circuits* **SC-17**: 28–33, 1982. (Some yield characterization numbers)

[Sakai 81] Y. Sakai, T. Hayashida, N. Hashimoto, O. Mimato, T. Masuhara, K. Nagasawa, T. Yasui, and N. Tanimura, "Advanced Hi-CMOS Device Technology," *IEEE International Electron Device Meeting*, 534–537, Washington, D.C., 1981.

[Sakamoto 80] M. Sakamoto and K. Hamano, "A New Self-Aligned Contact Technology," *IEEE International Electron Device Meeting*: 136–139, Washington, D.C., 1980.

[Samuelson 80] G. Samuelson, "Polyimide for Multilevel VLSI." *Division of Organic Coating and Plastics Chemistry, ACS Preprints* **43**: 1587–1594, 1980.

[Smith 81] R. T. Smith, J. D. Chlipala, J. F. M. Brindels, R. G. Nelson, F. H. Fisher, and T. F. Mantz, "Laser Programmable Redundancy and Yield Improvement in a 64K DRAM," *IEEE J. Solid-State Circuits* **SC-16**: 506–514, 1981.

[Straub 83] R. R. Straub, "Hydrofluoric Acid Burns of the Hand," *Orthopedics* **6**: 978–980, 1983.

[Sze 83] S. M. Sze, *VLSI Technology*, McGraw-Hill, New York, 1983.

[Theuwissen 82] A. J. P. Theuwissen, "Linear CCD-Imagers with Polyimide Insulation for Double Level Metallization," *IEEE Trans. Electron Devices Lett.* **EDL-3**: 308–309, 1982.

[Wollensen 79] D. L. Wollensen, "C-MOS LSI: comparing second-generation approaches," *Electronics* **52**: 116–123, September 13, 1979.

[Yu 81] K. Yu, R. J. C. Chwang, M. T. Bohr, P. A. Warkentin, S. Stern, and C. N. Berglund, "HMOS-CMOS—A Low-Power High-Performance Technology," *IEEE J. Solid-State Circuits* **SC-16**: 454–459, 1981.

4 THE LOGIC ABSTRACTION

The logic abstraction forms a link between the digital and electrical domains. In the digital domain, signals are represented by ones and zeros, and in the electrical domain, signals are analog in nature, contain noise, and are handled by nonideal transducers. In this chapter, we analyze the transformation from the electrical to the digital domain. We examine static and dynamic issues of noise, noise margins, and worst-case design, and present a quantitative analysis of many building block components.

4.1 Restoring logic

Valid logic signals are represented by ranges of voltages. We require that the voltages representing valid logic levels be restored as they propagate through a digital network. That is, we assume that the logic voltages are degraded as they propagate through a network so that logic signals tend to move outside the specified ranges. We wish to restore these signals to the valid ranges.

We define V_{IL} and V_{IH} as the valid low and high input logic voltages to a restoring gate. For positive logic, we have

$$V_{OH} \geq V_{IH} + NM_H \qquad (4.1)\star$$

and

$$V_{OL} \leq V_{IL} - NM_L, \qquad (4.2)\star$$

205

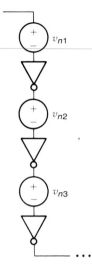

FIGURE 4.1 Voltage noise on logic gates.

FIGURE 4.2 A worst-case noise situation for a chain of inverters.

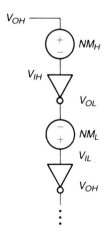

where V_{OL} and V_{OH} are the valid output logic voltages. The low and high noise margins NM_L and NM_H must be positive even under worst-case temperature and processing conditions.

Fig. 4.1 illustrates a chain of inverters with voltage noise sources between the gates. A noise margin specification allows one to say that if these noise voltages are below the limits specified by the noise margins, then the circuit will operate correctly. A worst-case noise scenario is illustrated in Fig. 4.2.

Due to the threshold drop incurred in nMOS pass transistor logic, there are typically two or more V_{IH} and V_{OH} levels in depletion-mode nMOS circuits. In our example of an nMOS process, there are three: one for a depletion-mode pullup, another for the enhancement-mode pullup, and a third for the zero threshold pullup. In CMOS there is typically only one value for V_{OL} and one value for V_{OH}. Nevertheless, depending on the pass transistor methodology, one can also have multiple logic levels in CMOS. Note that we have tacitly assumed in Eqs. (4.1) and (4.2) that there are no circumstances in which a high is too high nor a low too low.[1]

Fig. 4.3 illustrates a typical inverter transfer function. Regions of valid logic levels are superimposed on the plot. The shaded region represents the area in which the inverter transfer function must lie for it to be a legal, restoring inverter. When the input is below V_{IL}, the output must be above V_{OH}. When the input is between V_{IL} and V_{IH}, its logic value is undefined and the output may be anywhere. When the input is above V_{IH}, the output must be below V_{OL}. Since $V_{OH} - V_{OL}$ must be greater than $V_{IH} - V_{IL}$, the inverter must have differential gain somewhere in the middle region. Note also that within quite a large latitude, the assignment of the logic levels is fairly arbitrary. This has the advantage that many different logic gate implementations can be members of a given logic family. The valid voltage levels in a digital logic family are a contract or methodology in which one agrees that every gate in the family will meet or exceed certain specifications on their dc voltage transfer functions. That one or more gates may exceed this specification does not make the noise margin and voltage level assignments invalid. It may indicate that they may not have been the best choice, but that is a different issue, which we will discuss later. Noise margins help

---

[1] Usable logic families can possess many other properties. The input–output transfer characteristics of gates should be such that perturbations about valid logic voltages are attenuated. That is, $|dv_{\text{out}}/dv_{\text{in}}|$ is less than 1 for all v_{in} that correspond to valid logic input levels. While this property is nice, it is not strictly required. A second important characteristic is that gates should be nonreciprocal so that logic signals can propagate through a restoring gate in only one direction. Many physical phenomena can provide gain but not nonreciprocity. Lasers are one example. Reflections become a problem unless isolators are used. But, except in esoteric technologies, one usually does not have to worry about the restrictions mentioned in this footnote.

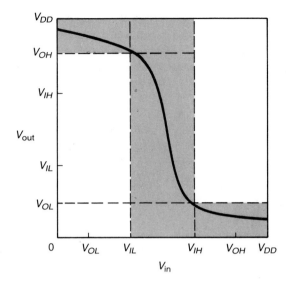

FIGURE 4.3   Inverter transfer function with valid logic voltage levels notated. The inverter transfer must lie within the shaded region to be a member of the logic family.

guarantee that it is valid to interpret certain signals in a network as representing ones and zeros. (A clocking methodology proscribes when to look at them.)

Because there are two or more legal V_{OH} levels in a depletion-mode nMOS integrated circuit, it is sometimes assumed that if a high output voltage can rise only to $V_{DD} - V_T$, the output is not restored, and if it can rise to V_{DD}, it is restored. This is not true. It is important to separate the concept of the level of the valid output voltage from the concept of restoring logic. In a restoring logic gate, there is a region of the input-to-output transfer function that has differential gain. The first two circuits in Fig. 4.4 are restoring but have valid V_{OH} levels of $V_{DD} - V_T$. Fig. 4.4(a) is an inverter with an enhancement-mode transistor load and Fig. 4.4(b) is steering logic implementation of an AND gate. The third circuit is a bootstrapped pass transistor. Its output can go to V_{DD}, but it is not a restoring circuit. Fig. 1.37(b) illustrates a CMOS exclusive OR circuit in which only one input is restoring.

To differentiate between the concepts of restored logic voltage and output voltage level, we refer to the closeness of the output to a power supply rail as the "strength" of that signal. That is, an output that rises to V_{DD} is a strong output, whereas one that rises only to $V_{DD} - V_T$ is a weak output. This consideration is orthogonal to whether the output is restored. Similarly, we can have strong and weak low signals. A weak low signal might be 300 mV, whereas a strong low signal might be ground. This second case is appropriate, for instance, for the input to the gate of a pass transistor that controls dynamic charge storage. Because of subthreshold leakage, one must lower v_{GS} of a transistor significantly

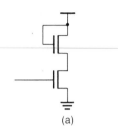

(a)

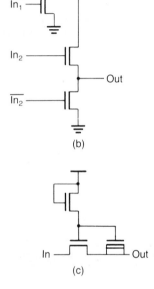

(b)

(c)

In ———— Out

FIGURE 4.4 Logic gates (a) restoring with weak high, (b) restoring with weak high, and (c) nonrestoring with strong high.

below threshold to keep the slow leakage of charge through the transistor on the same order as the reverse p–n diode junction leakage.

In our logic gate family, all primitive gates (such as NOT, NAND, and NOR) invert. We define the static transfer function between any input and the output as

$$V_{\text{out}} \equiv H(V_{\text{in}}).\qquad(4.3)$$

Using Eqs. (4.1) and (4.2) with an inverter, we have

$$H(H(V_{OH} - NM_H) + NM_L) \geq V_{OH}\qquad(4.4)$$

and

$$H(H(V_{OL} + NM_L) - NM_H) \leq V_{OL}.\qquad(4.5)$$

Fig. 4.2 illustrates this point. Equations (4.4) and (4.5) can be generalized to handle NOR and NAND gates. For a NAND gate, the output must be less than or equal to V_{OL} if all of the inputs are greater than or equal to V_{IH}. If any of the inputs is less than or equal to V_{IL}, then the output must be at least V_{OH}. A multidimensional geometric interpretation similar to Fig. 4.3 is possible. Fig. 4.5 illustrates a worst-case noise situation for a network of NAND and NOR gates.

Given the basic noise margin concept, we can look at some derived properties. One can choose the assignment of the four logic voltages in such a way that the sum of the noise margins is maximized. This objective function can be justified by looking at a long chain of identical inverters. Since a high-going transition into one stage produces a low-going transition going into the next stage, we can maximize the total "system" noise margin by maximizing the sum of the high and low noise margins. We will discuss cases where it is advantageous to do this and cases where it is not. For the inverter, the sum of the noise margins is given by

$$NM_L + NM_H = V_{IL} - V_{IH} + H(V_{IL}) - H(V_{IH}).\qquad(4.6)$$

Solving for the optimum assignment of V_{IL} and V_{IH}, we obtain

$$-1 = \frac{dH(V_{IL})}{dV_{IL}}\qquad(4.7)$$

and

$$-1 = \frac{dH(V_{IH})}{dV_{IH}}.\qquad(4.8)$$

Note that Eqs. (4.7) and (4.8) were derived under certain restricted assumptions. They are not part of the basic definition of noise margins. To translate this result to logic gates with a fan-in greater than unity, what Eqs. (4.7) and (4.8) show is that if *all* of the inputs of a NOR gate are connected together and raised to V_{IL}, perturbations of this collected input voltage will cause a perturbation on the output of equal magnitude and opposite sign. This means that the small signal gain from

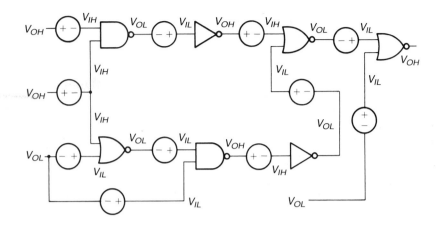

FIGURE 4.5 Worst-case noise scenario for a collection of logic gates. The circuit is guaranteed to compute correctly providing that the noise is below these limits.

any individual logic input to the output must be less than 1 in magnitude when all of the inputs are low. If there are M identical inputs to, say, a NOR gate, then at the optimal V_{IL} the differential gain is $-M^{-1}$. This makes sense because if the small signal gain of several inputs were unity, then that gate would act as a noise summer. Note that this derivation of the optimum logic voltage assignments assumes that when their inputs are tied together, the transfer functions of each logic gate in the system reduce to that of a single inverter with transfer function H. We will later see an example of this in a CMOS methodology.

The noise margins are defined such that the entire system could be plagued by noise, all the way to the noise margin limit, on every wire, and the system should still work. That is, if the noise in the system is below the noise margins, then the system is guaranteed to function. We can ask the complementary question. At what level of noise is the system guaranteed not to function? Suppose we have only one noise pulse in the system, on only one wire. If that noise does not cause the output of the affected logic gate to cross V_{inv}, then the system will eventually recover. (Again assuming, for simplicity, identical transfer functions for all logic gates.) If the noise transition does cause the output to cross V_{inv}, an error will occur. V_{inv} is the voltage at which the output of an inverter is equal to its input. That is

$$V_{inv} \equiv H(V_{inv}). \tag{4.9}$$

The voltage between V_{inv} and V_{OL} or V_{OH} is a partial measure of the robustness of the design to low probability noise bursts. The noise robustness measures are given by

$$NR_L \equiv V_{inv} - V_{OL} \tag{4.10}$$

and

$$NR_H \equiv V_{OH} - V_{inv}. \tag{4.11}$$

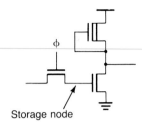

FIGURE 4.6   A dynamic node that is sensitive to coupled noise.

Voltage is not the only parameter that relates to the noise in digital systems. Impedance, loosely defined, is an important issue. Capacitive coupling noise of all sorts produces larger noise voltages on high impedance nodes than on low impedance nodes. Because they are isolated, the most noise-susceptible nodes are ones that store dynamic charge. Precharged nodes fall into this class. It is no accident that in nMOS we precharge nodes high, and that NM_H is much larger than NM_L. This is because we can anticipate large noise voltages in precharged circuits. In CMOS circuit design, nodes can be precharged either high or low, but precharged nodes are usually followed by an inverter that can restore the large noise voltages seen on precharged nodes.

In these cases, the noise margins and impedance levels are consistent. Other cases may have a mismatch, and extreme care must be taken in the layout in order to minimize noise problems. An example of such a circuit is the dynamic storage node in an nMOS shift register cell, illustrated schematically in Fig. 4.6. The low capacitance of the storage node, coupled with the fact that it is isolated, is good reason not to, for instance, run a metal signal line over the top of the storage node. The noise coupling due to gate-to-source capacitance of the pass transistor must also be considered. A particularly sensitive structure, which can occur either in nMOS or CMOS, is shown in Fig. 4.7 with a CMOS implementation. In this circuit, if the high-impedance node moves by one threshold, the output will flip. Using a fully complementary inverter instead of the precharged one means the high-impedance node must move nearly all the way to V_{inv} to flip the output.

Noise margins in nMOS circuits

In MOS integrated circuits we try to take advantage of the tracking of parameters on a chip and define at least one of the logic levels in relation to a transistor threshold voltage rather than some fixed voltage. We define V_{IL} in relation to V_{TE0} (V_{TE} at $v_S = 0$) in nMOS.[2]

FIGURE 4.7   A precharged CMOS OR gate that is very sensitive to noise. If the high impedance precharged bus moves by as much as $|V_{TP0}|$, the output will flip.

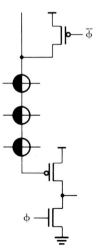

---

[2] In Chapter 2, we referred all voltages to the substrate. That was appropriate for looking at the transistor from a device physics point of view. But when thinking about circuits, we want to reference our voltages to ground—signal ground, not the substrate. In many nMOS technologies the substrate is biased at a negative voltage with respect to signal ground. We therefore change coordinates so that the substrate is at the negative voltage V_{BB}. Voltages are referred to signal ground. And threshold voltages are taken at the appropriate substrate bias. For $V_{BB} = 0$, we have $V_{TX} = V_{TE0}$. But for $V_{BB} < 0$, we have

$$V_{TE0} = V_{TX} + \gamma \left(\sqrt{-V_{BB} + 2\Phi_{Fp}} - \sqrt{2\Phi_{Fp}} \right).$$

V_{TE0} is a function of geometry and is defined at $V_{DS} = 0$. Typically, $V_{BB} = 0$ for CMOS technologies, and we use V_{TN0} to denote the nMOS enhancement-mode device threshold and V_{TP0} for the enhancement-mode pMOS device threshold when $v_{SB} = 0$ for both the n- and p-channel CMOS devices.

In an nMOS circuit the key to this problem is the NOR gate. By definition, if any of the input voltages to a NOR gate are at least V_{IH}, then the output voltage of the NOR gate must be less than or equal to V_{OL}. This constrains the ratios of the shape factors S to each of the pulldown transistors and the pullup to a value greater than some constant called the β-ratio. We define β as

$$\beta \equiv \frac{S_{pd}}{S_{pu}}, \qquad (4.12)$$

where S_{pd} is the shape factor of a pulldown and S_{pu} is the shape factor of the pullup. This V_{OL} constraint is independent of the fan-in of the NOR gate (including the special case of the inverter). We also have the constraint that the output voltage of the NOR gate must be at least V_{OH} if all of the inputs are less than or equal to V_{IL}. The β-ratio constraint can be independent of fan-in only if all of the pulldown devices are completely off. This implies[3]

$$V_{IL} \leq V_{TE0} - \sigma V_{OH}. \qquad (4.13)$$

Note that in Eq. (4.13) we have included the degradation of the threshold voltage due to drain-induced barrier lowering. For depletion load nMOS, we have $V_{OH} = V_{DD}$. Typical noise margins for 5 volt nMOS are 100 to 200 mV for NM_L and 800 to 1000 mV for NM_H.

We can relate Eq. (4.13) to the derived optimal V_{IL}. If one takes the limit of the -1 slope points as the fan-in of the NOR gate increases to infinity, one discovers that these limits coincide with the values of V_{IL} and V_{OL} discussed above. By choosing V_{IL} as the enhancement-mode pulldown threshold, we have chosen V_{IL} as the minimum of the optimal V_{IL} assignments corresponding to different fan-ins. For a precharged gate, the noise margin is $V_{TE0} - \sigma V_{OH} - V_{OL}$, which corresponds to the ratioed case when $\beta = \infty$. Fig. 4.8 illustrates the transfer function of an nMOS depletion load inverter as a function of the width of the pulldown device.

Noise margins in CMOS circuits

For CMOS circuits, the most straightforward way of defining the valid logic levels is to let $V_{OH} = V_{DD}$ and $V_{OL} = 0$. Then the n- and p-channel enhancement-mode thresholds (shifted to include drain-induced barrier lowering) will be the low and high noise margins, respectively. This is sometimes done, but it has the problem that the sum of the noise margins is not very large because the value of V_{IL} or V_{IH} needed

---

[3] In large PLAs, where the pulldown transistors are small and numerous, the leakage through even supposedly off devices becomes a problem.

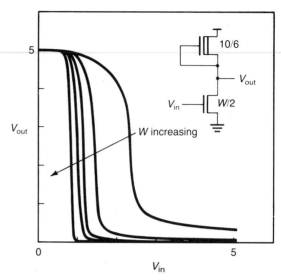

FIGURE 4.8 Transfer characteristics of a depletion load nMOS inverter as a function of increasing pulldown width.

to achieve output levels of V_{DD} or ground is close to the ground or V_{DD} when the thresholds are low.

For performance reasons, the fan-in on classical CMOS circuits is usually small. A classical CMOS circuit is one in which there is one p-type device for each n-type device. Whereas in the nMOS case we took the limit of fan-in $M \to \infty$ to set V_{IL}, we do not need to do that for CMOS. For instance, we could set V_{IL} and V_{IH} by taking the minimum and maximum, respectively, of the optimum V_{IL} and V_{IH} values of NAND and NOR gates with $M = 1$, 2, and 3.

The push–pull nature of the classical CMOS logic gate leads to high differential gain and hence a fairly ideal transfer function. Classical CMOS thus tends to have very good noise properties. When precharging or other advanced CMOS circuit techniques are used, good noise performance cannot be taken for granted.

4.2 The inverter

The MOS inverter consists of a pair of MOS transistors connected in such a way that they realize the transfer function specified by Eqs. (4.4) and (4.5). We will analyze four of the many possible configurations and develop a number of formulas for simple MOS circuits. Many of these results are used in later sections and chapters. Besides being a useful reference, the derivations of these results serve two other purposes. First, they illustrate circuit analysis techniques, and second, they point out sometimes surprising sensitivities and dependencies. The two basic types of MOS inverters are those in which both devices are active and those in which one of the devices acts as a passive load. We will look first at the nMOS inverter with an ideal resistive load.

nMOS inverter with a resistive load

The nMOS inverter with a resistive load illustrated in Fig. 4.9 is the simplest of MOS circuits. This configuration closely models the case of a depletion load device with high body effect or an enhancement-mode pullup with its gate connected to a voltage much greater than V_{DD}. It is, of course, a good model for devices with polysilicon load resistors. The current through the pullup resistor is given by

$$i_{pu} = (V_{DD} - v_{\text{out}})G. \tag{4.14}$$

The current through the enhancement-mode pulldown transistor is

$$i_{pd} = \begin{cases} 0 & \text{for } v_{\text{in}} \leq V_{TE0} \\ \dfrac{K'S}{1+\delta}(v_{\text{in}} - V_{TE0})^2 & \text{for } 0 < v_{\text{in}} - V_{TE0} < (1+\delta)v_{\text{out}} \\ K'S\left(2(v_{\text{in}} - V_{TE0})v_{\text{out}} - (1+\delta)v_{\text{out}}^2\right) & \text{for } v_{\text{in}} - V_{TE0} \geq (1+\delta)v_{\text{out}} \end{cases} \tag{4.15}$$

Equating the two currents yields the steady state transfer function for the device. We then have

$$V_{\text{out}} = \begin{cases} V_{DD} & \text{for } V_{\text{in}} \leq V_{TE0} \\ V_{DD} - \dfrac{1}{B}\left(\dfrac{V_{\text{in}} - V_{TE0}}{1+\delta}\right)^2 & \text{for } 0 < V_{\text{in}} - V_{TE0} < (1+\delta)V_{\text{out}} \\ \dfrac{V_{\text{in}} - V_{TE0}}{1+\delta} + \dfrac{B}{2} - \left(\left(\dfrac{V_{\text{in}} - V_{TE0}}{1+\delta} + \dfrac{B}{2}\right)^2 - BV_{DD}\right)^{1/2} & \text{for } V_{\text{in}} - V_{TE0} > (1+\delta)V_{\text{out}}, \end{cases} \tag{4.16}$$

where

$$B \equiv \frac{G}{K'S(1+\delta)}. \tag{4.17}$$

The inverse voltage V_{inv} is defined as the voltage at which $V_{\text{in}} = V_{\text{out}}$. This occurs when the transistor is in saturation. Solving, we have

$$V_{\text{inv}} = V_{TE0} - \frac{B(1+\delta)^2}{2} + \frac{(1+\delta)}{2}\sqrt{\frac{B^2(1+\delta)^2}{4} + B(V_{DD} - V_{TE0})}. \tag{4.18}$$

The point of maximum small signal gain also occurs when the transistor is in saturation. There are two ways of solving this problem.

Differentiating V_{out} with respect to V_{in}, we have

$$\frac{\partial V_{\text{out}}}{\partial V_{\text{in}}} = a_0 = -\frac{2K'S}{(1+\delta)G}(V_{\text{in}} - V_{TE0})$$

$$= -2\sqrt{\frac{K'S}{(1+\delta)G}} \sqrt{V_{DD} - V_{\text{out}}}$$

(4.19)

for $V_{\text{in}} - V_{TE0} \leq (1+\delta)V_{\text{out}}$. Note that the magnitude of the differential gain a_0 increases as K' and S increase and as δ and G decrease. The maximum absolute gain occurs at V_{inv}. We have

$$a_{0\text{max}} = 1 - \sqrt{\frac{1}{4} + \frac{V_{DD} - V_{TE0}}{(1+\delta)^2 B}}$$

$$= 1 - \sqrt{\frac{1}{4} + \frac{(V_{DD} - V_{TE0})K'S}{(1+\delta)G}}.$$

(4.20)

For the inverter to be restoring, $|a_0|$ must be greater than one. Note that $|a_{0\text{max}}|$ depends only on the square root of all major parameters. This is one of the reasons it is difficult to obtain high differential gain in MOS circuits.

A second technique for looking at the small signal gain is to develop a small signal model of the transistor. We define the transconductance g_m of the transistor as

$$g_m \equiv \left.\frac{\partial i_{DS}}{\partial v_{GS}}\right|_{V_{DS}}.$$

(4.21)

We have

$$g_m = \begin{cases} 0 & \text{for } V_{GS} \leq V_{TE0} \\ \dfrac{2K'S}{1+\delta}(V_{GS} - V_{TE0}) & \text{for } 0 < V_{GS} - V_{TE0} < (1+\delta)V_{DS} \\ 2K'SV_{DS} & \text{for } V_{GS} - V_{TE0} \geq (1+\delta)V_{DS}. \end{cases}$$

(4.22)

We define the output conductance g_D as

$$g_D \equiv \left.\frac{\partial i_{DS}}{\partial v_{DS}}\right|_{V_{GS}}.$$

(4.23)

Solving,[4] we have

$$g_D = \begin{cases} 0 & \text{for } V_{GS} - V_{TE0} \leq (1+\delta)V_{DS} \\ 2K'S(V_{GS} - V_{TE0} - (1+\delta)V_{DS}) & \text{for } V_{GS} - V_{TE0} \geq (1+\delta)V_{DS}. \end{cases}$$

(4.24)

---

[4] The output conductance g_D is not actually zero in saturation. Its nonideal value is due to drain-induced barrier lowering, channel length modulation, and series drain and source resistance.

Using elementary circuit analysis techniques, the gain of the inverter is given by

$$a_0 = -\frac{g_m}{g_D + G},\tag{4.25}$$

which reduces to Eq. (4.19) when the proper substitutions are made.

By differentiating the steady state transfer function, we can find the small signal gain for all inputs. We have

$$a_0 = \begin{cases} 0 & \text{for } V_{\text{in}} < V_{TE0} \\[2mm] -\dfrac{2(V_{\text{in}} - V_{TE0})}{(1+\delta)^2 B} & \text{for } 0 < V_{\text{in}} - V_{TE0} < (1+\delta)V_{\text{out}} \\[4mm] \dfrac{1}{1+\delta}\left(1 - \left(\left(\dfrac{V_{\text{in}} - V_{TE0}}{1+\delta} + \dfrac{B}{2}\right)^2 - BV_{DD}\right)^{-1/2}\left(\dfrac{V_{\text{in}} - V_{TE0}}{1+\delta} + \dfrac{B}{2}\right)\right) & \text{for } V_{\text{in}} - V_{TE0} > (1+\delta)V_{\text{out}}. \end{cases}\tag{4.26}$$

There are either zero, one, or two -1 slope points. If the inverter is restoring, there are two solutions. They are found at

$$V_{\text{in}} = \frac{(1+\delta)^2 B}{2} + V_{TE0}\tag{4.27}$$

and

$$V_{\text{in}} = V_{TE0} + (1+\delta)\left(\left(\frac{(2+\delta)^2 BV_{DD}}{\delta^2 + 4\delta + 3}\right)^{1/2} - \frac{B}{2}\right).\tag{4.28}$$

We may relate the parameters of the resistive load inverter to the noise margins developed in Section 4.1. Using the definitions $V_{IL} \equiv V_{TE0}$ and $V_{OH} \equiv V_{DD}$, we then have

$$V_{OL} = \frac{V_{IH} - V_{TE0}}{1+\delta} + \frac{B}{2} - \left(\left(\frac{V_{IH} - V_{TE0}}{1+\delta} + \frac{B}{2}\right)^2 - BV_{DD}\right)^{1/2},\tag{4.29}$$

where V_{IH} is given by Eq. (4.1). This analysis has one subtlety. Recall that V_{TE0} actually depends on V_{DS} because of drain-induced barrier lowering. V_{TE0} is also a function of device width and length. Noise margins are a concept that applies to one logic gate driving not itself but another logic gate. If the widths and lengths of the devices in both logic gates are the same, then the dependence of V_{TE0} on these parameters need not be considered (but if they are different, then of course one must take them into account). Looking at the equations we have just written, we see that if Eq. (4.16) refers to the gate under study, then $V_{IL} = V_{TE0}$ refers to the next gate. Because the output of the next gate is high when its input is V_{IL}, we may more accurately write $V_{IL} = V_{TE0} - \sigma V_{DD}$, which agrees with Eq. (4.13). This subtlety can have a significant effect on circuit design.

The transient behavior of the resistive load inverter is asymmetric. For simplicity, assume that the input to the gate is either a voltage step from zero to V_{DD}, or from V_{DD} to zero. The transient behavior of a logic gate is almost always determined by extrinsic rather than intrinsic capacitances.[5] We analyze the case of an inverter driving a linear load capacitor of value C_L. For a low-going input transition at time $t = 0$, the rising output transient is determined by the load conductance G. We have

$$v_{\text{out}}(t) = (V_{DD} - V_{OL})\left(1 - e^{-tG/C_L}\right) + V_{OL}. \tag{4.30}$$

If the input transition is from zero to V_{DD}, the situation is more complicated. The time to reach some particular voltage v_{out} is given by

$$t = \int_{V_{DD}}^{V_{\text{out}}} \frac{C}{i_{pu} - i_{pd}}\, dv_{\text{out}}, \tag{4.31}$$

where i_{pu} and i_{pd} are given by Eqs. (4.14) and (4.15), with $v_{\text{in}} = V_{DD}$. (See Appendix B.) The shape of $v_{\text{out}}(t)$ is linear at the beginning, as behooves the voltage on a capacitor being driven by a current source, and exponential near the end.

In reality, the output voltage of the inverter can drop slightly below ground in response to a falling step input because of the charge pumped through the Miller capacitor C_{GD}. While Miller capacitance can have a considerable detailed effect on the transient response, its integrated effect is limited by the fact that since this is a digital logic application, the change in charge on C_{GD} is limited to $2C_{GD}V_{DD}$. For logic gates driven from nonzero impedance sources, the Miller capacitance can cause notches in the voltage transients.

In the extreme, the effective input capacitance for the resistive load inverter can vary from about 30% of WLC_{OX} to about 130% of WLC_{OX}, depending on the speed and direction of the input transition. W and L are the width and length of the inverter's pulldown transistor, respectively. For a high-going input, the intrinsic capacitance of the MOS pulldown transistor starts near zero and rises abruptly at threshold to about $\frac{2}{3}WLC_{OX}$. As the output of the inverter falls, the Miller effect boosts the input capacitance. After the transition is mostly over, the input capacitance of the inverter is WLC_{OX}. Note that depending on the relationship between the output and input transition times, the effective input capacitance of the inverter can range from less than $\frac{2}{3}WLC_{OX}$ (very fast inputs) to a much larger value as determined by the Miller capacitances, and so forth (slower inputs). For falling input transients there is a different sequence of capacitance values. All of this seems very grim for a designer's ability to "eyeball" the delay through a chain of devices. Luckily for a circuit designer's ability to estimate,

---

[5] Intrinsic capacitances are those intimately associated with the transistor, that is, those in Eqs. (2.119) and (2.120).

but unluckily for performance, most circuit delays are dominated not by gate capacitances but by interconnect capacitances. One can typically use simple delay models to estimate delays within 30% or so, but the effects discussed here do limit the accuracy of these simple estimates.

Most of the effects discussed so far become worse as devices get shorter and shorter. So why do we push for short channel devices? The answer is found by looking at a pair of ratioed logic gates. The power dissipation in the logic gate is primarily determined by the pullup device. We examine what happens to the delay for a given size pullup and varying lengths on the pulldown. The shape factor of the pulldown S_{pd} must remain constant in order for the logic gate to satisfy noise margin constraints. The input capacitance of the logic gate is proportional to $W_{pd}L_{pd}$, and hence the delay and power delay product for a fixed size pullup driving the logic gate is also proportional to $W_{pd}L_{pd}$. (The power is constant.) But because S_{pd} is constant, $W_{pd} = S_{pd}L_{pd}$. Hence the delay and power delay product are both proportional to L_{pd}^2. This is a very strong motivation to make L_{pd} small. In CMOS technology, one also develops an L^2 dependence of delay, given a fixed transistor width.

The depletion load nMOS inverter

In ratioed nMOS, the load (which connects the output to V_{DD}) is as much, if not more, a determinant of the logic gate's performance than the pulldown transistor. Assuming that the load is a two-terminal resistive element without hysteresis, there are several constraints on its current versus voltage characteristics. Given that the pullup is passive, the current through the load should be zero when the output is at V_{DD}. In addition, when the input is at V_{IH}, the output should be at V_{OL}, and the current through the pullup must equal the current through the pulldown. Fig. 4.10 illustrates four different pullup I–V curves that could be used with the same pulldown transistor. They all source the same current when the output of the logic gate is V_{OL}. The enhancement-mode pullup will not pull as high as the other three devices, but other than that, all are equivalent from a static viewpoint.

In terms of dynamics, there is a huge difference. If the output is at V_{OL} and the pulldown transistor is turned off, the output will rise in a time given by

$$\Delta T = C_{\text{LOAD}} \int_{V_{DD}-V_{OL}}^{V_{DD}-V_{\text{finish}}} I_{pu}^{-1} \, dV \qquad (4.32)$$

to a voltage V_{finish}. We can use $V_{\text{finish}} = V_{IH}$, or even a little lower. It is in the context of this equation that the pullups are very different. The enhancement load is by far the worst, followed by the linear resistor. The depletion load is the best of the conventional loads—and the lower the body effect, the better it is. The tunnel diode is the best from a theoretical standpoint, but not very practical. Since it has negative differential resistance, it is potentially unstable unless compensated by

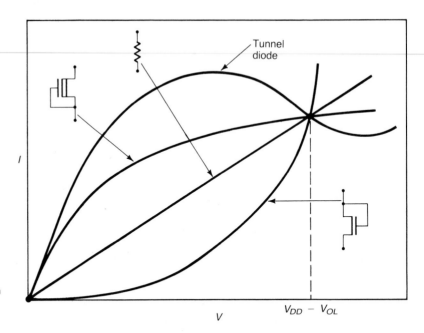

FIGURE 4.10 Comparison of various $I-V$ curves of load devices.

the pulldown's positive resistance. We include it in this discussion only to point out that there is still room for new ideas in this area. In ratioed nMOS, the majority of the power dissipation occurs when the outputs of the logic gates are low. Dynamic switching power is typically less than 5% of this quasi-dc value. If we neglect dynamic power, then each of the pullups we have illustrated draws about the same wattage. Therefore the delay given in Eq. (4.32) is proportional to the power delay product. Depletion load logic families have a better power delay product than enhancement or resistive load logic families.

Modern depletion load nMOS transistors have a very low body effect because the substrate doping is light and the gate oxides are thin. Thus δ can typically be ignored in these technologies. The current in the pullup is given by

$$i_{pu} = \begin{cases} K'_D S_D (-2V_{TD} - V_{DD} + v_{\text{out}})(V_{DD} - v_{\text{out}}) & \text{for } V_{DD} - v_{\text{out}} \leq V_{TD} \\ K'_D S_D V_{TD}^2 & \text{for } V_{DD} - v_{\text{out}} \geq V_{TD}. \end{cases}$$
(**4.33**)

V_{TD0} is normally around -2.5 volts. Due to body effect, V_{TD} is a function of v_{out}. Typical v_{out} versus v_{in} transfer functions are plotted in Fig. 4.8. There are two regimes of operation that warrant closer attention. When both devices are in saturation, the differential gain is largest. Neglecting δ and σ, the small signal characteristics of the saturated pulldown are

$$\frac{\partial i_{pd}}{\partial v_{\text{in}}} = 2\sqrt{K'_E S_E I_{pd}}.$$
(**4.34**)

For the pullup, we have

$$\frac{\partial i_{pu}}{\partial v_{\text{out}}} = -2b\sqrt{K'_D S_D I_{pu}}, \qquad (4.35)$$

where we have

$$b = \frac{\partial V_{TD}}{\partial v_{\text{out}}}. \qquad (4.36)$$

Evaluating b, we obtain[6]

$$b = \frac{\gamma}{2\sqrt{V_{SB} + 2\Phi_{Fp}}}. \qquad (4.37)$$

In terms of the load line characteristics seen before, b should be as low as possible in order to achieve the best power delay product. This implies that we want a low value of γ and a large value of V_{SB}. This is one of the reasons a negative V_{BB} bias is used in high performance nMOS technologies. For processes with a low value of γ, it also pays to keep $|V_{TD}|$ low so that the pullup is in the saturation regime for as long as possible. This serves to keep i_{pu} constant for the maximum range of pullup voltages. It also keeps the pullup capacitance low. Note that $b = \delta$ in our model. This is an artifact of our particular modeling assumptions and should not be construed as an eternal truth. The small signal gain of the depletion load inverter is obtained by equating i_{pu} and i_{pd}. We obtain

$$a_o = \frac{\partial v_{\text{out}}}{\partial v_{\text{in}}} = -\frac{1}{b}\sqrt{\frac{K'_E S_E}{K'_D S_D}} \approx -\frac{\sqrt{\beta}}{b}. \qquad (4.38)$$

For very high gains with short channel devices, drain-induced barrier lowering and channel length modulation can become limiting. For instance, even if the load were a perfect current source of value I_{pu}, the magnitude of the small signal gain would be limited to σ^{-1}. We have

$$I_{pu} = K'_E S_E (v_{\text{in}} - V_{TE0})^2. \qquad (4.39)$$

Differentiating with respect to v_{in}, we have

$$0 = 2K'_E S_E (V_{\text{in}} - V_{TE0})\frac{\partial}{\partial v_{\text{in}}}(v_{\text{in}} - V_{TE0})$$
$$= 1 - \frac{\partial v_{\text{out}}}{\partial v_{\text{in}}}\frac{\partial V_{TE0}}{\partial v_{\text{out}}} \qquad (4.40)$$
$$= 1 + a_o \sigma,$$

where we have identified

$$a_o = \frac{\partial v_{\text{out}}}{\partial v_{\text{in}}} \qquad (4.41)$$

---

[6] It should be pointed out that most MOS device model parameters are extracted from measured data by curve fitting to the large signal parameters. Small signal parameters that typically refer to differentials can be far from their true value.

and

$$\sigma = -\frac{\partial V_{TE0}}{\partial v_{\text{out}}}. \tag{4.42}$$

Solving, we obtain

$$a_o = -\frac{1}{\sigma}. \tag{4.43}$$

When the gain of a depletion load inverter is limited by this mechanism, it can be increased by actually lengthening the pulldown device to get it out of the short channel regime!

The second important regime is when the output is near V_{OL}. Assuming reasonable thresholds, the pullup device is in saturation. Equating the pullup and pulldown currents, we obtain

$$K'_D S_D V_{TD}^2 = K'_E S_E \left(2(v_{\text{in}} - V_{TE0})v_{\text{out}} - v_{\text{out}}^2 \right), \tag{4.44}$$

where the pulldown has been assumed to be in the nonsaturation regime. In this regime, $V_{TD} \approx V_{TD0}$. Equation (4.44) defines the constraint on the β-ratio when $v_{\text{in}} = V_{IH}$ and $v_{\text{out}} = V_{OL}$. We have

$$\beta \geq \frac{K'_D}{K'_E} \frac{V_{TD0}^2}{\left(2(V_{IH} - V_{TE0})V_{OL} - V_{OL}^2 \right)}. \tag{4.45}$$

The dynamics of the depletion load inverter are quite complicated. Let us concentrate on the important initial 40% or so of the output waveform. Assume a step input. The enhancement-mode transistor determines the pulldown transient. If the input goes from zero to V_{DD}, then the pulldown current starts at

$$i_{pd} = K'_E S_E (V_{DD} - V_{TE0})^2 \tag{4.46}$$

and the initial slew rate at the output is given by

$$\frac{dv_{\text{out}}}{dt} \approx \frac{K'_E S_E (V_{DD} - V_{TE0})^2}{C_{\text{LOAD}}}. \tag{4.47}$$

For the pullup transient, the pulldown is off. The pullup transistor starts in saturation. We have

$$I_{pu} = K'_D S_D V_{TD}^2 \tag{4.48}$$

or

$$\frac{dv_{\text{out}}}{dt} \approx \frac{K'_D S_D V_{TD}^2}{C_{\text{LOAD}}}. \tag{4.49}$$

Note the trade-off in choosing $|V_{TD0}|$. If $|V_{TD0}|$ is made small, the pullup will lay out compactly. But since uncertainties in thresholds tend not to depend on the threshold magnitude, the delay is more controlled when $|V_{TD0}|$ is large. Solving for the ratio of the pullup to pulldown transition rate β_τ, we have

$$\beta_\tau \equiv \frac{\text{rising slew rate}}{\text{falling slew rate}} = \beta \frac{K_E'}{K_D'} \frac{(V_{DD} - V_{TE0})^2}{V_{TD}^2}. \qquad (4.50)$$

Using β from Eq. (4.45), Eq. (4.50) reduces to

$$\beta_\tau = \frac{(V_{DD} - V_{TE0})^2}{(2(V_{IH} - V_{TE0})V_{OL} - V_{OL}^2)}. \qquad (4.51)$$

Note that to first order, β_τ does not depend on V_{TD} and increases with V_{DD}. For reasonable threshold choices, $\beta_\tau \approx 2\beta$. This tendency is influenced in detail by three facts: the transistors do not stay in saturation, the inputs are not step functions (this slows the fall time more than the rise time), and the switching point V_{inv} is generally closer to ground than to V_{DD} (lengthening the effective fall time relative to the rise time).

Fig. 1.28 illustrates the layout of two typical nMOS inverters. The load on the first inverter consists of self-loading capacitance, interconnect capacitance, and input capacitance of the next stage. The self-loading capacitance has several components, including the channel-to-substrate capacitance of the pullup, the capacitance of the buried contact, the poly extension capacitance (proportional to the length of the pullup), the diffusion capacitance (proportional to the width of the pulldown), the gate-to-drain capacitance of the pullup (mostly overlap capacitance since the pullup is in saturation), and so forth. Of these self capacitances, the capacitance of the buried contact often dominates when the devices are small. Interconnect capacitances were discussed in Section 2.4.

For very slow ramping inputs, the transient response of the inverter is determined by the dc transfer function. Defining the transition time for a logical voltage as the time it takes the voltage to go from one valid input voltage to another, the transition time on the output of a logic gate is roughly the input transition time divided by the "gain" of the logic gate. Imagine a long string of identical inverters. If the first one starts with a very long transition time for the input waveform, as the signal propagates through the chain the transition gets faster and faster until it reaches a steady state. In the steady state, the transition time is limited by a combination of load capacitances and the dc transfer function. Consider ratioed nMOS, where the pullup resistance is usually about ten times larger than the pulldown resistance. The integrated "gain" of the inverter is also about ten. In this case, the load capacitance limits the pullup voltage transient, which is the slow case. The low-going voltage transient, on the other hand, is an ugly mixture of gain and load capacitance limited behavior. This case must be simulated to get an accurate delay transition time estimate. CMOS circuits are more ideal in this respect, yet they too go more slowly than one would expect simply

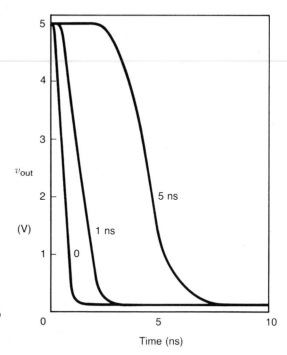

FIGURE 4.11 Inverter output waveforms for ramp inputs with 20%–80% rise times of 0, I, and 5 ns.

from a step response characterization. Fig. 4.11 illustrates the output waveforms of a depletion-load nMOS gate in response to a variety of input waveforms.

The input capacitance of a depletion load inverter is the same as that for the resistive load inverter discussed earlier. The input charging characteristics of several different circuits are shown in Fig. 4.12 for a rising input voltage transient.

FIGURE 4.12 The total charge into the gate of a MOS transistor used in various circuit configurations. The input is a ramp, and all of the devices are the same size. The source follower configurations are the easiest to charge.

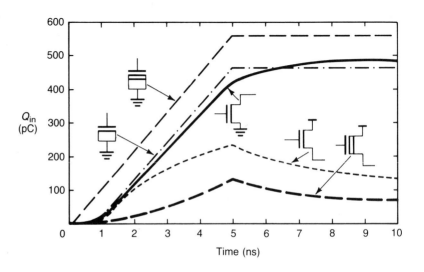

While we have qualitatively discussed the effective input capacitance for these next stages, we have not really defined it. The most reasonable definition is probably given by

$$C_{\text{in}} \equiv \frac{dQ_{\text{in}}}{dv_{\text{in}}}\bigg|_{\text{region of interest}} \tag{4.52}$$

The problem here is that $Q_{\text{in}}(t)$ can continue moving after $v_{\text{in}}(t)$ has stopped, rendering C_{in} less than useful. It is better to look at two quantities. One can examine the initial $dQ_{\text{in}}/dv_{\text{in}}$ because this must be well-behaved and is a good reflection of the load that the drive sees in the important initial nanoseconds of the transient. Thus Eq. (4.52) is useful when used over the "region of interest." The other physically meaningful quantity is the total charge placed on the node. This is a reasonable quantity since digital circuits do, in some sense, integrate over the entire waveform. Note, for instance, that the total charge into an enhancement-mode source follower is approximately zero due to the Miller effect. Source followers are indeed easy to drive.

Having determined the β-ratio for an inverter, the next step is to generalize these results to complex gates. To do this, note that regardless of the logic function to be performed, the total current through, and voltage across, the pulldown network must remain constant for a given V_{IH} and V_{OL}. Consider the case of the prototypical inverter with a W_0/L_0 pulldown, a drain voltage of V_{OL}, and a gate voltage of V_{IH}. Ignoring short and narrow channel effects, we would have exactly the same current and voltage if the pulldown width was χW_0 and the length χL_0, since S would remain the same. This is true whether the device is operating in the linear, nonsaturation, saturation, or even cutoff regime.[7] We can break the transistor into two series devices of width χW_0 and lengths L_0 and $(\chi - 1)L_0$. The voltage V_Z at the center node is the same as it would be at the length $(\chi - 1)L_0$ along the channel in the $\chi W_0/\chi L_0$ device. Since the bottom device has a width of χW_0 and a length $(\chi - 1)L_0$, we can transform this device to an equivalent device (keeping S constant) of width $\chi W_0/(\chi - 1)$ and length L_0. These transformations are illustrated in Fig. 4.13. By generalizing this argument, one can demonstrate that providing the number of squares of pulldown resistance remains constant, V_{OL} will be preserved as the output voltage. Note that this argument is independent of the voltage nonlinearity of transistors, and works for any value of γ. Though we assumed no short or narrow channel effects, L was kept constant at L_0 for all transistors. This eliminates the role of short channel effects from these calculations. When W is kept wide, narrow channel effects are also unimportant, and one need only count squares to size the pulldown network.

---

[7] This argument does not work when velocity saturation is important because then i_{DS} is not proportional to S, but for low V_{OL} this is usually not an issue.

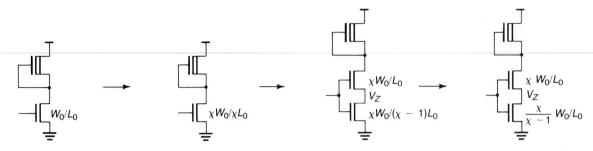

FIGURE 4.13 Sizing a NAND gate (V_{OL} is the same in all cases).

Because we know the length and width of a transistor vary with processing, we must determine which W and L to use. It certainly is not the drawn length and width, since the arguments above were in terms of effective lengths and widths. It turns out that over the set of all the length and width variations, one set is the worst in terms of noise. These worst-case noise parameters are the ones to use. The worst-case noise parameters are discussed in Section 4.4.

The previous development did not depend on the value of V_{IH}, but it did depend on V_{IH} being the same for all pulldown devices. If all pulldowns do not have the same input voltage, we estimate the required shapes by invoking linearity. This is less rigorous since the pulldown is not completely linear, but the approximation is not bad. Say we discover that for a given pullup the width of pulldown needed is W_D when the inverter input comes directly from depletion load inverter and W_P when the input comes through a pass device. That is, we are examining the case when there are both strong and weak high voltages. If we had a two-input NAND gate with both inputs driven strongly from depletion load gates, we would expect a required width of $2W_D$ for both pulldowns. If both inputs came through pass devices, we would expect both pulldowns to be of width $2W_P$. Assume the following model. Let R_D be the "resistance" of one square of pulldown transistor in the case where the input is driven directly. The choice of R_D is arbitrary, but because of the preceding scaling arguments, any pulldown network with a resistance of $R_D L_0/W_D$ will be legitimate. Thus for a two-input NAND with incommensurate input levels, if the pulldown resistance of the strongly driven device is αR_D, the pulldown resistance of the weakly driven device must be $((L_0/W_D) - \alpha) R_D$. The width of the strongly driven device must be L_0/α, and the width of the pass driven device must be $W_P/(1 - \alpha W_D/L_0)$.

It is often useful to have a simple first order model of the delay through MOS logic gates when reasoning about trade-offs, optimization strategies, and doing "back of the envelope" delay calculations. We said before that the time required by a MOS logic gate to charge a linear load capacitance can be written as

$$T = C \int \frac{1}{i} dv \qquad (4.53)$$

in the case of a step input voltage. For the purpose of obtaining simple delay estimates, we can identify

$$R = \int \frac{1}{i} dv, \qquad (4.54)$$

where R is the equivalent "resistance" of the logic gate for purpose of RC delay calculations. R is different for the pullup and pulldown. For a depletion load inverter, we have

$$R_{pu} \approx \beta_\tau R_{pd}. \qquad (4.55)$$

R is inversely proportional to the width W of the transistor. It is convenient to define a resistance R_τ of a pulldown transistor that has been drawn square. Thus, a pulldown transistor in our 2 μm nMOS depletion load process would have an equivalent dynamic resistance of $R_\tau/5$ if it was drawn 10 μm wide. This is obviously a gross, but useful, simplification. The pulldown transistor also has an effective input capacitance, and that capacitance is proportional to W. If we define C_τ as the effective input capacitance of a pulldown transistor that is of minimum length and as wide as it is long, then the 10/2 pulldown can be shown to have an approximate effective capacitance of $5C_\tau$. For any width pulldown, the product of R_τ and C_τ is constant. We define τ, a constant of the technology, as

$$\tau \equiv R_\tau C_\tau. \qquad (4.56)$$

To be as accurate as possible, R_τ should be extracted from delay simulations of typical circuit configurations. C_τ is usually taken as $L^2 C_{OX}$, where L is the minimum drawn length. Typical values of R_τ are 5 to 50 k$\Omega/\square$ for n-channel devices and 10 to 100 k$\Omega/\square$ for p-channel devices.

The CMOS inverter

The classical CMOS inverter illustrated in Fig. 1.34 consists of a p-channel pullup transistor connected to an n-channel pulldown transistor. Both transistors are enhancement-mode devices. The gates of the two transistors are wired together. The dc transfer function of this device has three regimes of operations. In the first regime one of the two transistors is off. The pullup is off when $v_{in} > V_{DD} + V_{TP0}$ (remember that V_{TP0} is

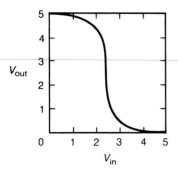

FIGURE 4.14 Transfer function of a CMOS inverter.

negative for an enhancement-type pMOS device) or when $v_{\text{in}} < V_{TN0}$. V_{TP0} and V_{TN0} are the thresholds of the p- and n-type devices when the source voltage on the devices are V_{DD} and 0, respectively.[8] In the second important regime of operation, both transistors are in saturation. The gain in this regime is very high, limited by channel length modulation, series resistance, and drain-induced barrier lowering. Fig. 4.14 illustrates the transfer function of a CMOS inverter. Note that the body effect is unimportant in this particular circuit configuration because the sources of both devices are pinned. We can compute the input voltage that causes the inverter to enter this regime. Equating the saturation currents, we have

$$V_{\text{inv}} = V_{TN0} + \frac{V_{DD} - V_{TN0} + V_{TP0}}{1 + \sqrt{\beta \frac{K'_N}{K'_P} \frac{1+\delta_P}{1+\delta_N}}}. \tag{4.57}$$

Note that V_{inv} is determined more by device thresholds than the β-ratio of the inverter. In typical CMOS processes the body effect is high and hence δ must be included in our calculations. At V_{inv}, the CMOS inverter consists of two current sources in series. Small changes in the input voltage will cause the output to swing widely. To first order, the gate has high gain between the points when one of the transistors goes out of saturation. These output voltages are

$$\frac{V_{\text{inv}} - V_{TN0}}{1 + \delta_N} \tag{4.58}$$

and

$$\frac{V_{\text{inv}} - V_{TP0}}{1 + \delta_P}. \tag{4.59}$$

In the third regime of operation, one transistor is in saturation and the other is in nonsaturation.

As mentioned in Section 4.1, there are at least two common ways of defining the voltage logic levels in CMOS circuits. Let us examine the easiest one first. Following the procedure we used in nMOS of taking a threshold as a reference voltage, we can define $V_{IL} \equiv V_{TN0}$ and $V_{IH} = V_{DD} + V_{TP0}$. In this case, the sizing of the transistors in the CMOS logic gate is completely arbitrary, from a noise margin standpoint. One must be careful, however, to have fairly high voltage thresholds to obtain a sufficient level of noise immunity. This type of methodology is appropriate for medium to low performance chips.

The other common style of CMOS circuit design uses the $-M^{-1}$ slope points of the logic gates to define the noise margins. While it

---

[8] In most CMOS processes, the body of the n-channel devices is tied to ground and the body of the p-channel devices is tied to V_{DD}. Special substrate-bias generators, common in nMOS, are not typically used.

is clear that one can design robust VLSI circuits with this technique, there is no rigorous theory. Present techniques involve a lot of art. The basic strategy is to keep V_{inv} close to $V_{DD}/2$ for most gates, and follow precharged stages by inverters.

Unlike static nMOS circuits, classical static CMOS circuits draw current only while switching. This current has two components. One component of the current does the useful work of charging and discharging external load capacitances. The other component is due to overlap currents when both the pullup and pulldown devices are on simultaneously. The power required to charge and discharge a capacitor C_{LOAD} at a switching frequency f_S is $f_S C_{\text{LOAD}} V^2$, where V is the difference between the high and low voltages on the capacitor. We have made the assumption that the energy on the capacitor is dissipated in each cycle. V usually equals V_{DD} in digital CMOS circuits. Fig. 4.15 illustrates an idealized logic gate. The integral of the charging current, per cycle, is given by

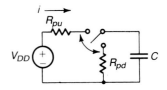

FIGURE 4.15   Idealized model of a CMOS inverter. Half the power is dissipated in R_{pu} and half in R_{pd}.

$$\int i\,dt = Q = CV_{DD} \qquad (4.60)$$

because all of the current goes to charge the capacitor. This current must flow out of the power supply. The energy drawn from the power supply is

$$\int (i\,V_{DD})\,dt = CV_{DD}^2. \qquad (4.61)$$

Half of the energy is dissipated in R_{pu} and half in R_{pd}, independent of their values.

Even if there were no output capacitance connected to the inverter, it would still, unfortunately, dissipate switching energy unless the input switched instantly. The slower the input transition, the longer the inverter is in the region where both transistors are on and the higher the excess power dissipation. The maximum excess current is given by

$$I_{\max} = \frac{K_N' S_N}{1 + \delta_N} \left(\frac{V_{DD} - V_{TN0} + V_{TP0}}{1 + \sqrt{\beta \frac{K_N'}{K_P'} \frac{1 + \delta_P}{1 + \delta_N}}} \right)^2. \qquad (4.62)$$

The excess power dissipation in CMOS circuits is very important and can easily be of the same order as the $f_S CV^2$ power dissipation. Both types of power dissipation are proportional to frequency. T_{out} is the output switching time. The switching energy E has two components. Concentrating only on the discharge transient, we have

$$E = E_{\text{excess}} + E_{\text{cap}} \qquad (4.63)$$

where E_{excess} is the energy due to overlap current. The energy stored

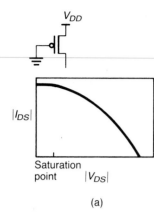

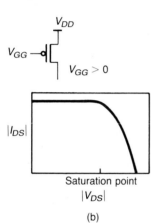

FIGURE 4.16 A p-type load device with (a) $V_{GG} = 0$ and (b) V_{GG} near $V_{DD} + V_{TP0}$. The load line in (b) is more ideal than the one in (a).

on the capacitor is

$$E_{\text{cap}} = 1 \le 2C_L V_{DD}^2 \qquad (4.64)$$

and

$$E_{\text{excess}} = (Q_T - V_{DD}C_L) V_{DD} . \qquad (4.65)$$

Solving, we have

$$E = Q_T V_{DD} - 1 \le 2C_L V_{DD}^2 . \qquad (4.66)$$

For fast inputs, $Q_T \approx C_L V_{DD}$, and hence we have

$$E \approx 1 \le 2C_L V_{DD}^2 = E_{\text{cap}} . \qquad (4.67)$$

For very slow inputs, C_L is unimportant and the inverter traces out its transfer function. This causes $T_{\text{in}}/T_{\text{out}}$ to be constant. The total charge Q_T is thus proportional to $I_{\max}T_{\text{in}}$. When V_{TN0} and $|V_{TP0}|$ are small, $I_{\max} \propto V_{DD}^2$, so

$$E \propto V_{DD}^3 \, T_{\text{in}} \qquad (4.68)$$

in this regime. ($E \propto V_{DD}^2 T_{\text{in}}$ for heavily velocity saturated devices). One can simulate the dynamic power dissipated in a CMOS circuit by using a current controlled current source driving a capacitor to keep track of the charge supplied by the power supply.

A style of static CMOS called "mostly nMOS" dissipates dc power. The philosophy is quite simple: n-channel devices are better than p-channel devices, so we will use mostly n-channel devices. Over the years the load for nMOS pulldowns has gotten better and better. One can view the p-channel transistor as just a better load. Fig. 4.16(a) illustrates a pMOS pullup with a grounded gate. This device does a good imitation of an ideal depletion load (no body effect). Such devices find application, for instance, in large static CMOS NOR gates.

We can improve on the grounded gate p-channel pullup in two ways. Fig. 4.16 illustrates two p-type loads. Note that as the voltage V_{GG} on the gate of the p-type device is increased, the load line becomes more ideal. (To maintain the same current, the pullup would need to be widened.) In addition, the current through the load can be made to track the strength of the pulldown by the use of a highly useful circuit technique called the current mirror. Fig. 4.17 illustrates a current mirror. Assuming that transistor M_1 is on, it must be in saturation. Its current is totally determined by $V_{DD} - V_{GG}$, the shape factor of the device, and intrinsic device parameters.[9] It is independent of V_{DS} to first order, especially if drawn long. Note that the gate-to-source voltage of transistor M_2 is identical to that of transistor M_1. Thus the current I_2, through transistor

---

[9] Remember that V_{DD} is an anachronism left over from nMOS design. V_{DD} is actually the source terminal of the p-channel pullups. Even more anachronistic is the use of V_{CC} ("C" as in "collector") for V_{DD}. V_{SS} is often used to denote ground.

M_2, is

$$I_2 = \frac{S_2}{S_1}I_1 \qquad (4.69)$$

as long as it stays in saturation. I_1 is the current through transistor M_1.

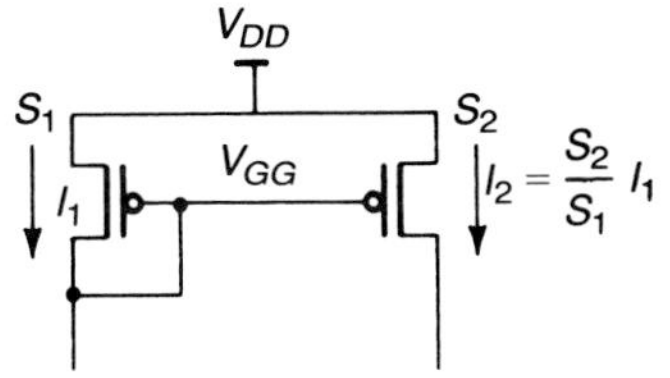

In the design of the perfect load for a MOS process in which transistor parameters vary from chip to chip, we want the load to be strong when the pulldown is strong, but we want the load to be weak when the pulldown is weak. In other words, we would like the ratio of the pulldown to pullup "resistance" to be constant, independent of fabrication uncertainties. The current mirror gives us a way to almost achieve this. The key is to build a circuit in which V_{GG} is generated as a reference voltage and pumped around the chip. Fig. 4.18 illustrates this. If the pulldown transistors in the circuit are strong, the reference transistor will also be strong. If the reference pulldown transistor is strong, V_{GG} is pulled lower and the pullups around the chip are turned on harder.

One of the important parameters in a circuit is its power dissipation. Normally, the current in a ratioed gate (assuming the output is pulled low) is proportional to the mobility of the pullup. For instance, doing a sensitivity analysis of the p-type pullup with the grounded gate shown in Fig. 4.16(a), we have

$$\frac{dI}{I} = \frac{d\mu_P}{\mu_P}. \qquad (4.70)$$

Let us examine the case of the current mirror pullup. We have

$$\frac{dI_{pu}}{d\mu_P} = \frac{I_{pu}}{\mu_P} - \frac{2I_{pu}}{(V_{DD} - V_{GG} + V_{TP0})}\frac{\partial V_{GG}}{\partial \mu_P}. \qquad (4.71)$$

The strategy is to use $\partial V_{GG}/\partial \mu_P$ to help lower the sensitivity. For the

FIGURE 4.17   A current mirror. This circuit works only as long as both devices are in saturation.

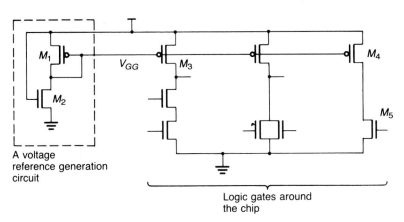

A voltage reference generation circuit

Logic gates around the chip

FIGURE 4.18   The voltage reference generator on the left provides a reference for ratioed logic gates about the chip.

pulldown, we have

$$\frac{\partial I_{pd}}{\partial \mu_P} = I_{pd} \left(\frac{1}{V_{GG}} - \frac{1 + \delta_N}{2(V_{DD} - V_{TN0}) + (1 + \delta_N)V_{GG}} \right) \frac{\partial V_{GG}}{\partial \mu_P} \tag{4.72}$$

or

$$\frac{\partial I_{pd}}{\partial \mu_P} \approx \frac{I_{pd}}{V_{GG}} \frac{\partial V_{GG}}{\partial \mu_P}. \tag{4.73}$$

Since we have

$$\frac{dI_{pd}}{d\mu_P} = \frac{dI_{pu}}{d\mu_P} = \frac{dI}{d\mu_P} \tag{4.74}$$

and

$$I_{pd} = I_{pu} = I, \tag{4.75}$$

we then have

$$\frac{dI}{d\mu_P} = \frac{I}{\mu} \left(\frac{V_{DD} - V_{GG} + V_{TP0}}{V_{DD} + V_{GG} + V_{TP0}} \right). \tag{4.76}$$

For $V_{GG} \approx 2, V_{TP0} \approx -1$, and $V_{DD} \approx 5$, we obtain

$$\frac{dI}{I} = \frac{d\mu_P}{3\mu_P} \tag{4.77}$$

or about one third the sensitivity of the normal case.

There are many variations on this technique. Fig. 4.19 shows another variant of this technique with the reference pulldown in saturation. In this case, the pulldown is biased further away from its actual use; however, the sensitivity of the reference voltage V_{GG} to changes in the n-type pulldown is better. Fig. 4.20 illustrates a clocked version of the circuit. This technique is also good for constructing bleeder devices. It should be noted that the p-type pullup is rarely made minimum size, so its current is more controlled. There is almost no performance penalty for this since the capacitance looking into the drain of this device is very low.

One problem with this technique is its high noise sensitivity. Since V_{GG} is pumped all around the chip, it can act as a noise conduit. One might expect, for instance, a lot of coupling to the clock lines. When V_{GG} is near $V_{DD} + V_{TP0}$, I_{pu} is extremely sensitive to V_{GG} noise, V_{DD} bounce, and on-chip threshold variations. The sensitivity of the current to V_T is

$$\frac{dI}{I} = -\frac{2dV_T}{V_{GS} - V_T}. \tag{4.78}$$

For instance, for $dV_T = 100$ mV and $V_{GS} - V_T = 1$ V, we see 20% variations in the pullup currents due just to this effect.

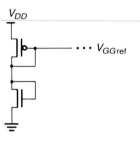

V_{DD}

$\cdots V_{GGref}$

FIGURE 4.19   Another voltage reference generator.

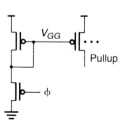

FIGURE 4.20   Combination current mirror and clocked reference generator circuit provides for lower power dissipation.

V_{GG}

Pullup

ϕ

4.3 Noise

The noise performance of a VLSI chip has three variables: the voltage noise margins, the impedance levels, and the characteristics of the noise sources. Several factors contribute to noise in integrated circuits. The physical noise sources such as thermal and $\frac{1}{f}$ [Stone 84] , which are familiar to the analog circuit designer, are not yet critical in digital MOS. Most noise in digital integrated circuits comes from phenomena that are in principle predictable but in practice too complicated to account for systematically.

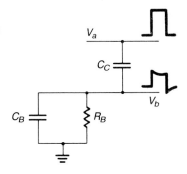

FIGURE 4.21   Capacitive coupling noise.

Capacitive coupling

Capacitive coupling is a significant cause of noise and will become even more important as MOS technology progresses to more numerous layers of interconnect and line geometries with greater aspect ratios. Consider the arbitrary nodes A and B in a VLSI circuit, with a coupling capacitor C_C between them, as illustrated in Fig. 4.21. We would like to determine the effect of a change in the voltage v_A at node A on the voltage v_B at node B. If the total small signal admittance (including C_C) between node B and all other nodes in the system is $Y(s)$, where s is complex frequency, then the transfer function between $V_a(s)$ and $V_b(s)$ is given by

$$V_b(s) = \frac{sC_C}{Y(s)}V_a(s). \tag{4.79}$$

Thus the greater the magnitude of $Y(s)$, the lower the coupling noise. There are two special cases of interest. The first is when node B is used for dynamic storage. If node B has a capacitance to ground of value C_B, then

$$v_B = \frac{C_C}{C_C + C_B}v_A. \tag{4.80}$$

If node B corresponds to the output of a restoring logic gate, then $Y(s)$ has a significant resistive component. For $R_B \ll \tau/C_C$, we have

$$V_b(s) = sC_C R_B V_a(s). \tag{4.81}$$

Because the small signal resistance of a pullup transistor is about β times larger than the corresponding pulldown resistance in nMOS technology, we need $NM_H \approx \beta NM_L$ in cases where this effect dominates. But there is an even more compelling reason to have NM_H greater than NM_L. In nMOS we typically precharge a line high. When the line is precharged high, its impedance is very large because the pullup shuts off. Thus we want the highest noise margin here that we can get. In some sense we can also afford a larger noise margin here because the high noise margin is some fraction of $V_{DD} - V_{TE0}$. The low noise margin is some fraction of V_{TE0}, which is a much smaller number.

Resistive coupling

Perhaps the most significant contribution to noise in MOS integrated circuits is from resistive feedback in ground and power lines. For packing reasons, and because the sources and drains of transistors must be in n+ or p+, power in integrated circuits is often distributed locally in diffused regions. Diffusion ordinarily has a resistance of tens of ohms per square. This resistance affects the operation of solitary gates by changing the pullup and pulldown resistances and hence the switching delay, V_{OL}, and power dissipation. Noise is caused by the effect of one gate on another. When one gate switches it can change the effective V_{DD} or ground voltage seen by a second logic gate. Fig. 4.22 illustrates a case where resistive coupling noise is important.

When two logic gates have a common parasitic resistance R_P that connects the logic gates to the voltage supply, the noise voltage v_P across that resistor is approximately equal to $i_P R_P$ if R_P is much smaller than the real part of the small signal impedance of the transistors. The current i_P is that current which flows through R_P due to one of the gates.

Resistive drops in power lines occur not only because of the resistive contribution of the diffusion but also because of the resistance of contact cuts, polycrystalline silicon, and metal. Since metal typically supplies power to many gates, we can say, with the possible exception of machines such as systolic arrays, that for statistical reasons the ac current component of the line is much smaller than the dc component in ratioed nMOS designs. It is not clear whether soft system failures occur because of the occasional simultaneous transitions of many gates. For chips with 100,000 or more transistors, the resistance of metal lines can be an important source of noise.

FIGURE 4.22   Resistive coupling noise. The noise voltage developed across R_P can cause the node that was precharged on ϕ_2 to discharge accidentally. This effect is seen in both CMOS and nMOS.

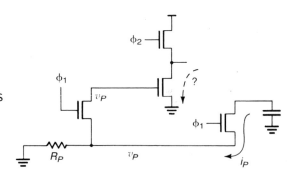

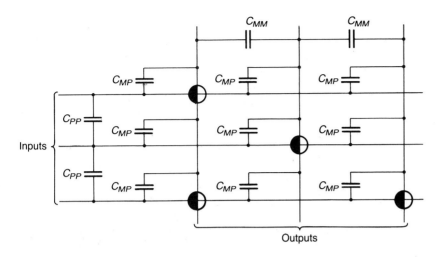

FIGURE 4.23   Capacitive coupling in an array.

Noise in array structures

Array structures are particularly sensitive to coupling noise due to the noise coherence and the tight packing density. Capacitive coupling occurs for crossing as well as adjacent wires. Fig. 4.23 illustrates a NOR plane with three of the coupling capacitances labeled. One technique for minimizing the detrimental effects of capacitive coupling is to keep voltage swings small. Another is to use a shielding conductor between the two coupled wires. This applies to wires that are side by side or ones that are stacked. For instance, a grounded sheet of first layer metal can isolate poly from a second layer of metal.

Resistive coupling due to common ground or V_{DD} currents can also cause noise problems. Another significant cause of noise in arrays is the overlap of signal or control lines. For instance, it is common to have a transient condition in which we simultaneously have two different word lines selected in a ROM or RAM array, despite the fact that this condition is prohibited in the ideal.

■■■■■■■ EXAMPLE 4.1

Noise coupling in an HROM
In Example 2.9, entitled SPICE Model of a ROM, we analyzed the delay through an nMOS HROM. We now extend the analysis to include coupling noise by solving the following problem. Assuming that the bit lines are precharged to V_{DD}, how far does an unprogrammed bit line fall when its two nearest neighbors move through 5 volts?

The worst-case is when the bit line capacitance is low and the coupling capacitance is high. Since the coupling capacitance between bit lines rises more rapidly than does the bit line capacitance to ground when the lines become closer together, the worst-case is when the bit lines are wide. Since the bit lines are drawn at 4 μm lines and spacings, if the worst-case metal variation is ± 0.4 μm/edge, then the final lines are 4.8 μm wide with 3.2 μm spacings.

We first find the average bit line capacitance to ground per micron of bit line length. The repetition interval is 16 μm, in which we have 14.4 μm^2 of C_{mdiff}, 19.2 μm^2 of C_{mpoly}, and 43.2 μm^2 of C_{mf} for the unprogrammed case. This results in an average bit line capacitance of 25.6×10^{-5} pF/μm. From Dang [Dang 81], we have discovered that the total coupling capacitance is 2.5×10^{-5} pF/μm, where we have assumed an average metal height above the ground plane of 1.1 μm and a metal thickness of 0.55 μm. The coupling is therefore down only about 21 dB, meaning that a 5 V change in the neighboring bit lines would result in a 460 mV change in the unprogrammed bit line. In the worst-case, even if the bit lines where heavily programmed, they would still move about 300 mV.   ■

Alpha particle noise

Alpha particles have been found to be a significant source of noise in dynamic memories with storage capacities of 4000 bits or above. Alpha particle-related soft errors in microprocessors, at enhanced radiation levels, have also been seen.

An alpha particle is a doubly ionized He atom with an energy of less than about 10 MeV. These particles penetrate the silicon, up to 20 or 30 μm, leaving a track of hole–electron pairs. The alpha particle will generate 25,000 to 80,000 hole–electron pairs per micron of track length. When these holes and electrons are collected they can form a noise current or, more important, a noise charge. Designing a dynamic RAM cell with a small collection efficiency is obviously important.

While there have been advances in packaging, it is unlikely that fluxes below 0.001 to 0.01 α/(cm^{2}h) will be achieved. Good circuit and system design is therefore important. Usually a critical charge Q_{crit} is determined for a memory cell. This gives a first order indication of the sensitivity of the cell to alpha particles. There are, however, many other factors involved, such as the collection area. Storage cells with N_{crit} above about 2×10^6 electrons seem to be fairly safe, regardless of the design details. Below this level, detailed design is required. For well-designed cells, Q_{crit} can be reduced by an order of magnitude. Assuming a square cell W by W, we have

$$Q = \Delta V_{\text{crit}} C_{OX} W^2 \tag{4.82}$$

or

$$W_{\text{crit}} = \sqrt{q N_{\text{crit}} \leq \Delta V_{\text{crit}} C_{OX}}. \tag{4.83}$$

For $\Delta V_{\text{crit}} = 1$ V, $C_{OX} = 10^{-3}$ pF/μm^2, and $N_{\text{crit}} = 2 \times 10^6$, we have $W_{\text{crit}} \approx 18\mu$m. While the value of N_{crit} we assumed was conservative, it is clear that alpha particle noise in dynamic logic circuits is not impossible. Looking at the minimum capacitance in the above example, we find $C_{\text{crit}} = 0.3$ pF—quite a large capacitance by VLSI circuit standards. (By careful design, C_{crit} in dRAMs is actually closer to 0.05 pF.) Looking at the problem another way, a 3 μm wide depletion layer can collect about 200,000 hole–electron pairs from a worst-case alpha particle. The resulting current pulse would have a peak current of about 1 mA [May 79]. As the number of dynamic storage nodes on a VLSI logic chip go up, soft errors due to alpha particles will clearly become an issue.

Processing noise

Parameter mismatch also produces a noise component. It is generally assumed that while device parameters vary from chip to chip and wafer to wafer, device parameters on any chip are exactly matched. When this is not true (and it is becoming less true as the number of devices on a chip increases), these mismatches cause noise. Consider, for instance, variations in V_{TE0}. These variations act as a voltage noise source in series with the gate of the pulldown transistor. That is, it is just like any other induced noise voltage, except that its noise spectrum consists of an impulse at dc.

To understand how on-chip processing variations in physical parameters, such as the width of poly lines, translate into noise voltages, we look at a simple example. In ratioed depletion mode nMOS, the output low voltage is determined by the ratio of the strengths of the pullup to the pulldown transistors. Let us consider only the first-order effects for an input high voltage near V_{DD}. Assume that $K'_E \approx K'_D$ and that the pulldown transistor is in its linear regime. We have

$$V_{\text{out}} \approx \frac{V_{TD0}^2}{2\beta(V_{\text{in}} - V_{TE0})}. \tag{4.84}$$

Focusing on variations in the length L of the pulldown transistor, examine dV_{out}/dL. We then have

$$\frac{dV_{\text{out}}}{dL} = \frac{\partial V_{\text{out}}}{\partial \beta}\frac{\partial \beta}{\partial L} + \frac{\partial V_{\text{out}}}{\partial V_{TE0}}\frac{\partial V_{TE0}}{\partial L} \tag{4.85}$$

or

$$\frac{dV_{\text{out}}}{dL} = V_{\text{out}}\left(\frac{1}{L} + \frac{1}{V_{\text{in}} - V_{TE0}}\frac{\partial V_{TE0}}{\partial L}\right), \tag{4.86}$$

where $\partial V_{TE0}/\partial L$ is found from the short channel model, or more safely from experimental data. We want to relate a change in L to an equivalent change in V_{in}. We have

$$\frac{dV_{\text{out}}}{dV_{\text{in}}} = -\frac{V_{\text{out}}}{V_{\text{in}} - V_{TE0}}. \qquad (4.87)$$

Solving for dV_{in}, we obtain

$$dV_{\text{in}} = -\left(\frac{V_{\text{in}} - V_{TE0}}{L} + \frac{\partial V_{TE0}}{\partial L}\right) dL. \qquad (4.88)$$

Note from Eq. (4.88) that for $V_{\text{in}} \approx 5$ volts and $V_{TE0} \approx 1$ volt, a 10% change in L can result in 400 millivolts of equivalent dc noise—not including the short channel effect contribution. This is a good advertisement for quality control.

■ EXAMPLE 4.2

Process variation noise

Assume, for the moment, that process variations in the enhancement-mode device threshold are the only noise source on the chip and a normal distribution adequately describes the statistics. We could ask, given a circuit with N pulldown transistors, what is the probability that the noise voltage will somewhere exceed the noise robustness? If the probability P requires that all noise voltages are less than the noise margin, then the probability that any particular noise voltage is below this limit must be $P^{1/N}$, assuming independence.

We can approximate this as

$$P^{1/N} \approx 1 - \frac{(1 - P)}{N}$$

for $P \approx 1$. The tail of the normal probability distribution $N(X)$ can be approximated as

$$(1 - N(X)) \equiv \int_X^{\infty} \frac{1}{\sqrt{2\pi}} e^{-(1/2)y^2} dy \approx \frac{1}{\sqrt{2\pi}X} e^{-(1/2)X^2},$$

where X is measured in units of σ, the standard deviation. We have

$$\frac{(1 - P)}{N} \approx \frac{1}{\sqrt{2\pi}X} e^{-(1/2)X^2}$$

or

$$X \approx \left(-2\ln\left(\sqrt{2\pi}X\frac{(1 - P)}{N}\right)\right)^{1/2}.$$

For $P = 0.99$ and $N = 10^5$, we have

$$X \approx 5.2$$

or

$$\sigma_{V_{TE0}} \leq \frac{NR_L}{5.2}$$

for this example. NR_L is the low-level noise robustness. For comparison, 95.6% of the transistors have threshold variations within $\pm 2\sigma_{VTE0}$ of nominal.   ∎

4.4 Worst-case design

We have examined the origins and effects of noise in integrated circuits. In some cases the noise was truly random in nature, while in others it was generated by phenomena that though predictable were nevertheless too complicated to contemplate—for example, signal coupling through parasitic resistances and capacitances.

Processing variations cause a plethora of additional dimensions for uncertainty. Device lengths, widths, impurity distributions, electron mobilities, and so forth all change from lot to lot, wafer to wafer, chip to chip, and even from transistor to transistor. Luckily, the control over the variation in these parameters is an order of magnitude better for transistors on a single chip than for transistors on different chips. It is helpful to divide the variation in process parameters into two components: a component that is constant over the extent of the chip (but not necessarily over a wafer) and a component that encompasses all parameter variations local to a single chip. The latter component is considered simply as another noise source (see Section 4.3).

We need to address the problem of designing an integrated circuit that meets all functional specifications and maintains all noise margins over a given temperature range. The parameters of the devices we have at our disposal have wide variability. However, on a single chip they are constant because we have factored out the nonconstant part so that it enters calculations through the noise margin. The parameters T_{OX}, V_T, ΔW, and ΔL should be assumed constant across the chip. Taking account of the correlations of device parameters is essential to building high performance circuits.

The choice of the proper design style is a function of the market the chip must serve. In general, the systems market demands parts that meet certain worst-case specifications while in the components market parts are often binned or graded—faster parts are sold for more money, but most slower parts are not thrown out. Asynchronous systems, under certain conditions, can also fall into the second category.

The way the processed chips are valued clearly influences the style of design needed to maximize this value. We will consider the case common to most system applications in which a chip that runs very fast but too hot is just as useless as one that runs very cool but too slow. Under these constraints, it is important to design to the statistically significant "worst" case.[10] That is, one should design so that the worst-case slow chip is fast enough while the worst-case fast chip is cool enough. In all cases, the logic must have positive noise margins.

To gain some intuition for this problem, we examine a trivial LSI circuit consisting of 1000 inverters. Nominally, each inverter can charge a 1 pF capacitor in 100 ns and draws 1 mW of dc power. The power delay product is 100 pJ. Let us further postulate that the only important processing parameter is the mobility, which has a nominal value of 1000 and varies from 670 to 1500. Thus, at one extreme of the allowed process variations, the chip draws 1.5 W and can charge the capacitors in 67 ns. At the other extreme, the chip draws 0.67 W and charges the capacitors in 150 ns. In both cases, the power delay product is 100 pJ. Note, however, what happens when one must meet worst-case specifications. Assume that we are told the chip must dissipate no more than 1 W. This means the shape factor of the transistors must be decreased so the worst-case power is 1 W rather than 1.5 W. The best-case power is now 0.44 W. The best-case speed is 100 ns, but the worst-case speed is 225 ns. Because one must simultaneously design to the worst-case power and the worst-case speed, it is the product of the worst-case power times the worst-case speed (225 pJ) that is important to worst-case design. This is true even though these two conditions never occur on the same chip. Not knowing which case will occur forces one to design for both cases almost as if they could occur simultaneously.

Because the process parameters evidence a continuous variation from the ridiculous to the sublime, one must decide the limit on the distribution curve where devices become unacceptable. This is the question of just what does "statistically significant" mean? The wider the parameter range the design can tolerate, the higher the yield, but the poorer the worst-case performance. Typical design practice is to accept device parameters out to the 2 or 3σ points. Since at 2σ, 95% of all devices will be considered good in any parameter, the maximum multiplicative factor by which the yield can be decreased by N independent parameters is 0.95^N. This is a lower bound because two independent parameters can be out of spec in such a way that the resulting circuit remains in spec. Note that N is a small number, which is independent of the size or number of transistors on the chip; it is perhaps more a measure of the complexity or maturity of the process. As we will see, the criteria for deciding how to specify the worst-case parameter ranges are very complex for three reasons: the statistics are complex, some parameters are more critical to the circuit design than

---

[10] The alternative is design centering, where one attempts to maximize the yield of a part based on the shape of the statistical distributions of the process parameters. This more sophisticated technique is appropriate to high-volume designs and is beyond the scope of this book.

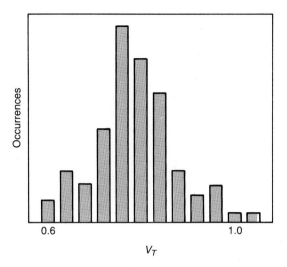

FIGURE 4.24   Bar graph showing the number of occurrences of various threshold voltages. Some wafers will have thresholds outside the acceptable limits.

others, and parameters are correlated. For instance, V_{TE0} depends on variations in transistor length as well as on implant dose. A bar graph of typical threshold variations is illustrated in Fig. 4.24. These variations are for a minimum length device. Fig. 4.25 illustrates a scatter plot showing the weak correlation of threshold voltage and device length.

The general problem of determining what combination of parameters yields the worst-case conditions for an arbitrary network is very difficult (the tolerancing problem). But ascertaining the worst-case conditions for digital MOS is not quite as complex because of the restrictions that the methodology places on the circuit forms. Worst-case design is a cooperative effort shared by the process engineer, device physicist, and circuit designer. The process engineer can give the circuit designer the

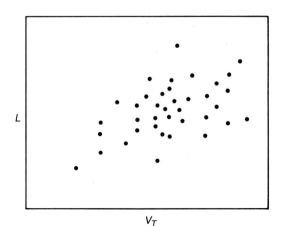

FIGURE 4.25   Scatter plot of threshold voltage versus channel length, showing a weak correlation between the two.

ranges and correlations of the device parameters. The process engineer cannot, in principle, give the circuit designer the worst-case parameters because the circuit forms must be chosen before we can interpret what "worst" means. In practice, the process engineer often gives the circuit designer the worst-case parameter lists, but this is because the circuit forms have reached a high level of maturity and stability. If the circuit designer were to invent a totally new circuit form, the process engineer and circuit designer would need to cooperatively evaluate what "worst" means in this new context. The worst-case parameter lists usually take the form of circuit simulation files. For instance, there might be a file for worst-case noise and another file for worst-case power. These files encompass process correlations and implicitly assume a set of circuit forms. Each worst-case file is said to represent a "corner" of the space of acceptable process variations.

The worst-case parameter lists can be thought of as a contract between the process engineer and the circuit designer. The process engineer agrees to deliver, with reasonable yield, wafers that fall within the corners of the worst-case specifications. The actual variations seen can be larger than those given by the worst-case specifications; but chips outside the specified range are, by mutual agreement, defective. On the other hand, the designer agrees not to take advantage of many hidden correlations, or at least does so at his or her own risk. We know, for instance, through Eq. (2.49), that a correlation exists between threshold voltage and gate oxide thickness. It would be foolish, however, for the circuit designer to try to take advantage of this correlation without first clearing it with the process engineer. There might be some other effect that is even more important; for instance, the implant dose and profile are also correlated to T_{OX} since the implant is done through the gate oxide. And, even if there were no other conditions, the process engineer might, and probably would, be counting on correlations to help meet the specifications with reasonable yield. Another important point is that the conservative circuit designer can count on conformance to the specifications only on parameters that are measured by the process line engineer.

nMOS

For the depletion load nMOS logic family that we have used in our examples, we have chosen $V_{IL} < V_{TE0}$, which implies $V_{OH} = V_{DD}$. For an input voltage of V_{IH}, V_{OL} must be less than or equal to $V_{TE0} - NM_L$ for all valid ranges of process parameters, temperature, and supply voltages. V_{OL} depends on many factors but, simply speaking, it is maximized for low ratios of pulldown to pullup conductances. The noise margin is minimized (considered one of the worst-case situations) when the threshold voltage is low at the same time that the effective β-ratio is low.

Worst-case power occurs under a different combination of parameters. It occurs when the conductance of the pullup and, to a lesser extent, the pulldown are maximized. Because most of the power dissipation in nMOS LSI is due to static phenomena, we need to consider the power dissipated in the charging and discharging of capacitors only in very special cases. Circuits typically run very fast under worst-case power conditions.

Worst-case speed occurs under almost the opposite conditions. Because the bulk of the delay in gates occurs during pullup transitions, the worst-case speed occurs when the pullup conductance is low. Worst-case speed is more complicated than worst-case power because the gate delay depends not only on the pullup transistor but also on the load capacitor. While metal and poly capacitances are largely independent of the effective transistor conductance, the gate and diffusion capacitances are correlated with the transistor parameters. Temperature is also a first-order effect. Because the electron mobility decreases with increasing temperature, the worst-case speed conditions occur at high temperature. Note that the subthreshold leakage also increases at high temperature.[11]

We now consider, parameter by parameter, the conditions leading to worst-case current and power dissipation in digital nMOS circuits. Under worst-case power conditions, both the depletion- and enhancement-mode devices are "fast." That is, both the pullup and pulldown are highly conductive. The majority of the power P is dissipated in the depletion load. In the depletion load pullup, the gate is shorted to the source. Approximating V_{DS} as V_{DD}, we have

$$P \approx K' S_{pu} V_{TD}^2 V_{DS} \approx K' S_{pu} V_{TD0}^2 V_{DD}. \qquad (4.89)$$

Under conditions of worst-case power, we have V_{DD}, K', and W maximum and L and V_{TD0} minimum. V_{TD0} is minimum because it is negative, and hence a minimum V_{TD0} maximizes V_{TD0}^2. K', in turn, is maximized when μ is maximum and T_{OX} is minimum. These are the first-order effects on power. Because the mobility decreases with temperature for temperatures above about 77 K, worst-case current occurs at low temperatures. Note that the supply voltage[12] V_{DD} and the temperature, while important in the calculation of the worst-case, are not part of the worst-case device models. We see in the temperature not only the first-order effect on the mobility but also a second-order effect on the threshold voltage. As the temperature decreases, the threshold voltage increases. This actually decreases the conductivity of the depletion- and enhancement-mode devices. This effect is in the opposite direction but weaker than the mobility contribution. It is an example of a mitigating

---

[11] One must also be careful at cryogenic temperatures because impurity freezeout can occur in the depletion-mode transistors.

[12] In a 5 volt process, typical V_{DD} specifications are 4.5 to 5.5 volts or 4.75 to 5.25 volts.

TABLE 4.1   Conditions leading to worst-case power and current

PARAMETER	VALUE AT WORST-CASE	REASON
τ	High (low)	
V_{DD}	High	
V_{TD}	More negative	Increased V_{TD}^2
μ	High	Large K'
T_{OX}	Low	Large K'
ΔW	Minimum	Wide devices
ΔL	Maximum	Short devices
L_{diff}	Maximum	Short devices
ΔL_{poly}	Maximum	Short devices
γ	Low	Low V_T
V_{BB}	Less negative	Low V_T
V_{TE0}	Low	

factor. The model for worst-case power is often called "FF" or "F" for "Fast depletion, Fast enhancement." Because we are interested in worst-case power when the chip is hot, worst-case power occurs at high temperature. That is, we are not concerned with the chip burning up when it is cold. Except for temperature, the parameters for worst-case power and worst-case current are the same. Worst-case power occurs at the maximum chip temperature, and worst-case current at the minimum. Since the mobility decreases with increasing temperature, the chip draws less current under worst-case power conditions than it does under worst-case current conditions. Table 4.1 lists the worst-case power dissipation conditions.

Worst-case power is only one of the three important corners of the design space. Worst-case speed occurs when both the depletion- and enhancement-mode transistors are "slow" (the "SS" or "S" model). Though the logic levels V_{IL} and V_{IH} do depend on V_{TE} and V_{TD}, the dominant effect of more positive threshold voltages is to lower the conductance of the pullup and pulldown transistors. For the same reasons that mobility variation causes worst-case current at low temperature, it causes worst-case speed at high temperature.

Dynamics are also important under conditions of worst-case speed. Interconnect capacitances all take on their maximum values. Gate capacitances, on the other hand, are low. This is because delay is proportional to capacitance and inversely proportional to drive. As the thickness of the gate oxide increases and the transistor width decreases, the drive available from a pullup or pulldown transistor decreases at almost[13] the same rate that the gate capacitance decreases. However, because the load capacitance is composed partially of gate capacitances and partially of interconnect capacitances, the decrease in

---

[13] "Almost" because of vertical field mobility degradation.

$K'S$ dominates. Under worst-case speed, gate and overlap capacitances are low; these are mitigating factors. The same argument applies to gate area. Under worst-case speed, ΔW is maximum and ΔL is minimum. If one were to build a string of inverters, one would find that the delay would increase with V_{DD}. This is because as V_{DD} increases, the voltage swings need to be higher and this takes more time given a saturated pullup transistor. On the other hand, if one looks at the total chip, precharged lines and all, the maximum clock frequency of the chip decreases as V_{DD} decreases. Therefore the worst-case is taken as V_{DD} low. Table 4.2 summarizes the conditions leading to worst-case speed.

The conditions for worst-case noise margin occur under the simultaneous conditions of a fast pullup, a slow pulldown, and a low V_{TE0} (the "FS" or "Noise" model). A low V_{TE0} contradicts a weak pulldown. From Eq. (4.45), the low noise margin is given by

$$NM_L = 2V_{TE0} - V_{IH} + \sqrt{(V_{IH} - V_{TE0})^2 - \frac{K'_D}{K'_E}\frac{V_{TD}^2}{\beta}}. \qquad (4.90)$$

In cases where the input is driven by enhancement-mode pullups, such as pass transistors on zero threshold superbuffers, V_{IH} in Eq. (4.90) is $V_{DD} - V_{TE} - NM_H$ or $V_{DD} - V_{TZ} - NM_H$, as appropriate. If the gate is driven by a strong high, we have $V_{IH} = V_{DD} - NM_H$. The dependence of the low noise margin on V_{TE0} is quite complex, as can be seen in Fig. 4.26.

Looking at the sensitivity of NM_L with respect to V_{TE0}, we have

$$\frac{\partial NM_L}{\partial V_{TE0}} = 2 - \frac{V_{IH} - V_{TE0}}{\sqrt{(V_{IH} - V_{TE0})^2 - \frac{K'_D}{K'_E}\frac{V_{TD}^2}{\beta}}}. \qquad (4.91)$$

TABLE 4.2   Conditions leading to worst-case speed

PARAMETER	VALUE AT WORST-CASE	REASON
τ	High	Low μ
V_{DD}	Low	Less drive
V_{TD}	Less negative	Less drive
μ	Low	Low K'
T_{OX}	Thick	Low K'
ΔW	Maximum	Narrow devices
ΔL	Minimum	Long devices
L_{diff}	Minimum	Long devices
ΔL_{poly}	Minimum	Long devices
γ	High	High V_T
V_{BB}	Less negative	More capacitance
V_{TE0}	High	

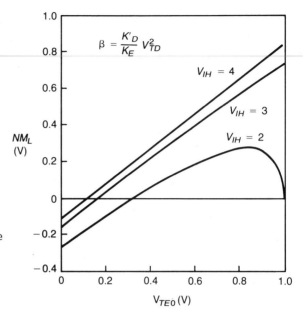

FIGURE 4.26 Plot of the low-level noise margin for various input high levels. Note the nonmonotonic behavior for low V_{IH} logic levels.

This sensitivity changes sign at

$$V_{IH} - V_{TE0} = \sqrt{\frac{4K_D'V_{TD}^2}{3K_E'\beta}}. \tag{4.92}$$

Since the minimum β-ratio is given by Eq. (4.45), we have

$$V_{OH} = V_{TE0} + NM_H + \frac{4(V_{TE0} - NM_L)^2}{8(V_{TE0} - NM_L) - 3} \tag{4.93}$$

as the voltage at which the sensitivity of the noise margin with respect to V_{TE0} equals zero.

Equation (4.93) implies that for the usual case of a moderately strong high input voltage, the sensitivity of the low noise margin to V_{TE0} is such that the worst case occurs for a low V_{TE0}. In the rarer case of a very low V_{IH}, such as one might have when converting from TTL to MOS voltage levels, the worst-case V_{TE0} actually occurs when V_{TE0} is high. For very low V_{IH}, the drive on the pulldown transistor is very low and this becomes the dominant effect in determining the worst case. In the worst case, the mobility of the enhancement-mode pulldown is low and that of the depletion-mode pullup is high. The correlation between these two quantities is generally reflected in the worst-case noise margin file.

To determine the proper worst-case values of ΔL and ΔW, we need to say more about the design methodology. All else being equal, we have the time constant of the process τ proportional to L_{pd}^2. This makes it very important to minimize the length of the pulldown transistor. The length of the depletion-mode pullup is not as critical to the performance because, since it spends most of the time in saturation, the capacitance that determines the logic gate speed is the channel-to-substrate capacitance, which is generally an order of magnitude lower than the gate-to-channel capacitance of the enhancement-mode pulldown. (The gate is shorted to the channel, via the source, in the depletion-mode pullup.)

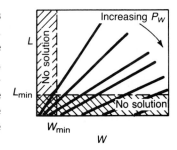

FIGURE 4.27   Contours of constant power dissipation.

We know that both the speed and power dissipation of a ratioed logic gate are primarily determined by the depletion load. We examine what happens to the worst-case delay as we keep worst-case power constant and vary the drawn width of the pullup transistor. For a nominal length L and a nominal width W, assume that there is a variation in the length of $\pm\delta L$ and in the width $\pm\delta W$. Ignoring short channel effects, in order to keep the worst-case power constant at P_W, the true shape factor of the pullup must be kept constant. Worst-case power occurs when the actual width is $W + \delta W$ and the length is $L - \delta L$. We have

$$P_W = P_o \frac{W + \delta W}{L - \delta L}, \tag{4.94}$$

where P_o is a proportionality constant. Solving for L, we obtain

$$L = \delta L + \frac{P_o}{P_W}(W + \delta W). \tag{4.95}$$

Fig. 4.27 shows contours of constant power in the W–L plane. Also shown in the figure are the design rule constraints. Fig. 4.28 illustrates the region of possible solutions to a "worst-case power must be less than ..." type constraint.

The worst-case speed situation is a little more complex because of the self-capacitance of the pullup. There are four important terms in the capacitance equation. There is a component that is independent of W and L. This includes the load and the buried contact. There is a component that is proportional to the actual device width, as well as a component that is proportional to the device length, and a channel-to-substrate component that is proportional to the actual device area. The current through the pullup is proportional to the actual shape factor. The minimum current occurs when the actual width is $W - \delta W$ and the actual length is $L + \delta L$. Summing all of these effects, we find

FIGURE 4.28   Region of possible solutions to a maximum power dissipation constraint.

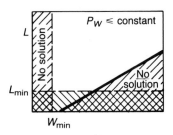

$$T_W = T_o \frac{L + \delta L}{W - \delta W}\left(1 + a_1(W - \delta W) + a_2(L + \delta L) + a_3(L + \delta L)(W - \delta W)\right), \tag{4.96}$$

where the a_i represent the constants in the self-capacitance equation and T_o is a constant of proportionality. Ignoring a_1 through a_3 for the

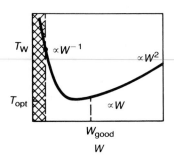

FIGURE 4.29 Delay versus pullup width for constant power dissipation. At low values of W the line-width control is a problem, and at high values of W the self-capacitance is the issue.

moment, and using Eq. (4.95) for L, we find that for a constant-worst case power, it pays to make W and L as large as possible. This is because a large W and L maximize the control of the current. When the self-capacitance terms corresponding to a_1 through a_3 are included, the worst-case delay T_W goes through a minimum. For very small W, we have that L, and hence the delay, must be made larger than one would like because of uncertainties in the fabrication process. For very large W, fabrication uncertainties are minimized but the self-capacitance begins to adversely impact the delay. Fig. 4.29 illustrates a typical plot of worst-case delay T_W versus W. Illustrated on the plot is a good choice of W.

In the design of a VLSI chip, one can not usually afford the time to optimize each gate to the level of detail discussed here. Indeed, we haven't even quite analyzed the "right" problem since power is a globally scarce resource, not a local constraint as discussed above. The real problem to solve pertains to an ensemble of thousands of gates. Nevertheless, the general behavior of Eq. (4.96) is the general behavior of the ensemble. It pays not to make the pullup transistor too small in either length or width so that the device is operating away from the steep W^{-1} part of the curve. As a general rule, then, we always make the pullup transistors in a depletion load nMOS technology longer than the pulldown devices, which are kept at minimum length. For instance, in our example 2 μm nMOS process we never (well, hardly ever) design depletion pullups less than 6 μm/6 μm.

We can now return to the original problem of determining the worst-case ΔL and ΔW for worst-case noise margin. Because the pullup is longer than the pulldown and worst-case noise margin occurs for fast pullups and slow pulldowns, we have ΔL minimum (L maximum). With reference to percentage, this has a larger effect on the pulldown. Typically, we also constrain the pulldown to be as wide, or wider, than the pullup. This is convenient because $\beta \gg 1$. Given this constraint on the design methodology, we have ΔW minimum (W maximum) for worst-case noise margin. The worst-case noise conditions for depletion load nMOS are listed in Table 4.3.

███████████████████ EXAMPLE 4.3

Inverter sizing for noise margins
Find the widths of the pulldown transistors required to meet noise margin constraints of 10/10 and 6/6 drawn pullup devices.

The specifications are: $NM_H = 800$ mV, $NM_L = 200$ mV, 4.5 V $\leq V_{DD} \leq 5.5$ V, -3 V $\leq V_{BB} \leq -2$ V, and $0°$ C $\leq T \leq 100°$ C.

At the worst-case noise corner, $V_{DD} = 4.5$ V, $V_{BB} = -2$ V, and $T = 100°$ C, and the worst-case file is the "N" or "FS" case. We can run SPICE with the N model and have it print out the threshold information.

TABLE 4.3   Conditions leading to worst-case noise margin.

PARAMETER	VALUE AT WORST-CASE	REASON
τ	High	V_{TE0} low
V_{DD}	Low	V_{OL} high
V_{TD}	More negative	Increased V_{TD}^2
μ_D	High	
μ_E	Low	
T_{OX}		
ΔW	Minimum	Wide pullup
ΔL	Minimum	Long pulldown
L_{diff}	Minimum	Long pulldown
ΔL_{poly}	Minimum	Long pulldown
γ	Low	Low V_T
V_{BB}	Less negative	Low V_T
V_{TE0}	Low (usually)	$V_{TE} - V_{OL}$ low

We need to comment out the NFS parameter because this is a flag to SPICE that turns on the subthreshold model and causes the output to lie about the device threshold. We find that at $\tau = 27°$ C, V_{TX} for a 20 μm-long device is 0.576 V. Increasing the temperature to 100° C drops V_{TX} to 0.480 V. For a 2.25 μm-long device, an 8 mV short channel effect lowers V_{TE} to 0.472 V. Applying $V_{BB} = -2$ V raises V_{TE0} to 0.503 V. And finally, drain-induced barrier lowering reduces V_{TE} by 109 mV, resulting in 0.394 V.

From Eq. (4.45), we have

$$W_{pd} = \frac{L_{pd} S_{pu} \mu_D V_{TD0}^2}{\mu_E \left(2 \left(V_{IH} - V_{TE0}\right) V_{OL} - V_{OL}^2\right)},$$

where $V_{IH} = 4.5 - 0.8 = 3.7$ V and $V_{OL} = 0.394 - 0.200 = 0.194$ V. At room temperature, $\mu_D/\mu_E = 730/670 = 1.09$. This ratio will be the same at 100° C. $L_{pd} = 2 + 0.25 - 0.38 = 1.87$ μm. (The term 0.38 is due to lateral diffusion.) For the shape factors of the pullups, the drawn 10/10 is 8.8/9.87 and the 6/6 is 4.8/5.87. Therefore in the 10/10 case, we have

$$W = \frac{1.87 \times 0.89 \times 1.09 \times (-2.3)^2}{(2 \times (3.7 - 0.5) \times 0.19 - 0.19^2)} = 8.13$$

or

$$W_{\text{drawn}} = 9.3.$$

For the 6/6 pullup, $W_{\text{drawn}} = 8.6$. Note the two different values of V_{TE}: 0.503 for $V_{DS} = 0$ and 0.394 for $V_{DS} = 5$. The $V_{DS} = 5$ value is used to calculate the necessary V_{OL}.

These numbers are only first guesses, since second-order effects, such as vertical field degradation, were not taken into account. From SPICE, we find that an 11/2 drawn pulldown yields $NM_L = 210$ mV for a 10/10 pullup, and a 10/2 pulldown yields $NM_L = 211$ mV for a 6/6 pullup.

Continuing the analysis, we look at the worst-case power. Setting $V_{DD} = 5.5$ V, changing the device lengths, models, and so on to the worst-case corner, we find that the 10/10 pullup dissipates 0.82 mW and the 6/6 pullup dissipates 0.76 mW. Looking at the time it takes to charge a 0.3 pF capacitor to 4 V, we find a worst-case delay of 29 ns in the 10/10 case and 33 ns in the 6/6 case. The 10/10 pullup has about a 5% better power delay product than the 6/6 pullup. ∎

FIGURE 4.30 Several CMOS circuits with nonobvious worst-case corners: (a) a ratioed NOR gate; (b) a mostly nMOS XOR gate using an n-channel selector; (c) half a shift register cell without complementary pass transistors; (d) a ratioed S–R latch.

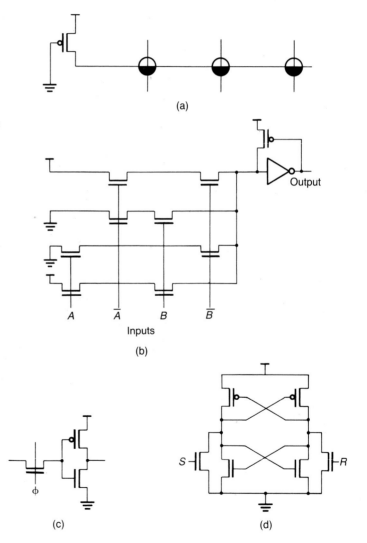

CMOS

The most commonly used CMOS worst-case corner is SS (slow p-type, slow n-type). As in the nMOS case, C_{OX} is low in the slow case, and all other capacitances are high.

Noise is often worst in the FF case because device thresholds are low and iR drops are large. When the clocking methodology has race conditions, FF can be the worst case because it is in this corner that the gates are fastest in relation to the clock edges, especially if poly is used for the clock distribution. (R_{poly} maximum leads to worst-case clock skew.)

Fig. 4.30 illustrates a number of ratioed CMOS circuits. Worst-case conditions for these circuits can occur under SF or FS conditions.

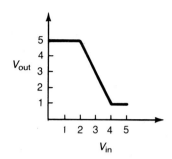

FIGURE 4.31

4.5 Perspective

Noise and processing variations are facts of life in VLSI. Noise principally stems from resistive and capacitive coupling. Digital circuits can be built to be reliable in spite of this. Noise margins teach us how to handle noise, and worst-case design teaches us how to handle processing variations. A key to successful worst-case design is to account for the correlations among process parameters on a chip. The worst-case corners are a function of the design methodology.

We used the inverter to study fundamental circuit issues. In the course of this study we saw device physics issues surface repeatedly.

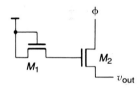

FIGURE 4.32

Problems

4.1 An idealized inverter has a voltage transfer function as illustrated in Fig. 4.31. What are the values of V_{OL}, V_{OH}, NM_L, NM_H, NR_L, and NR_H for $V_{IL} = 2.5$ V and $V_{IH} = 3.5$ V? What are the noise margins that optimize $NM_L + NM_H$?

4.2 We want the circuit in Fig. 4.32 to bootstrap. What are the worst-case conditions on ΔL_{poly}, ΔW, and T_{OX}? What parameter file, if any, do these conditions correspond to? Lay out the circuit using the 2 μm nMOS process from the appendixes. How wide must M_2 be for v_{out} to rise all the way to V_{DD}? Simulate with SPICE and compare.

4.3 Explain the operation of the circuit in Fig. 4.33. What are this circuit's advantages?

4.4 Derive closed-form expressions for the time-dependent output voltages of the three circuits shown in Figs. 4.34 through 4.36. $\triangle$

4.5 For an nMOS inverter with a 6/6 pullup and the example 2 μm process, what is the minimum width of the pulldown, assuming that the

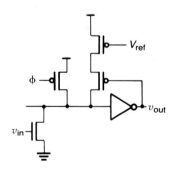

FIGURE 4.33

FIGURE 4.34

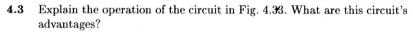

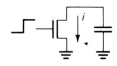

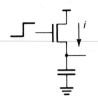

FIGURE 4.35

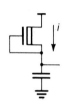

FIGURE 4.36

input comes through a 6/2 enhancement-mode pass transistor. Assume $NM_H = 600$ mV plus all of the other specifications from Example 4.3. It is better to turn off the subthreshold flag in SPICE when simulating the maximum pullup voltage on the input. Is the worst-case V_{TE} high or low? △

4.6 At what input voltages is the differential gain of a depletion load nMOS inverter equal to -1? What are the output voltages at these points?

4.7 Determine the β-ratio requirements on a depletion-mode pullup/enhancement-mode pulldown gate family without pass transistors, given that $V_{DD} = 2.5 \pm 0.2$ V, $NM_H = 600$ mV, and $NM_L = 200$ mV. Is the worst-case V_{TE0} high or low? △

4.8 The case of metal is unique because metal migration limits the allowable average current density $J_{\max}$. The maximum average voltage drop per unit length of line is thus equal to $\rho J_{\max}$, where ρ is the resistivity of the metal. What is this voltage drop for aluminum?

4.9 Rideout [Rideout 81] has proposed measuring process parameters at various points in the fabrication process and using the results of these measurements to control later steps. What steps would have the greatest leverage? What are some of the practical considerations? How would this affect worst-case design? △△

4.10 The large capacitance of the power and ground lines, in conjunction with the resistance of the metal layer, constitute a filter configuration that one might imagine to be effective in filtering out power supply noise. Show that this filter is actually not very effective. △

4.11 When the input of a CMOS inverter switches very slowly, the dissipated energy is almost totally due to the excess current. The switching energy can be approximated as QV_{DD}. Assume that the input voltage slews at a constant rate of s volts/s. Show that, neglecting δ, Q is given by

$$Q = \frac{K_N' S_N}{s(1+\alpha)^3}(V_{DD} + V_{TP} - V_{TN})^3,$$

where

$$\alpha \equiv \sqrt{\frac{S_N K_N'}{S_P K_P'}}. \quad △$$

4.12 We want to compare the dc performance of a p-type and n-type pullup transistor. Given $L = 2$ μm, $T_{OX} = 300$ Å, $\mu_p = 250$ cm^2/(V·s), $\mu_n = 600$ cm^2/(V·s), $\gamma = 0.2$ V$^{0.5}$, $V_{TN} = 0.8$ V, and $V_{TP} = -0.8$ V, determine the width W of the transistors required to pull a 20 mA current source load to 3 V. Assume $V_{DD} = 5$ V. Decide which device is wider—the n-type device with an input voltage of 5 V, or the p-type device with an input voltage of zero. Use hand calculation. Derive an expression for the output voltage at which the widths are equal. △

4.13 In the introduction to Section 4.4, we did an example of a 1000 inverter LSI chip. Redo the analysis assuming the capacitors have a variation of ±30%.

4.14 Consider the following situation. The drain, source, and substrate of an n-type enhancement-mode transistor are tied to ground and its gate is also initially low. A voltage pulse applied to the gate of the device will charge the gate and other intrinsic capacitances. Charges must be suplied by the drain, source, and substrate to mirror the charges placed on the gate terminal.

We want to investigate how SPICE models this process. To measure the charge pumped into or out of the various terminals, without affecting the measurement conditions, use the following setup. Tie the drain, source, and substrate to ground with separate 0 V voltage sources. Use current-controlled current sources to generate currents proportional to the current flowing through the four voltage sources associated with each of the four device terminals. If these currents are then fed into capacitors, the voltage developed across each capacitor will be a measure of the total amount of charge that has flowed into the associated terminal. You may have to adjust the current proportionality constant and the capacitor size to get the voltages in a reasonable range. Or you can try thinking first. Also, SPICE will not let you tie a capacitor in parallel with a current source without some dc path to ground, so add a very large resistor (1 GΩ, say) in parallel with the capacitor.

With this set up, apply a voltage pulse to the the gate and see what happens. Does SPICE seem to do reasonable things? Is charge conserved? Exactly? What happens when the gate voltage is brought back to its original low voltage? What values for X_{PN} does SPICE use, under the channel, above and below threshold?

References

[Bode 45] H. W. Bode, *Network Analysis and Feedback Amplifier Design*, Van Nostrand, New York, 1945. (The gain-bandwidth principle, among other wonderful things.)

[Burns 64] J. R. Burns, "Switching Response of Complementary Symmetry MOS Transistor Logic Circuits," *RCA Review*: 627–661, 1964.

[Dang 81] R. L. M. Dang and N. Shigyo, "Coupling Capacitances for Two-Dimensional Wires," *IEEE Trans. Electron Devices Lett.* **EDL-2**: 196–197, 1981.

[Demoulin 79] E. Demoulin, J. A. Greenfield, R. W. Dutton, P. K. Chatterjee, and A. F. Tasch, Jr., "Process Statistics of Submicron MOSFETs," *IEEE International Electron Device Meeting*, pp. 34–37, Washington, D.C., 1979.

[Fang 75] F. F. Fang and H. S. Rupprecht, "High Performance MOS Integrated Circuits Using Ion-implantation Techniques," *IEEE J. Solid-State Circuits* **SC-10**: 205–211, 1975.

[Forbes 73] L. Forbes, "n-Channel Ion-implanted Enhancement/Depletion FET Circuit and Fabrication Technology," *IEEE International Solid-State Circuits Conf.* **SC-8**: 226–230, New York, 1973.

[Glasser 81] L. A. Glasser, "The Analog Behavior of Digital Integrated Circuits," *Proceedings of the Eighteenth Design Automation Conference*, pp. 603–612, Nashville, Tenn., 1981.

[Haviland 80] G. L. Haviland and A. A. Tuszynski, "A CORDIC Arithmetic Processor Chip," *IEEE J. Solid-State Circuits* **SC-15**: 4–15, 1980.

[Hill 68] C. F. Hill, "Noise Margin and Noise Immunity in Logic Circuits," *Microelectronics* **1**: 16–21, 1968.

[Lo 61] A. W. Lo, "Some Thoughts on Digital Components and Circuit Techniques," *IRE Trans. Electronic Computers* **EC-10**: 416–425, 1961. (Includes discussion of nonreciprocal logic elements.)

[Lohstroh 79] J. Lohstroh, "Static and Dynamic Noise Margins of Logic Circuits," *IEEE J. Solid-State Circuits* **SC-14**: 591–598, 1979.

[Lohstroh 83] J. Lohstroh, E. Seevinck, and J. De Groot, "Worst-Case Static Noise Margin Criteria for Logic Circuits and Their Mathematical Equivalence," *IEEE J. Solid-State Circuits* **SC-18**: 803–807, 1983.

[Maly 82] W. Maly and A. J. Strojwas, "Statistical Simulation of the IC Manufacturing Process," *IEEE Trans. Computer-Aided Design* **CAD-1**: 120–130, 1982.

[Masuhara 72] T. Masuhara, M. Nagata, and N. Hashimoto, "A High-Performance n-Channel MOS LSI Using Depletion-Type Load Elements" *IEEE J. Solid-State Circuits* **SC-7**: 224–231, 1972.

[May 79] T. C. May, "Soft Errors in VLSI: Present and Future," *IEEE Trans. Components, Hybrids, Manufacturing Technology* **CHMT-2**: 377–387, 1979.

[Polkinghorn 70] R. W. Polkinghorn et al., "FET Driver Using Capacitor Feedback," U. S. Patent 3.506.851, 1970.

[Rideout 81] V. L. Rideout, "Trends in Silicon Processing," *Proc. Second Caltech Conf. Very Large Scale Integration*, C. L. Seitz (ed.): pp. 65–110, Pasadena, Calif., 1981.

[Sai-Halasy 83] G. A. Sai-Halasy, "Cosmic Ray Induced Soft Error Rate in VLSI Circuits," *IEEE Electron Device Lett.* **EDL-4**: 172–174, 1983.

[Stein 77] K.-U. Stein, "Noise-Induced Error Rate as Limiting Factor for Energy per Operation in Digital IC's," *IEEE J. Solid-State Circuits* **SC-12**: 527–530, 1977.

[Stone 84] D. C. Stone, J. E. Schroeder, R. Kaplan, A. R. Smith, "Analog CMOS Building Blocks for Custom and Semicustom Applications," *IEEE Trans. Electron Devices* **ED-31**: 189–195, 1984. (Noise sources in analog circuits.)

[Tokuda 83] T. Tokuda, K. Okazaki, K. Sakashita, I. Ohkura, and T. Enomoto, "Delay-Time Modeling for ED MOS Logic LSI," *IEEE Trans. Computer-Aided Design* **CAD-2**: 129–134, 1983.

5 CIRCUIT TECHNIQUES

In Chapters 2 through 4 we studied the issues defining the framework within which a VLSI circuit must be designed in order to be functional. We examined the physics and modeling of MOS structures, the constraints imposed by both the logic abstraction and technology, and the costs of yield and power dissipation. Beyond these basics, we would like to understand the circuit forms and idioms used to solve specific circuit problems. Most of these specialized circuit techniques exemplify fundamental circuit issues that transcend the specific implementation. In this chapter, we examine a collection of digital MOS circuits, together with the conceptual and analytical tools necessary to study them. These circuits provide a source of problem-solving circuit techniques as well as a basis for innovation and further study.

5.1 Circuit optimization techniques for simple logic gates

The first problem we examine is the optimization of simple logic gates. Transistor width is the primary variable available for the optimization of a VLSI circuit. In this section, we examine, analytically, the problem of optimally sizing the transistors in a critical path [Glasser 84]. This will lead to some simple formulas and inequalities that can provide general guidance in more complex situations. For this analysis, we will use the $R_\tau C_\tau$ model of MOS logic gates derived in Section 4.2. Writing the total delay T through a string of logic gates as a sum of RCs, we have

$$T = \sum_{i=0}^{N} t_i,$$ (5.1)

where

$$t_i \equiv R_i \left(C_{i+1} + Y_i + \frac{\alpha_i C_i}{n_i} \right) n_i$$ (5.2)

and where R_i is the equivalent pulldown resistance of the the ith stage, C_{i+1} is the effective input capacitance of the $i+1$st stage, and Y_i is the total effective interconnect capacitance of the wire connecting the ith and $i+1$st stages. Y_i also includes the equivalent input capacitances of stages not in the critical path. The self-capacitance loading of a stage is characterized by α_i. As the width of a transistor increases, its parasitics also increase. To first order, these increases are captured in α, where α_i is usually constant (that is, $\alpha_i = \alpha_0$ for all i). Figure 5.1 illustrates these parameters. The parameter n_i, called the fan-out parameter, has many distinct interpretations, depending on the problem being solved. Its basic purpose is to account for differences between pulldown networks. Figure 5.2 illustrates a simple cascade of logic gates. In this example, we have

$$T = R_0 \frac{\tau}{R_1} + R_1 \left(Y_1 + \frac{\tau}{R_2} \right) + 2\beta_\tau R_2 \left(Y_2 + \frac{\tau}{R_3} \right) + R_3 C_{\text{load}}$$ (5.3)

or

$$T = 5 R_0 C_\tau + \frac{R_\tau}{5} (Y_1 + 10 C_\tau) + \frac{\beta_\tau R_\tau}{5} (Y_2 + 10 C_\tau) + \frac{R_\tau}{10} C_{\text{load}}.$$ (5.4)

FIGURE 5.1   Important capacitances and "resistances" in a collection of logic gates. $\tau = R_i C_i$.

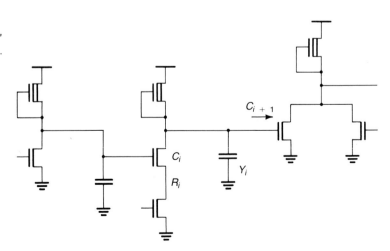

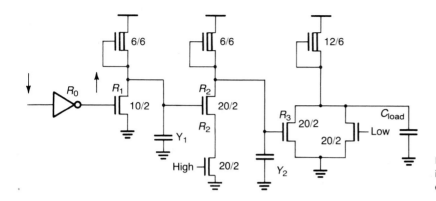

FIGURE 5.2 An example illustrating the components of the $\tau = RC$ delay model.

Using Eq. (4.36), we can rewrite Eq. (5.1) as

$$T = \tau \sum_{i=0}^{N} \left(R_i \left(\frac{1}{R_{i+1}} + \frac{Y_i}{\tau} \right) + \frac{\alpha_i}{n_i} \right) n_i, \qquad (5.5)$$

where $n_0 \equiv 1$. R_0 is the source equivalent resistance and $C_{N+1} = 0$. In Eq. (5.5), we have defined $Y_N \equiv 0$ and $R_{N+1} \equiv \tau/C_{\text{load}}$ where C_{load} is the output capacitance. This is purely a notational convenience.

Equation (5.5) is the most important equation in this section. We first solve it in closed form in a number of special cases. The most important special case is where all the wire capacitances Y_i are much, much smaller than the respective input capacitances $C_{i+1} = \tau/R_{i+1}$ (with the exception of C_{load}). This case is likely to occur under conditions of minimum delay, where the drive transistors are very wide. We find the conditions of optimum speed by taking the gradient of T with respect to R_i and setting it equal to zero. We have

$$0 = \nabla_{R_i} T = -\frac{R_{i-1} n_{i-1}}{R_i^2} + \frac{n_i}{R_{i+1}} \qquad \text{for } i = 1 \text{ to } N. \qquad (5.6)$$

Because in this case we have

$$t_i = n_i \tau \left(\frac{R_i}{R_{i+1}} + \frac{\alpha_i}{n_i} \right), \qquad (5.7)$$

we can see that in the special case of $\alpha_i = \alpha_0$, Eq. (5.6) leads to all of the t_i being equal. We then have

$$t_i = t_{i+1} \equiv t_0. \qquad (5.8)$$

In this special case, the delays per stage are all equal. For instance, they are independent of whether some stages are NOR gates and others are NAND gates. We define t_{00} as

$$t_{00} \equiv t_0 - \alpha_0 \tau. \qquad (5.9)$$

Equation (5.6) can be solved in closed form. The results are

$$R_i = \tau \frac{n_{i-1}}{t_0} R_{i-1}, \tag{5.10}$$

$$R_i = R_0 \left(\frac{t_{00}}{\tau} \right)^{-i} \Pi_{j=i}^{i-1} n_j, \tag{5.11}$$

$$t_{00} = \tau \left(\frac{R_0}{R_{N+1}} \Pi_{j=1}^{N} n_j \right)^{\frac{1}{N+1}}, \tag{5.12}$$

and

$$T_{\min} = (N+1)t_0. \tag{5.13}$$

These formulas can provide a simple first guess of the optimum device sizes and delays for a string of simple gates. For the simple case of all $n_i = 1$, these results reduce to the familiar geometric device scaling in which each device is a constant times the size of the preceding device.

We have not been able to find closed form solutions in cases where wiring capacitance cannot be neglected, or in cases where we want to consider power dissipation or area as an additional metric. Let us write the total power or area in the form

$$B = \sum_{i=1}^{N} b_i. \tag{5.14}$$

That is, we associate with each stage a quantity that captures the resources used by that stage. For the case of power dissipation, we interpret the b_i as the average power dissipation of each stage. In the case of active area, we interpret b_i as the area of each stage. For both the cases of power dissipation and active area, b_i is of the form

$$b_i = m_i C_i, \tag{5.15}$$

where m_i is a constant that depends on the topology of the logic gate. For instance, the active area of a CMOS logic gate is proportional to the number of transistors in the logic gate times the area of each transistor. Similarly, the power dissipated in an nMOS logic gate depends on the width of the pulldown transistors (which is directly proportional to the capacitance C_i) times the β-ratio times the number of pulldown transistors connected in series. We therefore have

$$B = \sum_{i=1}^{N} m_i C_i = \tau \sum_{i=1}^{N} \frac{m_i}{R_i}. \tag{5.16}$$

Given T, we now want to minimize B. That is, given a delay specification, minimize either the area or power dissipation of the total circuit. To do this, we invoke the marginal utility argument, which states that to minimize B (the globally scarce resource) we require that locally each stage must make the best use of that scarce resource. This will be true if the sensitivity of the total delay to the global resources used at each stage is the same for all stages. To understand this argument, imagine that the sensitivity of the total delay to the power used in the 5th stage is larger than the same sensitivity in the 7th stage. It would then be advantageous to remove power from the 7th stage, slowing down the circuit slightly, and use a part of that power to speed up the 5th stage to compensate for the slowdown of the 7th. The circuit would now operate at the original speed, but with less power dissipation. This marginal utility argument is equivalent to a Lagrange multiplier technique formulation of the problem. We have

$$0 = \nabla_{b_i} B + \lambda \nabla_{b_i} T, \tag{5.17}$$

which reduces to

$$-\lambda^{-1} = \nabla_{b_i} T. \tag{5.18}$$

Solving, we obtain

$$\lambda^{-1} = \frac{R_i}{\tau m_i} \left(\frac{n_i R_i}{R_{i+1}} + \frac{n_i R_i Y_i}{\tau} - \frac{n_{i-1} R_{i-1}}{R_i} \right) \quad \text{for } i = 1 \text{ to } N, \tag{5.19}$$

where λ is the Lagrange multiplier.

Many useful observations can be made. By setting $\lambda^{-1} = 0$, Eq. (5.19) reduces to Eq. (5.6) in the special case of optimum speed. The negative of the derivative of delay with respect to area or power is represented by λ^{-1}. Optimum solutions will always occur for nonnegative λ because a negative λ implies that we are operating in a region where we can simultaneously decrease the area and decrease the delay. We are hardly at an optimum if we are in a region where such happy circumstances exist. In terms of the t_i, we can rewrite Eq. (5.19) as

$$t_i = t_{i-1} + \frac{\tau^2 m_i}{\lambda R_i} - n_{i-1} R_{i-1} Y_{i-1} + \tau(\alpha_{i-1} - \alpha_i), \tag{5.20}$$

where τ, n_i, m_i, R_i, Y_i, and λ are all nonnegative. When α is constant, this equation teaches us that for optimum speed ($\lambda^{-1} = 0$) in the presence of interconnect wiring parasitics, the stage delay is a monotonically decreasing function of position in the chain. That is, the greatest delay occurs at the beginning. Physically, this occurs because the drive capability at the beginning of a chain of gates is limited, but by the end of the chain one has had a chance to telescope that drive to a higher level and hence make the interconnect capacitance more negligible.

On the other hand, if there are no interconnect capacitances, but power or area is important, then the stage delay is a monotonically increasing function of position. We can understand this by considering the problem of driving a large load capacitance through many stages. If λ^{-1} were equal to zero, the delay in every stage would be the same and the sensitivities of delay to power would also be the same (zero). Nevertheless, the last stages would draw the greatest percentage of the power. Thus if we wanted to remove large quantities of power from the design and have the chip run slower, it would be most advantageous to draw the power from the last stages, slowing them down, because that is the only place that large quantities of power are found.

A final caveat: In a VLSI design, it is not always obvious which of the many paths is the critical path. Moreover, as that path is optimized, other paths become critical. Since the optimization of one path can influence the speed of another path, large cycles can occur if one tries the straightforward algorithm of optimizing the most critical path, then the second most critical, and so forth. The techniques presented here can be extended to the multipath problem. Matson [Matson 83] has investigated the solution of the multipath problem, including more accurate macromodels.

5.2 Large fan-out circuits and clock drivers

In either CMOS or depletion load nMOS technology, when one logic gate is driving another nearby logic gate, there is very little one can do to optimize or improve the delay other than size the transistors and manipulate the layouts. On the other hand, when there is a large mismatch between the size of the drive transistors and the load, there is ample opportunity, and need, for circuit creativity. Such situations are common because large fan-out conditions and regular structures go hand in hand. Large fan-out situations are those in which a small transistor must drive a very large capacitance.

Staged buffers and repeaters

When the capacitive load is a lot bigger than the gate capacitance of the drive transistor, it is sometimes appropriate to put several stages of buffering between the drive device and the load. Each buffer should be a little larger than the one before [Lin 75]. This is illustrated in Fig. 5.3. If we assume M stages of buffering, a given source resistance R_0, and a given load capacitance C_L, we then have

$$T = \sum_{i=0}^{M} R_i C_{i+1} = \tau \sum_{i=0}^{M} \frac{C_{i+1}}{C_i}, \qquad (5.21)$$

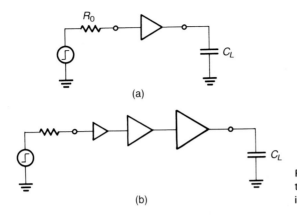

(a)

(b)

FIGURE 5.3   Transformation of a single buffer (a) into a staged buffer (b).

where $C_{M+1} = C_L$. From Section 5.1, we know that the minimum delay occurs when $t_i = t_{i+1} = t_0$. Therefore we know that there is some constant g such that

$$C_{i+1} = gC_i. \tag{5.22}$$

We have

$$T = \tau(M + 1)g \tag{5.23}$$

and

$$C_L = g^{M+1}C_0, \tag{5.24}$$

where $C_0 \equiv \tau/R_0$. Combining Eqs. (5.22) and (5.24), we obtain

$$T = \tau(M + 1)\left(\frac{C_L}{C_0}\right)^{\frac{1}{M+1}} \tag{5.25}$$

Optimizing with respect to M, we obtain

$$g = e \tag{5.26}$$

and

$$M + 1 = \ln\frac{C_L}{C_0}. \tag{5.27}$$

The minimum total delay is

$$T = e\tau \ln\frac{C_L}{C_0}. \tag{5.28}$$

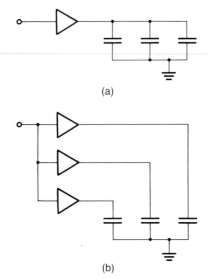

(a)

FIGURE 5.4   Trans-
formation of a single
buffer (a) into different sized
buffers for different critical
paths (b).

(b)

FIGURE 5.5   A model of an
RC line with repeaters.

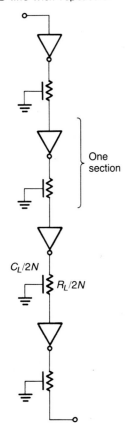

One
section

$C_L/2N$

$R_L/2N$

Note that the optimum number of stages and the total delay increase
only logarithmically as the ratio of the mismatch between the drive and
the load. When multiple loads are involved, we can decouple them and
optimize each path separately. By this technique, one can, for instance,
get away with expending power driving only loads in the critical path.
This technique is illustrated in Fig. 5.4.

There is one last bit of intuition that we can squeeze out of the simple
resistance model before moving on to more complicated viewpoints. Let
us examine the problem of breaking up a long resistive line with repeaters
to improve its delay. We know that as the line gets longer its delay
increases quadratically. If we install a repeater, the repeater introduces
some extra delay in the line, but also restores the signal rise time. Assume
that we break up a long RC line into $2N$ sections. If the total resistance
of the line is R_L and the total capacitance is C_L, then each section has
resistance $R_L/2N$ and capacitance $C_L/2N$. We define τ_L as

$$\tau_L \equiv \frac{R_L C_L}{2}, \tag{5.29}$$

where the τ_L is the "time constant" of the unperturbed line.[1] As shown
in Fig. 5.5, each section of the optimized line starts with a MOS inverter.

Let us assume nMOS depletion load technology. Each section of RC
line is terminated by a load capacitance of SC_τ, where S is the shape
factor of the pulldown. The zero-order pullup and pulldown resistances
are $\beta_\tau R_\tau/S$ and R_τ/S, respectively. The simplest way to estimate the

---

[1] From Section 2.4 we know better ways to estimate the delay, but recall that this
derivation borders on the heuristic anyhow.

behavior of a pair of gates is to sum the capacitances and resistances. We multiply each resistance by the sum of all the capacitances to the right, as illustrated in Fig. 5.6. *RC* lines are modeled by a T network as shown in Fig. 5.7. We have

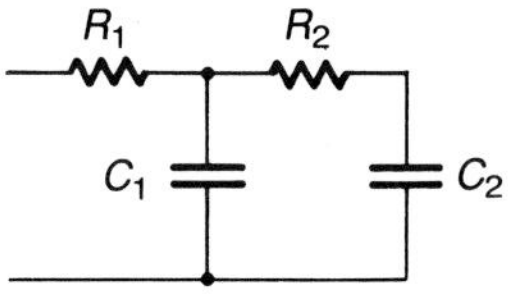

$$t_{\text{pair}} = \left(\frac{(1+\beta_\tau)R_\tau}{S} + \frac{R_L}{2N} \right) \left(SC_\tau + \frac{C_L}{2N} \right) + \frac{R_L C_L}{4N^2} \tag{5.30}$$

and the total delay for the chain sums to

$$T = N t_{\text{pair}}. \tag{5.31}$$

We can minimize T by setting its gradient equal to zero. We have

$$0 = \nabla_{S,N} T, \tag{5.32}$$

which reduces to the following simultaneous equations:

$$0 = (1+\beta_\tau)R_\tau C_\tau - \frac{R_L C_L}{2N^2} \tag{5.33}$$

and

$$0 = \frac{R_L C_\tau}{2} - \frac{(1+\beta_\tau)R_\tau C_L}{2S^2}. \tag{5.34}$$

Identifying $\tau_L = R_L C_L / 2$ and $\tau = R_\tau C_\tau$, we obtain

$$N = \sqrt{\frac{\tau_L}{2(1+\beta_\tau)\tau}} \tag{5.35}$$

and

$$S = \sqrt{(1+\beta_\tau)\frac{R_\tau}{R_L}\frac{C_L}{C_\tau}}. \tag{5.36}$$

Solving for the optimum delay, we obtain

$$T_{\text{opt}} = (1+\sqrt{2})\sqrt{1+\beta_\tau}\sqrt{\tau \tau_L}. \tag{5.37}$$

The optimum delay is some proportionality constant times the geometric mean between the intrinsic speed of the technology τ and the speed of the interconnect technology τ_L. As a technology improves, τ generally gets better but τ_L generally does not. From Eqs. (5.35) and (5.37), we can conclude that repeaters will become more common in advanced technologies.

Superbuffers

It takes a lot of power and area to drive a large capacitive load. When the load is sufficiently large, it pays to employ specialized circuit techniques

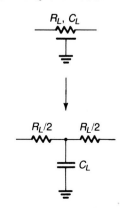

FIGURE 5.6   Estimation of the delay in a two-time constant system: $T \approx R_1(C_1 + C_2) + R_2 C_2$.

FIGURE 5.7   Modeling of *RC* line by a π model.

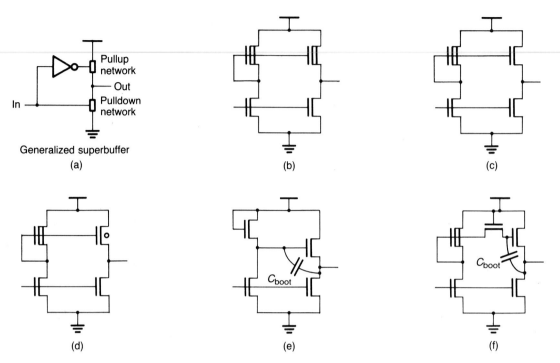

Generalized superbuffer
(a)

(b)   (c)

(d)   (e)   (f)

FIGURE 5.8   A drive of superbuffers.

to try to achieve a more efficient, if more complex, solution. The problem is worst in nMOS where the static power dissipation is so fierce. Two basic techniques are used. To save static power a push–pull output stage is used. The object is to make nMOS look like CMOS. In the steady state, the drive pullup and pulldown are never on simultaneously. The second basic technique is to overdrive[2] the transistors so that during the switching transients, the current per micron of transistor width is larger than normal.

Figure 5.8 shows a generalized superbuffer and a handful of possible realizations. Of these, version (b), with two depletion-mode pullups and two enhancement-mode pulldowns, is easiest to analyze, so we will use it as an example. The basic trade-offs in the other versions are similar, though the details are certainly different.

We do an example we can handle analytically. Assume $V_{TD} = -V_{DD}/2$ and that the body effect is negligible. This example also ignores the delay and power in the predriver stage. We examine the time it takes the buffer to charge a load capacitance C from ground to $V_{DD}/2$. Assume a step input from V_{DD} to zero. For the case of a standard style depletion

---

[2] In highly optimized technologies, these overdrive voltages can represent a reliability problem. In CMOS, bootstrapped voltages can cause latchup.

load inverter, the load is in saturation. The current through the pullup is

$$I_{pu} \doteq K'SV_{TD}^2. \tag{5.38}$$

The output voltage rises as a ramp to $V_{DD}/2$ in a delay time t_D given by

$$t_D = \frac{C}{-2K'SV_{TD}} \left(\frac{V_{DD}}{-V_{TD}} \right). \tag{5.39}$$

In the case of the superbuffer we have two possible situations. If the load C is small, the output responds very quickly compared with the predriver. In this case, V_{GS} of the depletion-mode drive transistor is approximately zero and the superbuffer is essentially equivalent to an inverter, except that its self-capacitance and area are a little worse. On the other hand, if C is large, the predriver responds much more quickly than the drive device. In the limit, V_{GD} of the pullup drive transistor goes immediately to zero. Since the pullup is in the nonsaturation regime, we have

$$I_{pu} = K'S(V_{DS} - 2V_{TD})V_{DS}. \tag{5.40}$$

Integrating the pullup current, we have

$$t_D = \frac{C}{K'S} \int_{V_{DD}}^{V_{DD}/2} \frac{dV_{DS}}{V_{DS}(V_{DS} - 2V_{TD})}. \tag{5.41}$$

Solving, we obtain

$$t_D = \frac{C}{-2K'SV_{TD}} \ln \frac{V_{DD} - 4V_{TD}}{V_{DD} - 2V_{TD}}. \tag{5.42}$$

With $V_{TD} = -V_{DD}/2$, the superbuffer is five times faster than the depletion load inverter of the same size. This means that the power delay product is approximately five times better with the superbuffer because both load currents, under steady state conditions, are the same. This comparison is more accurate if we adjust both the input capacitance and dc power of the superbuffer so that they are equal. Illustrated in Fig. 5.9 are layouts of both circuits. Shown in the Fig. 5.10 is a plot of rise time delay t_D to $V_{DD}/2$ versus load capacitance C. These simulated results include layout capacitances. The layout of the superbuffer is about 10% larger in this example, and this percentage decreases as the buffer gets bigger. The minimum size superbuffer can be over 50% larger than the minimum size inverter.

The depletion drive transistor superbuffer was unique in one respect. It was possible to keep both the input capacitance and power dissipation constant. In the other buffer designs, these parameters are only loosely

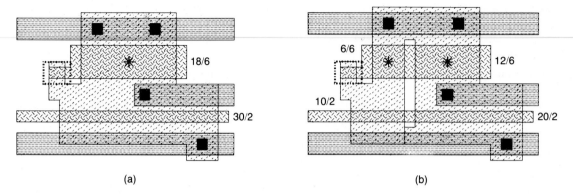

(a)   (b)

FIGURE 5.9   Two devices with equivalent input capacitance and dc power dissipation: (a) a superbuffer and (b) an inverter.

FIGURE 5.10   Performance comparison of the superbuffer and inverter in Fig. 5.9 as a function of load capacitance.

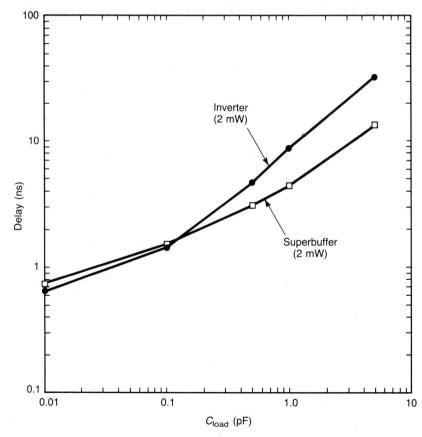

correlated. For instance, for the same input capacitance, the circuit represented in Fig. 5.8(d) uses less power but is slower than the design in Fig. 5.8(b). This makes comparisons difficult.

Recall the discussion of the various load line I–V characteristics in Section 4.2. The superbuffers illustrated in Fig. 5.8 have time-dependent load lines. Figure 5.11 shows two of these load line characteristics.

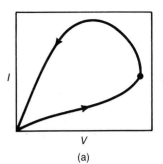

I

V

(a)

Clocked buffers

Ground and V_{DD} are not the only low impedance source of power on a synchronous MOS chip. The clocks, whether they are brought in externally, or carefully generated internal to the chip, are excellent sources of dynamic power. In the design of large TTL machines, it is common to enforce a discipline in which the clocks are not gated. This is not so in MOS VLSI. Consider a common situation in which a control signal becomes available at the end of one clock phase and then, at the beginning of the next clock phase, must drive many loads. We can use superbuffers, but there is a technique that takes almost no power and area—and is perfectly synchronized to the clock.

Figure 5.12 illustrates a variety of clocked buffers. All of these designs have three elements in common. The gated clock is separated from the output by a pullup pass transistor, which is usually wide. This pass transistor is overdriven to minimize the skew between the clock and the output. The output swings from ground to the most positive clock voltage (typically V_{DD}) and thus has both a strong low and a strong high. The second element is an isolation network that couples the input to the gate of the pullup pass transistor, yet allows the gate node to boot to a voltage greater than V_{DD}. The third element, a pulldown transistor, ensures that the output goes all the way to ground when it is low.

The detailed timing of these circuits varies quite a bit. For instance, referring to Fig. 5.12, circuits (a), (b), (c), and (f) require the input to stay high for all of ϕ_2, or the pullup pass transistor will shut off and the output will be pulled low. In circuits (d) and (e), a low going input after ϕ_1 has fallen will have no effect. However, in circuit (b), there will be a conflict in such a case. Some of these circuits, such as (e), are very sensitive to layout because of a high sensitivity to layout capacitance. Circuit (c) uses the dynamic depletion technique where the depletion transistor is actually shut off.[3] The advantage of this circuit is that it allows a potentially higher bootstrap voltage.

Circuit (a) shows two coupling capacitors, labeled C_{boot} and C_{yank}, which can generate the overdrive voltage. C_{yank} varies from

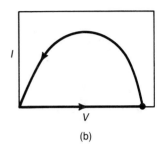

I

V

(b)

FIGURE 5.11   Time-dependent load lines for the superbuffers: (a) from Fig. 5.8(b); (b) from Fig. 5.8(d).

---

[3] Recall from Chapter 2 that depletion devices, because of the deep implant, are hard to shut off. Since $V_S = V_{DD} - V_{BB}$ in this case, the chances that the depletion device can not be shut off are minimal. Nevertheless, one should characterize a new technology with regard to this application before it is used.

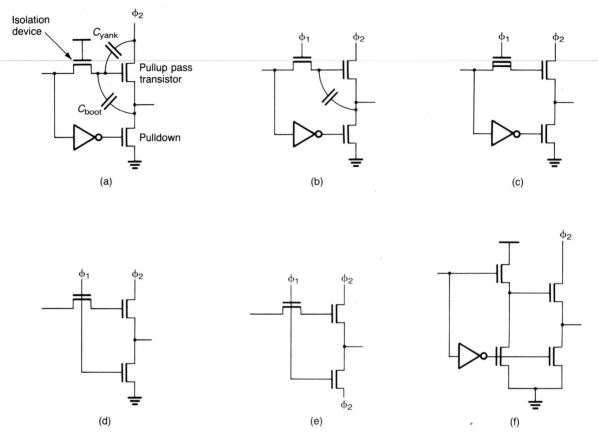

FIGURE 5.12 A drive of clocked buffers. These circuits differ in timing, low-impedance output, number of clocks required, and length of time the input must be held valid.

C_{GDO} to $C_{GDO} + WLC_{OX}/2$. C_{boot} varies from $C_{GSO} + WLC_{OX}/2$ to $C_{GSO} + 2WLC_{OX}/3$. To use C_{yank}, one must take advantage of the varactor effect. When the gate of a transistor is above threshold, the gate-to-source capacitance is large, but when $V_{GS} < V_T$, the capacitance is low. Thus C_{yank} and the parasitics must be designed so that the gate voltage is yanked high only when $V_G > V_T$, but not when $V_G < V_T$. This is a very fast circuit. The bootstrap capacitance mechanism is much less tricky. As the output rises, it boots the gate node. If the pass transistor pullup is completely off, the output will never rise and no bootstrapping action will occur.

In circuit environments with a lot of coupling noise, circuits (d) and (e) may be problematic. During ϕ_2, an output high voltage has a low impedance, but an output low has a high impedance. On the other hand, these designs are relatively compact. By attaching a weak latch to the

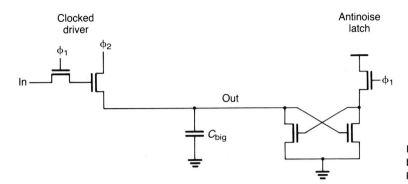

FIGURE 5.13   A clocked buffer with a hold down latch to combat noise.

output line, we can hold it low against noise pulses—and the latch can be at the other end of the line from the driver. This increases the layout freedom. Figure 5.13 illustrates a typical configuration.

Clock drivers

In the design of a clock driver, unlike a buffer, one sometimes has the opportunity to trade delay for rise time. In a buffer, the objective is to get data from one place to another as quickly as possible, whereas the quality of the clock edge is preeminently important in many clocking applications. (When buffering an external clock, the techniques we are about to discuss are not quite as applicable.) The best way to generate a sharp clock edge is to take a "wind it up and let it snap" approach.

We will examine five different nMOS clock drivers, each employing increasingly more sophisticated (and tricky) circuitry. These are only a few of the many possible variations on this theme. Case I, illustrated in Fig. 5.14, shows the simplest buffer. This buffer drives all the way to V_{DD} and all the way to ground, a property important for clocks. The buffer is driving a 20 pF load. Clock loads vary from under 10 pF to over

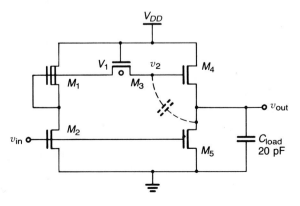

FIGURE 5.14   A bootstrapped superbuffer, Case I.

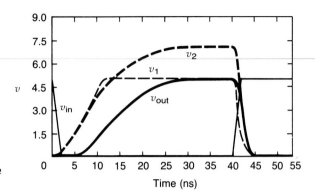

FIGURE 5.15 Voltage waveforms for Case I.

FIGURE 5.16 A boot-strapped superbuffer with delayed boot, Case II.

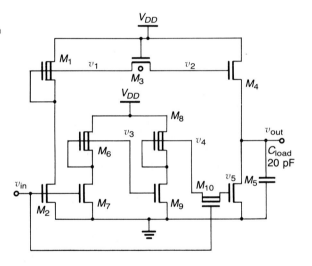

FIGURE 5.17 Voltage waveforms for Case II.

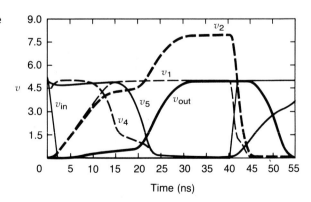

100 pF, depending on the technology and circuit style. For instance, the load will be higher when the clock must drive the sources and drains of transistors as well as their gates. In Case I, the transistor M_3 is designed to isolate the bootstrapping of v_2 from v_1. M_4 is sufficiently wide that an extrinsic bootstrap capacitor is not needed. The proper action of this circuit depends on v_{out} not being able to follow v_2 quickly. (If v_2 did follow quickly, v_2 would not go above V_{DD}.) Figure 5.15 shows a SPICE simulation of this circuit using typical parameters. Because v_{out} is rising, it is difficult to place as large a bootstrap voltage on v_2 as one would like.

The Case II circuit in Fig. 5.16 helps solve the bootstrap problem by delaying the fall of the voltage on the gate of M_5 [Schroeder 75]. This holds the output voltage v_{out} low for a longer time, enabling v_2 to charge to a higher voltage before v_{out} begins to rise. Figure 5.17 illustrates the simulated response of the Case II circuit. The longer the delay, the sharper the rising output voltage. To first order, one must trade delay for rise time because it takes delay to charge the bootstrap capacitor, and the better charged the bootstrap capacitor, the faster the output transition. The transistors M_6 through M_{10} provide the delay.

Case III in Fig. 5.18 illustrates a different way of controlling the isolation transistor needed for bootstrapping. In this case, M_3 can be a driven depletion device allowing v_2 to charge to V_{DD} before booting. This is particularly important if zero threshold transistors are not available or if they have a large body effect parameter. The performance of this circuit is similar to the Case II circuit.

In Cases I through III, the bootstrapping action is accomplished at the speed of the load device. In Case IV, illustrated in Fig. 5.19, a separate "output" stage $(M_{11}$ and $M_{12})$ is used to drive an extrinsic

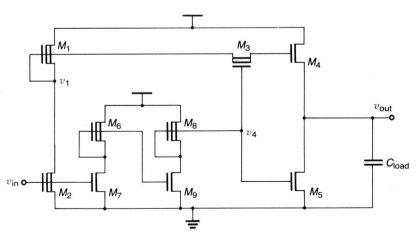

FIGURE 5.18   A bootstrapped superbuffer with dynamic depletion isolation transistor, Case III.

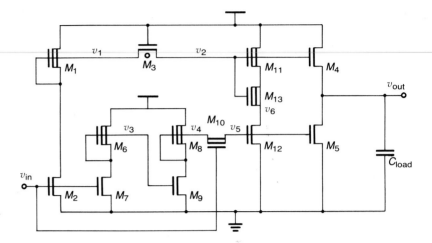

FIGURE 5.19 A bootstrapped superbuffer with predriven bootstrap capacitor, Case IV.

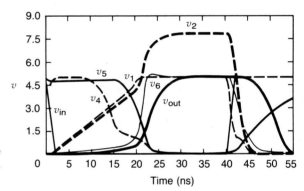

FIGURE 5.20 Voltage waveforms for Case IV.

FIGURE 5.21 Comparison of the output waveforms for the different schemes. Note the trade-off between rise time and the time when the output first starts moving.

FIGURE 5.22 A dynamic depletion bootstrapped superbuffer, Case V.

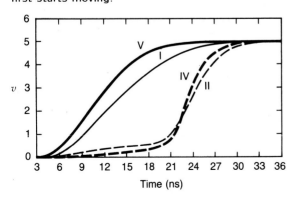

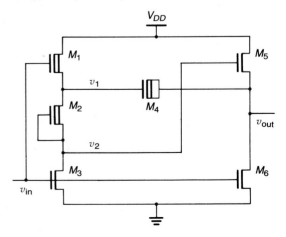

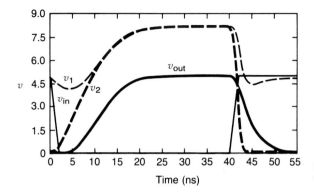

FIGURE 5.23 Voltage waveforms for Case V.

bootstrap capacitor M_{13} [Proebsting 75]. This allows the drive voltage V_6 on the bootstrap capacitor to snap faster than in the previous cases. This is particularly important with extremely large loads. Figure 5.20 shows the voltage waveforms of Case IV.

Figure 5.21 compares the output voltages of the four cases. It clearly illustrates the delay versus rise time trade-off. Care must be taken when driving these circuits since circuits with delay lines are sensitive to glitches on their inputs.

One might ask why it is necessary to wait until a signal arrives to start to charge the bootstrap capacitor. This is, after all, where much of the delay is incurred. In the next technique, the bootstrap capacitor is always charged, ready to be used. Case V, illustrated in Fig. 5.22, is more convoluted than the previous four cases [Knepper 78]. When the input is high, all of the transistors except for M_5 are on. The stack of transistors M_1 through M_3 is ratioed so that v_1 is high and v_2 is low. M_3 is wide. This means that the voltage across the capacitor M_4 is about V_{DD}. It turns out that to first order, the voltage across this capacitor remains constant. When the input voltage falls, transistors M_1, M_3, and M_6 are turned off.[4] We then get charge sharing between the capacitor M_4 and the gate of the drive transistor M_5. This sharing occurs through the "resistor" M_2. Assuming that the capacitor M_4 is large (it is typically larger than a normal bootstrap capacitor), then V_{DD} is placed on the gate of M_5. As the output rises, v_2 and v_1, which are now at the same potential, rise toward $2V_{DD}$. Figure 5.23 illustrates some simulated waveforms for this circuit. We may enhance this circuit with the technique for decoupling the load from the bootstrap in the manner illustrated in Case IV.

---

[4] M_3 is a dynamic depletion device in this realization. An enhancement-mode device could be used instead.

5.3 Large fan-in circuits

When designing circuits to drive large capacitive loads, most of the sophistication is in designing these circuits to dissipate a minimum of power or generate good edges. Until one runs into the limits of the technology, one can always increase the dissipated power to achieve a faster circuit. As we showed in Eq. (5.28), the total delay can be made to increase only logarithmically as the load capacitance increases, for a slightly worse than linearly increasing cost in power. The situation for circuits with high fan-in is quite different. An example of a high fan-in situation is the large OR-plane NOR gate in a PLA. In a high fan-in situation, the load capacitance (or self-capacitance) of a logic gate is very large and the charging or discharging current is limited. We have

$$dt = \frac{C}{I}dv. \tag{5.43}$$

In the PLA NOR gate example, the discharge current I is fixed by the maximum width of the pulldown transistor that will fit in the array. The capacitance C is determined by the number of words in the PLA. Looking at Eq. (5.43), the only way to make dt smaller, assuming C and I are fixed, is to detect a smaller change in dv. This implies the use of a different class of circuit techniques.

The problem divides into two tasks. One task is to design a circuit to detect a small change in voltage, and the other is to design a circuit that keeps the load in the region where the detection circuit is highly sensitive. In this section, we specialize in single-ended detection techniques. In Section 5.5, we will look at double-ended or "differential" detection techniques. In single-ended techniques we try to detect an absolute change in some voltage, whereas in double-ended sensing techniques we get to compare a signal either to a reference voltage or to a signal moving in the opposite direction.

One way we will detect a small change in voltage is to use a charge sharing amplifier (a cascode connection). Figure 5.24 illustrates the configuration. In Fig. 5.25 the canonical charge sharing amplifier is illustrated. We assume that the voltage v_β on C_{big} and the voltage v_σ on C_{small} are both greater than V_{ref}. The trigger voltage $V_{\beta\text{trigger}}$, at which C_{big} and C_{small} become connected by M_1, is

$$V_{\beta\text{trigger}} = V_{\text{ref}} - V_{T1}. \tag{5.44}$$

The gain a_0 of the amplifier is given by

$$a_0 = \frac{V_{\sigma0} - V_{\beta\text{trigger}}}{V_{\beta0} - V_{\beta\text{trigger}}}, \tag{5.45}$$

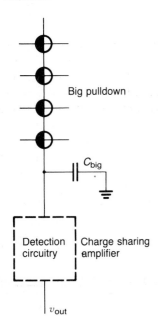

FIGURE 5.24 Large fan-in single-ended circuit with charge-sharing sense amplifier.

Big pulldown

C_{big}

Detection circuitry | Charge sharing amplifier

v_{out}

where $V_{\sigma 0}$ and $V_{\beta 0}$ are the initial conditions on v_σ and v_β, respectively. We have assumed that $V_{\beta 0}$ is unperturbed by the charge from C_{small}. We can identify $V_{\beta 0} - V_{\beta\text{trigger}}$ as the high noise margin of the circuit. Equation (5.45) becomes

$$a_0 = \frac{V_{\sigma 0} - V_{\beta\text{trigger}}}{NM_H}. \qquad (5.46)$$

Typical values for NM_H range between 300 mV and 800 mV for a 5 volt process. Assuming $V_{\beta 0}$ is less than V_{DD}, the maximum gain $a_{0\text{max}}$ becomes

$$a_{0\text{max}} = \frac{V_{DD}}{NM_H}. \qquad (5.47)$$

Since, from Eq. (5.43), the delay is proportional to the noise margin, we see the fundamental trade-off between noise margin and speed. By giving up noise margin, the speed of the detection circuitry can be improved. Figure 5.26 illustrates a typical plot of v_α and v_β as a function of time.

Figure 5.27 illustrates one of many methods for setting up $V_{\beta 0}$ and $V_{\sigma 0}$. We illustrate this circuit to show the importance of second-order device effects. Note that all of the voltages on transistors M_1 and M_2 are identical. The difference between $V_{\beta 0}$ and $V_{\beta\text{trigger}}$ is obtained in this case by use of the short channel effect. There are, of course, more straightforward ways of precharging $V_{\sigma 0}$ and $V_{\beta 0}$, but note that this method does track very well with process parameter variations. There are trade-offs in the placement of V_{ref}. If we keep V_{ref} low, then the receiver that looks at v_σ can be simple and the gain a_0 is high. On the other hand, in a PLA, I_{pd} is maximized when $V_{\beta 0}$ is high.

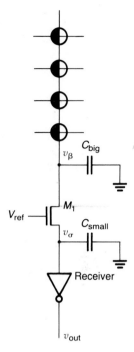

FIGURE 5.25 Charge-sharing sense amplifier implementation.

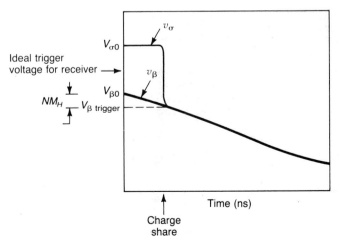

FIGURE 5.26 Voltage waveforms for the charge-sharing sense amplifier.

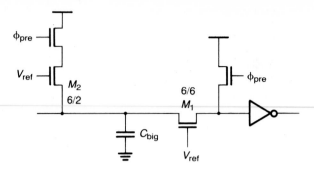

FIGURE 5.27 Implementation that exploits the short-channel effect. The noise margin of this circuit depends on the amount that the short-channel effect lowers the threshold of the 6/2 device. The worst-case speed corner of this circuit can be counterintuitive.

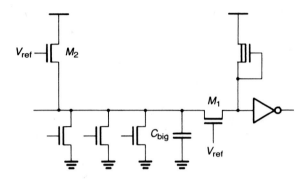

FIGURE 5.28 A static implementation with a transistor M_2 to limit the swing on C_{big}.

FIGURE 5.29 A second static implementation.

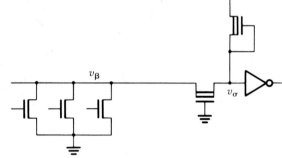

FIGURE 5.30 Voltage waveforms for the circuit in Fig. 5.29.

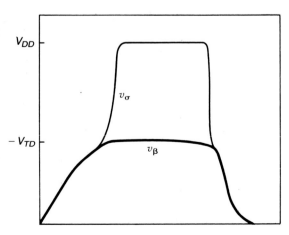

If the NOR gate is used in a ROM, that is, if only one pulldown can be on at a time, we can design our circuitry such that M_2 applies the "brakes" to v_β. This will keep the bit line from pulling too low. This is important in the design of a static circuit where we cannot use a precharge transistor to pull v_β back up because it decreases the time required between successive reads. Keeping the voltage swing small also minimizes capacitively coupled noise. Figure 5.28 illustrates this design. The current through M_2 is a quadratic function of $(V_{DD} - v_\beta)$. This tends to keep v_β from going very low, providing only one pulldown is on at a time.

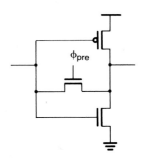

FIGURE 5.31   Circuit balanced to V_{inv}.

Another useful variant on this circuit is shown in Fig. 5.29. Figure 5.30 illustrates typical switching waveforms.

Another single-ended technique for making circuits fast is the technique of precharging a circuit to V_{inv}. This is illustrated simply in Fig. 5.31. This technique is particularly fast if dual rail logic is used. In dual rail logic, every time one generates a signal, its complement is also generated. Noise must be considered carefully in these circuits.

■■■■■■ EXAMPLE 5.1

The design of a reference voltage
Good reference voltages in MOS are extremely difficult to design. Often it is good enough to develop a voltage reference by dividing down the power supply voltage. Fig. 5.32 illustrates two circuits for developing a voltage reference off V_{DD}. The circuit in Fig. 5.32(a) is less sensitive to variations in V_{TD} than the circuit in Fig. 5.32(b). For a depletion load

FIGURE 5.32   Reference voltage generators, (a) robust and (b) sensitive.

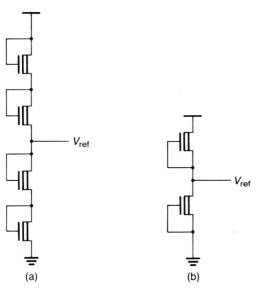

(a)   (b)

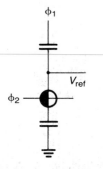

FIGURE 5.33 Reference voltage generator that works by capacitive division.

device with gate and source connected, we have, assuming the transistor is in saturation,

$$I = K'SV_{TD}^2.$$

For the same device, with drain and gate connected,

$$I = K'S\left(V_{\text{out}}^2 - 2V_{\text{out}}V_{TD}\right),$$

where V_{out} is V_{DS}. Let us look at an example with $V_{\text{out}} = 2$ V, and $-4 \leq V_{TD} \leq -3$. In the first case, I varies by 44% and in the second case, I varies by only 20%.

A second class of techniques is useful in generating reference voltages for dynamic circuits. In this case, one can use capacitive division as illustrated in Fig. 5.33. ∎

5.4 Bistability

Bistability can occur anywhere there is gain and positive feedback. It is used in many places in a VLSI circuit, including the static latch, the flip-flop, and the Schmitt trigger.[5]

The static latch

The purpose of a latch is to remember. There are two basic types of latches—static and dynamic. In the absence of noise above a fixed amplitude, a static latch will remember one bit of state for as long as the power supply is on. This is because the static latch contains elements with power gain that continually restore the integrity of the signals representing the state. Power gain is needed to compensate for the nonideal nature of the circuit elements and environment. The dynamic latch does not restore the signals representing the state— there is no source of power. In some technologies, one can nevertheless build memory elements that can store information for long periods of time—years in the case of superconductors. In MOS, dynamic storage is accomplished by storing charge on a capacitor. Depending on the temperature and other effects, the ratio of the dynamic storage time to τ is between 10^6 and 10^9 for typical digital MOS processes. The dynamic latch is analogous to pass transistor logic in that perturbations

---

[5] It also occurs in some places where it is less than useful—CMOS latchup, for example.

about valid logic levels are not restored. The static latch is analogous to restoring logic.

The simplest form of static latch consists of a pair of cross-coupled inverters, as shown in Fig. 5.34. The bistable latch has three quasi-static solutions, as shown in Fig. 5.35. In this figure, the solutions are found graphically. By symmetry, there is a static solution of the network in which the voltages on both drains are at the same voltage V. If the latch is bistable, then the circuit will be unstable at this symmetric operating point. Because both M_1 and M_2 (assumed identical) are enhancement-mode devices with their gates at the same voltage as their drains, we know that M_1 and M_2 are in saturation. Because the current through G_L equals the current through the transistor, we have

$$(V_{DD} - V)G_L = \frac{K'S}{1+\delta}(V - V_{TE0})^2. \qquad (5.48)$$

Solving the quadratic, we obtain

$$V - V_{TE0} = \frac{G_L(1+\delta)}{2K'S}\left(\left(1 + 4(V_{DD} - V_{TE0})\frac{K'S}{G_L(1+\delta)}\right)^{1/2} - 1\right). \qquad (5.49)$$

FIGURE 5.34   Bistable circuit.

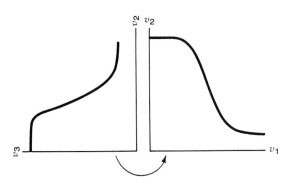

FIGURE 5.35   Graphical determination of the three static solutions, only two of which are stable.

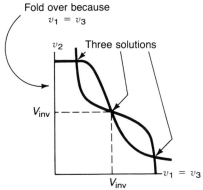

Fold over because $v_1 = v_3$

Three solutions

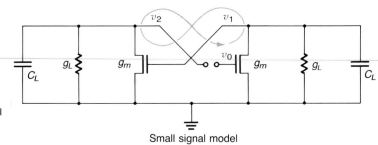

FIGURE 5.36 Small-signal model of a cross-coupled latch with a passive load.

Small signal model

The latch is bistable if the small signal loop gain shown in Fig. 5.36 is greater than one. The loop is composed of two identically biased inverters; the gain of each must be greater than one in magnitude for the latch to be bistable. Therefore, for bistability, we have

$$g_m > g_L, \tag{5.50}$$

where g_L is the small signal conductance of the (possibly nonlinear) load and

$$g_m = \frac{2K'S}{1+\delta}(V - V_{TE0}) . \tag{5.51}$$

Combining Eqs. (5.49), (5.50), and (5.51), we have

$$g_L < G_L \left(\left(1 + 4(V_{DD} - V_{TE0})\frac{K'S}{G_L(1+\delta)} \right)^{1/2} - 1 \right) . \tag{5.52}$$

Simplifying, we have

$$g_L \left(2 + \frac{g_L}{G_L} \right) < 4\,(V_{DD} - V_{TE0})\frac{K'S}{1+\delta} \tag{5.53}$$

as the condition for bistability. For resistors, we have

$$1 = \frac{g_L}{G_L} . \tag{5.54}$$

FIGURE 5.37 Correspondence between the large- and small-signal conductances of a nonlinear load device.

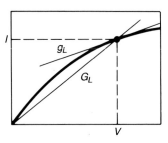

Equation (5.49) can be used to relate V to G_L, and therefore g_L to G_L, by determining the operating point. Figure 5.37 illustrates the relationship between g_L and G_L.

There are standard techniques for analyzing both the small and large signal dynamics of the cross-coupled latch. We look first at the small signal analysis. Referring to Fig. 5.36, we have

$$g_m v_1 = - \left(g_L + C_L \frac{d}{dt} \right) v_2 \tag{5.55}$$

and

$$g_m v_2 = -\left(g_L + C_L \frac{d}{dt}\right) v_1. \tag{5.56}$$

We are looking for the amplification of a voltage difference $v_2 - v_1$. We define Δv as

$$\Delta v \equiv v_2 - v_1. \tag{5.57}$$

Subtracting Eq. (5.55) from Eq. (5.56), we obtain

$$\frac{d}{dt}\Delta v = \left(\frac{g_m - g_L}{C_L}\right) \Delta v \tag{5.58}$$

or

$$\Delta v(t) = \Delta V \exp\left(\frac{g_m - g_L}{C_L} t\right), \tag{5.59}$$

where ΔV is the initial voltage difference. Equation (5.59) corroborates our previous conclusion that the cross-coupled latch is bistable if $g_m > g_L$. Note in Eq. (5.59) that Δv grows if $g_m > g_L$ and decays if $g_m < g_L$. The time for $\Delta v(t)$ to reach a particular value ΔV_{switch} is given by

$$t_{\text{switch}} = \frac{C_L}{g_m - g_L} \ln \frac{\Delta V_{\text{switch}}}{\Delta V}. \tag{5.60}$$

Note in (5.60) that t_{switch} tends to infinity as ΔV tends towards zero. This inability for the latch to "decide" in a bounded amount of time causes the synchronizer problem we will discuss in Chapter 6. For fast switching, we want C_L small and the transconductance g_m large. We define $C_L/(g_m - g_L)$ as the characteristic time constant τ_{switch} of the latch.[6] The minimum value of τ_{switch} is found when C_L is dominated by the transistor capacitance and when $g_m \gg g_L$. For this case, we have

$$\tau_{\text{switch}} \geq \frac{2}{3} \frac{L^2}{\mu(V_{DD} - V_{TE})}. \tag{5.61}$$

This small signal analysis yields most of the information about the static latch that we might want.

For a view of the latch that takes into account large signal effects, phase plane analysis techniques can be used. We have

$$C_L \frac{dv_2}{dt} = I(v_1, v_2) \tag{5.62}$$

and

$$C_L \frac{dv_1}{dt} = I(v_2, v_1), \tag{5.63}$$

---

[6] Miller capacitance is actually important in this calculation.

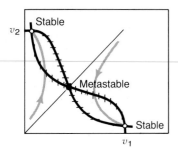

FIGURE 5.38 State–space solutions of the large-signal characteristics of a bistable latch.

where

$$I(v_1, v_2) = I_{pu}(v_1, v_2) - I_{pd}(v_1, v_2) \tag{5.64}$$

and

$$I_{pu}(v_1, v_2) = (V_{DD} - v_2)G_L. \tag{5.65}$$

To use the phase plane technique, we take the ratios of the derivatives of the voltages. We have

$$\frac{dv_2}{dv_1} = \frac{I(v_1, v_2)}{I(v_2, v_1)}. \tag{5.66}$$

The numerator and denominator of the right-hand side of Eq. (5.66) describe the static transfer characteristics of the two inverters when $dv = 0$. Thus in phase space, illustrated in Fig. 5.38, we can easily discover where dv_2/dv_1 is zero or infinity. By symmetry, $dv_2/dv_1 = 1$ when $v_1 = v_2$. Some possible v_1, v_2 trajectories, with time as a free parameter, are shown in Fig. 5.38.

When driving large capacitive loads, as with register files or where the output of the latch is taken through a pass transistor, it is possible to inadvertently flip the latch. The problem can be seen by briefly explaining the operation of the six-transistor RAM cell shown in Fig. 5.39. To write a 1 into the cell, the word line (connected to the two pass devices) is raised simultaneously with the raising of the left bit line and the lowering of the right bit line. To write a 0, the left bit line is lowered while the right bit line is raised. The value is stored when the word line is lowered. To perform a read, both bit lines are precharged high and then the word line is activated. The cell then forces one of the bit lines low. This is when the cell is in danger of loosing its value. When the pass transistors M_3 and M_4 are turned on, nodes A and B

FIGURE 5.39 A static RAM cell showing (a) the access transistors and (b) the small-signal model.

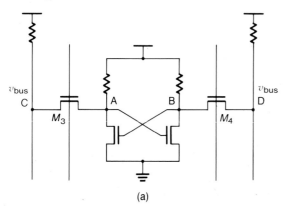

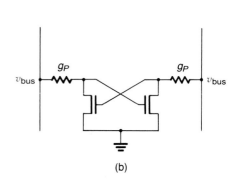

(a)

(b)

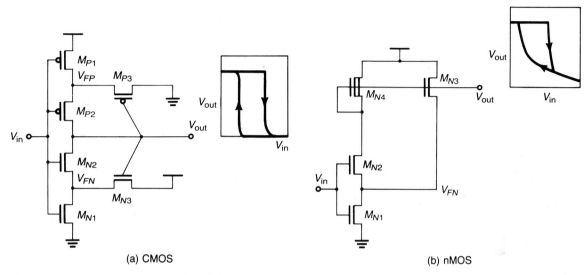

FIGURE 5.40   The Schmitt trigger: (a) CMOS implementation and static transfer curve; (b) nMOS implementation and static transfer curve.

are connected to nodes C and D, respectively. The voltage on nodes C and D can be quite a bit lower than V_{DD} because of a patterning effect where zeros read in previous register accesses pull nodes C and D low faster than the bus pullup transistors can respond. When this happens, it is important for the latch not to forget its state. We have conflicting requirements on the size of the pass transistors M_3 and M_4. We want them to be very wide so the large bus capacitances on nodes C and D get charged or discharged quickly. On the other hand, when transistors M_3 and M_4 first turn on hard, the equivalent load and supply voltages are determined by the small signal conductance of the pass transistors g_P and voltage on the bus v_{bus}. (We assume $g_P \gg g_L$.) In this case, if the loop gain is not greater than one, all is lost. Equation (5.53) can be used to determine the necessary conditions with the substitution of g_P and G_P for g_L, and G_L and v_{bus} for V_{DD}. Another important effect is charge coupling into the cell when M_3 and M_4 are first enabled. The theoretical considerations here are followed, in the design of a practical circuit, by numerous circuit simulations. These simulations are directed by the insights gained through analysis.

The Schmitt trigger

The Schmitt trigger is often used to turn a signal with a very slow or sloppy transition into a signal with a sharp transition. They are often found on chip inputs. Both an nMOS and a CMOS version of the Schmitt trigger are illustrated schematically in Fig. 5.40. The analysis of the trigger point for a high going input transient to both the CMOS and nMOS triggers is the same. To first order, the trigger point occurs when M_{N2} first turns on. We analyze the point when the current through M_{N2} is zero. If the input trigger voltage is V_{trigger}, the feedback voltage V_{FN} is

$$V_{FN} = V_{\text{trigger}} - V_{TN}, \tag{5.67}$$

where V_{TN} is the threshold voltage of the n-channel transistor.[7] For an input voltage at the trigger point, the pulldown transistor M_{N1} is on the edge of saturation. V_{FN} is determined by the ratio of M_{N3} to M_{N1}. Solving for the trigger voltage, we have

$$V_{\text{trigger}} = \frac{V_{DD} + \sqrt{\beta}V_{TN}}{1 + \sqrt{\beta}}, \tag{5.68}$$

where

$$\beta = \frac{S_{TN1}}{S_{TN3}}. \tag{5.69}$$

Clearly, $V_{TN} < V_{\text{trigger}} < V_{DD}$. The important point to note about Eq. (5.68) is that the trigger voltage is insensitive to process parameters. When β is low, even V_{TN} is not critical. It is also independent of the depletion (or p-channel) threshold, which does not even appear in these first-order equations.

Above the trigger voltage, the gain around the feedback loop is greater than one. This causes the circuit to switch abruptly to a new stable state. By symmetry, the trigger voltage for a low-going signal in a CMOS Schmitt trigger is also sharp. Note, however, that the trigger voltage for the low-going input is determined by the p-channel rather than the n-channel devices.

The Schmitt trigger evidences hysteresis as shown in the dc transfer function plots in Fig. 5.40. The Schmitt trigger can be used for level conversion, but one must be careful in the nMOS version to watch out for the soft switching response to low-going inputs. The output voltage of the nMOS Schmitt trigger can rise above V_{OL} before the switching action is initiated. Schmitt triggers can also be built with latches and resistive dividers. Note that, because of the gain-bandwidth trade-off, a Schmitt trigger is generally slower than an inverter of the same power. The Schmitt trigger is considered a current hog in micropower CMOS because several wide devices are simultaneously on for a relatively long time. A layout of an nMOS Schmitt trigger is illustrated in Fig. 5.41.

---

[7] Since this is a rough analysis, we are ignoring the body effect. It could easily be included.

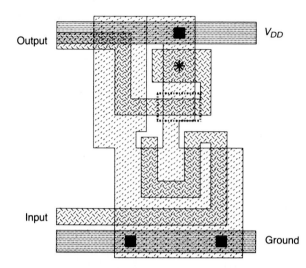

Output

V_{DD}

Input

Ground

FIGURE 5.41   Layout of a
Schmitt trigger.

■■■■■■■■■■■■■■■■ EXAMPLE 5.2

A timing generator
Since the Schmitt trigger has such a stable trigger voltage, it can be
used to generate a fairly stable delay. If we set the trigger point high,
its only important variation is with V_{DD}. If we use a resistor connected
to V_{DD} to charge a capacitor, and use the Schmitt trigger to detect
when the voltage on the capacitor exceeds some threshold value, then
the V_{DD} variations cancel out of the delay determination. We are left
with only the variations in the value of the capacitor and resistor. The
gate capacitance, on a given fabrication line, is closely controlled. To get
a good resistance, we can use either an external resistor or an on-chip
transistor. We can form a reasonable "resistor" in nMOS by hooking
the gate of a depletion-mode device to V_{DD}. With this configuration,
one can approach resistance variations on the order of the mobility
variation by drawing the devices large. A schematic of this sort of circuit
is shown in Fig. 5.42. Note that gate capacitance variations cancel out.
In CMOS, one should use the better controlled device in the linear mode
to determine the charging current. Current mirrors can be used if one
ends up needing the "wrong" device.   ■

FIGURE 5.42   Delay
generator. Main source
of imprecision is mobility
variations.

5.5   Sense amplifiers

Because of its central role in memory technology, there is perhaps no
MOS circuit that has been more carefully studied and optimized than
the sense amplifier. The sense amplifier detects and amplifies small

Reset

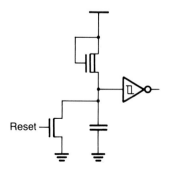

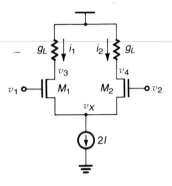

FIGURE 5.43 Model of a static differential amplifier.

input voltage differences. We examine three of the most important sense amplifier circuits.

The static differential amplifier

The static differential amplifier is the basic front-end stage used in analog operational amplifiers. Figure 5.43 illustrates the basic circuit configuration. The current source is typically built out of a long channel transistor in saturation, biased by a current mirror to track process variations. When the inputs v_1 and v_2 are equal, the current through M_1 and M_2 is I by symmetry.

Imbalances in the input voltage cause one of the two transistors to hog the current. In the limit, one transistor is off, with its drain at V_{DD}, while the other transistor is sinking $2I$, with its drain voltage at $V_{DD} - 2I/G_L$. We want M_1 and M_2 to be in saturation over as much of the operating range as possible because this maximizes the amplifier gain. We may write the current through the input transistors as

$$I_1 = K'S(v_1 - v_X - V_T)^2 \tag{5.70}$$

and

$$I_2 = K'S(v_2 - v_X - V_T)^2 \tag{5.71}$$

as long as they are in saturation. Solving for $v_2 - v_1$, we obtain

$$v_2 - v_1 = \sqrt{\frac{I_2}{K'S}} - \sqrt{\frac{I_1}{K'S}}. \tag{5.72}$$

We can see from Eq. (5.72) that the input dynamic range is given by

$$V_2 - V_1 = \pm\sqrt{\frac{2I}{K'S}}. \tag{5.73}$$

When the input voltages are equal, and if M_1 and M_2 are in saturation, then the transconductance of these transistors is given by

$$g_m = 2\sqrt{K'SI}. \tag{5.74}$$

To find the gain from input to output, we linearize the circuit and study its response to the difference of the two input voltages, as well as its response to the average input voltage. We define Δv_{in} as

$$\Delta v_{in} \equiv v_2 - v_1 \tag{5.75}$$

and Δv_{out} as

$$\Delta v_{out} \equiv v_4 - v_3. \tag{5.76}$$

The sum of the small signal currents into the current source must be zero. We have

$$0 = i_1 + i_2, \qquad (5.77)$$

where

$$i_1 = g_m(v_1 - v_X) \qquad (5.78)$$

and

$$i_2 = g_m(v_2 - v_X). \qquad (5.79)$$

We also have

$$g_L v_3 = -i_1 \qquad (5.80)$$

and

$$g_L v_4 = -i_2. \qquad (5.81)$$

Combining Eqs. (5.78) and (5.80), and Eqs. (5.79) and (5.81), we obtain

$$g_L v_3 = -g_m(v_1 - v_X) \qquad (5.82)$$

and

$$g_L v_4 = -g_m(v_2 - v_X). \qquad (5.83)$$

Subtracting Eq. (5.82) from Eq. (5.83), we obtain

$$\Delta v_{\text{out}} = -\frac{g_m}{g_L} \Delta v_{\text{in}}. \qquad (5.84)$$

Note that the gain does not depend on the exact value of the input voltages, providing M_1 and M_2 are in saturation, because g_m is set by the current source. In the large signal case, the gain is obviously not constant. The maximum input range over which the gain can be large is given by Eq. (5.73).

Adding Eq. (5.82) to Eq. (5.83), we obtain

$$(v_3 + v_4) = -\frac{g_m}{g_L}(v_1 + v_2 - 2v_X). \qquad (5.85)$$

But from Eq. (5.77) through Eq. (5.79), we have

$$0 = v_1 + v_2 - 2v_X. \qquad (5.86)$$

Combining Eq. (5.85) and Eq. (5.86), we have

$$0 = v_3 + v_4, \qquad (5.87)$$

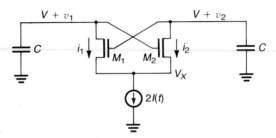

FIGURE 5.44   Model of a
clocked differential amplifier.

which tells us that the "common mode" gain is zero. The current source
keeps the average output voltage constant.

The body effect on transistors M_1 and M_2 and other nonideal
transistor characteristics prevents the common mode gain from being
zero. A more important limitation on this circuit is the difficulty in
keeping all of the transistors in saturation. This limits, among other
things, the magnitude of the output voltage swings.

The clocked differential amplifier

The clocked differential amplifier, like the static differential amplifier,
uses a current source [Stein 72]. In this circuit, the current source is
time varying. Figure 5.44 shows the clocked differential amplifier. This
circuit does not have pullup devices. A difference between the voltages
on the capacitors is amplified by the cross-coupled transistor structure.
An inequality in the input voltages will cause one of the cross-coupled
transistors to draw more current than the other. Because of the positive
feedback, the input node with the lower voltage will be pulled even lower.
Our objective is to design this circuit so there is essentially no movement
of the higher-input voltage, while the lower-input voltage is yanked to
ground.

We first examine the case in which the difference voltage is zero.
This is the common mode response of the network. We have

$$I(t) = -C\frac{dV(t)}{dt} \tag{5.88}$$

and

$$I(t) = K'S\left(V(t) - V_{TE} - V_X(t)\right)^2. \tag{5.89}$$

Solving for $V(t)$, we have

$$V(t) = V_0 - \frac{1}{C}\int I(t)dt \tag{5.90}$$

and for $V_X(t)$, we have

$$V_X(t) = V_0 - V_{TE} - \sqrt{\frac{I}{K'S}} - \frac{1}{C}\int I dt. \tag{5.91}$$

These equations are valid as long as $V - V_{TE} \geq V_X$. V and V_X both decay as $I(t)$ rises. The larger the current, the faster $V(t)$ falls.

To understand how this circuit functions as a sense amplifier, we examine the circuit's response when a difference perturbation is superimposed on the large signal common mode response. Assume perturbation voltages $v_1 = -\Delta v/2$ and $v_2 = \Delta v/2$. The sum of the small signal currents must be zero. We have

$$i_1 + i_2 = 0, \tag{5.92}$$

where

$$i_1 = -\frac{C}{2}\frac{d}{dt}\Delta v \tag{5.93}$$

and

$$i_2 = \frac{C}{2}\frac{d}{dt}\Delta v. \tag{5.94}$$

We also have

$$i_1 = -g_m \left(\frac{\Delta v}{2} + v_X\right) \tag{5.95}$$

and

$$i_2 = g_m \left(\frac{\Delta v}{2} - v_X\right). \tag{5.96}$$

Combining Eqs. (5.93) through (5.96), we obtain

$$C\frac{d}{dt}\Delta v = g_m \Delta v. \tag{5.97}$$

Unlike in the static differential amplifier, g_m is a function of time. We can rewrite Eq. (5.97) as

$$\frac{d}{dt}\ln \Delta v = \frac{g_m}{C} = \frac{2}{C}\sqrt{IK'S} \tag{5.98}$$

or

$$\Delta v = \Delta v_0 \exp\left(\frac{2}{C}\sqrt{K'S}\int \sqrt{I(t)}dt\right). \tag{5.99}$$

Equation (5.99) is, of course, valid only for small Δv. Eventually, one transistor will be cut off and the other will go out of saturation.

As circuit designers, one of our problems is to choose an appropriate current waveform $I(t)$. We know that $V(t)$ is generally falling and $\Delta v(t)$ is generally growing. We want the voltage on the capacitor with the higher initial voltage to remain constant while the voltage on the capacitor with the lower initial voltage is driven to ground. In terms of our common and differential mode signals, this translates to $\Delta v(t)$ growing twice as fast as $V(t)$ falls. In this way, $v_2(t)$ will stay constant

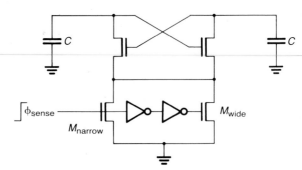

FIGURE 5.45 Clocking technique for dual slope sensing.

and $v_1(t)$ will fall at a rate of $d\Delta v/dt$. The average will fall at half this rate. Mathematically, we want

$$\frac{\frac{d}{dt}\Delta v}{\frac{d}{dt}V} \approx -2.$$

(5.100)

From Eq. (5.88), we have

$$\frac{d}{dt}V = -\frac{I}{C}$$

(5.101)

and from Eqs. (5.97) and (5.74), we have

$$\frac{d}{dt}\Delta v = \frac{2\Delta v}{C}\sqrt{IK'S}.$$

(5.102)

Combining Eqs. (5.100) through (5.102), we obtain

$$I(t) = K'S(\Delta v)^2.$$

(5.103)

Since $|\Delta v|$ is (we hope) increasing as a function of time, Eq. (5.103) implies that I should increase slowly at first and quickly later. If $I(t)$ is jammed on, Δv will not have a chance to grow and $V(t)$ will collapse before the difference is amplified. The speed at which Δv grows (see Eq. 5.98) is related to the strength of the current. Therefore we do want $I(t)$ as large as possible within the constraint that both sides of the sense amplifier do not collapse to zero. The ideal current waveform can be approximated with the circuit illustrated in Fig. 5.45, or with similar circuit tricks. If all sense amplifiers see approximately the same average input voltage, then v_X should be the same for all amplifiers, whether they are reading a 1 or a 0. This allows us to get good results by connecting all of the v_X nodes together since they are at the same potential anyhow. In this way, the clock pulldown devices can be shared by all amplifiers, and removed from the cells. While it is possible to make

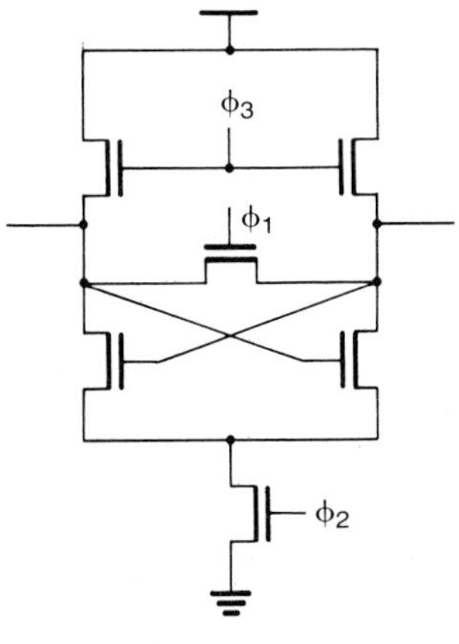

I too big, making it too small only slows the circuit down—it does not kill the functionality. For optimum speed, Lynch and Boll [Lynch 74] have shown that allowing the high line to also fall part way toward ground results in a speed improvement over the ideal case we derived. The general shape of the current waveform remains the same. In this case, a circuit may be required to restore or "boost" the bus. Either the bus booster circuit of Example 2.2 or clock pullups such as those in Fig. 5.46 could be used.

The first two sense amplifiers discussed in this section have very dissimilar characteristics. The gain of the static amplifier does not, to first order, depend on Δv. It does not need a clock. On the other hand, it is fairly tricky to design. It is a very hard to design a biasing circuit that keeps all of the devices biased in saturation under all processing variations. It is also hard to get good voltage swings. The gain of the clocked sense amplifier, on the other hand, depends on Δv. While there are no biasing problems, the circuit must be equilibrated, and care must be taken not to jam the current on too fast. In addition, one must take care to avoid capacitive steering. This circuit will amplify differences in the capacitive load C as effectively as voltage differences. Great care must be taken in the layout. Indeed, the speed of the clocked sense amplifiers is determined, in part, by the size of C. Therefore decoupling pass devices are often used to isolate the amplifier from the load after Δv is large enough to be sensed effectively. The isolation transistors can be gated by a clock, connected as resistors, or gated by the output of the sense amplifier.

FIGURE 5.46 Level restoring with the clocked differential amplifier: ϕ_1 equilibrate; ϕ_2 sense; ϕ_3 restore/boost.

The differential charge amplifier

We can break apart the common current source of the clocked sense amplifier, as shown in Fig. 5.47. If initially $v_1 = v_2$, then instead of amplifying differences in the drain voltages, this circuit will amplify differences in the source voltages v_{X1} and v_{X2}. The transistor with the lower source voltage will draw a larger current, decreasing v_{GS} of the other device. The cross-coupled topology provides positive feedback. Eventually, v_{GS} of the transistor with the higher source voltage will reach V_{TE} and that device will shut off. The other device will continue to conduct, and v_{DS} of that device will go to zero. Thus a final difference voltage of at least V_{TE} will be generated.

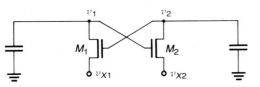

FIGURE 5.47 Another differential amplifier.

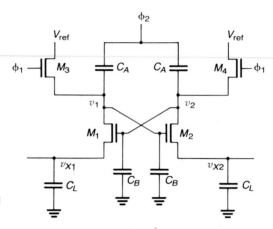

FIGURE 5.48 The differential charge preamplifier.

Figure 5.48 illustrates a simple sense amplifier [Heller 79] that uses this topology. The precharge clock is ϕ_1. V_{ref} is chosen such that M_3 and M_4 operate in the nonsaturation regime during precharge, and hence v_1 and v_2 are precharged all the way to V_{ref}. Thus the thresholds of M_3 and M_4 are unimportant. The source voltages v_{X1} and v_{X2} are precharged to one threshold drop below V_{ref}. One beautiful aspect of this circuit is that the precharge is accomplished through the transistors which will later be used for sensing the voltage. Thus threshold voltage differences in M_1 and M_2 cancel out. Voltage differences as low as a few millivolts have been detected with this circuit [Yee 78].

After ϕ_1 is lowered, ϕ_2 is raised. As with the clocked differential sense amplifier in Section 5.5, ϕ_2 must be ramped up slowly for maximum sensitivity.

The analysis of this circuit follows closely the analysis of the clocked sense amplifier. We will do only a small signal analysis. Equating the capacitor charging currents, we have

$$-v_{X2}C_L = v_2 C \tag{5.104}$$

and

$$-v_{X1}C_L = v_1 C, \tag{5.105}$$

where C, the total capacitance drain nodes, is defined as

$$C \equiv C_A + C_B. \tag{5.106}$$

We also have

$$C\frac{dv_1}{dt} = -g_m(v_2 - v_{X1}) \tag{5.107}$$

and

$$C\frac{dv_2}{dt} = -g_m(v_1 - v_{X2}). \tag{5.108}$$

Combining, we obtain

$$\tau_A \frac{d\Delta v}{dt} = \Delta v, \tag{5.109}$$

where

$$\Delta v \equiv v_2 - v_1 \tag{5.110}$$

and

$$\tau_A = \frac{C}{g_m \left(1 - \frac{C}{C_L}\right)}. \tag{5.111}$$

Note that Eq. (5.109) approaches Eq. (5.97) for $C_L \gg C$. The initial voltage gain from $v_{X2} - v_{X1}$ to $v_2 - v_1$ is $-C_L/C$.

If the inputs v_{X1} and v_{X2} are close in voltage, the maximum final difference voltage is on the order of V_{TE}. For this reason, this circuit is generally viewed as a preamplifier. It is typically combined with the clocked sense amplifier to provide a larger final voltage separation. The sequence of operations is thus

ϕ_1	HIGH	Precharge v_{X1} and v_{X2}.
ϕ_1	LOW	Observe some external circuit imbalances v_{X1} and v_{X2}.
ϕ_2	HIGH	Preamplify voltage difference.
ϕ_3	HIGH	Amplify voltage difference.

The final sense amplifier and timing diagram is illustrated in Fig. 5.49.

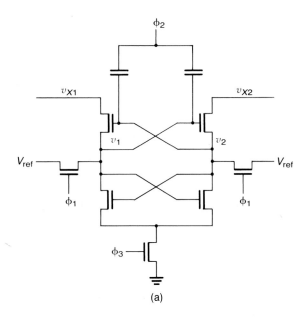

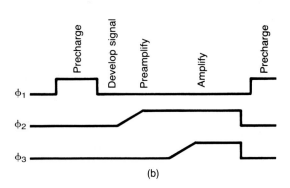

FIGURE 5.49   Schematic (a) and waveforms (b) of combined differential charge and clocked sense amplifier.

(a)

(b)

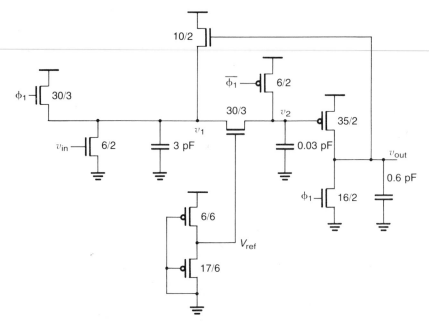

FIGURE 5.50   CMOS charge-sharing sense amplifier with fast recovery feedback transistor.

EXAMPLE 5.3

Design of a ROM sense amplifier

Figure 5.50 illustrates a single-ended CMOS charge sharing sense amplifier,[8] in the style of Section 5.3. It is executed in the example CMOS process. It might be used, for instance, on the output of a large ROM. The noise margin of this circuit was designed to be 0.5 volts. In the layout of this circuit one would need to carefully minimize the noise coupling to the voltage reference. Figure 5.51 illustrates simulated voltage waveforms. Note that shutting off the p-channel precharge transistor pumps the precharge node a few tenths of a volt above V_{DD}. Since this sort of precharge circuit is so common, processes must be designed so that latchup does not occur under these conditions.

There are many other ways to build a ROM sense amplifier, including the use of the differential sensing techniques of Section 5.5. There are two basic alternatives for generating the reference voltage. One could use a fixed voltage reference as was done in the circuit in Fig. 5.50 (a good idea), or one might place an extra column (bit line) in the ROM. The theory is that by making this extra line fully populated, it could be made to fall slower than any of the others. The voltage

---

[8] Designed by C. Bamji.

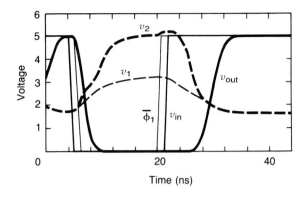

FIGURE 5.51   Voltage waveforms of circuit in Fig. 5.50.

waveform might look like those illustrated in Fig. 5.52. Both slow and fast cases are shown. If we examine the slow (solid) waveforms, we see that at time T it is quite reasonable to sense whether the selected bit line is reading a 1 (v_{bit1}) or a 0 (v_{bit0}). The limitation of this technique is that, for fast chips, performing the sense operation at time T is a poor strategy since both v'_{bit0} and v'_{ref} will have collapsed to ground.

There are certainly ways to work a little harder (a lot harder?) and make this scheme work, but there is a lesson to be learned here. In this application, if we use a fixed reference, we need concentrate only on the slow process corner. If, on the other hand, we use a dynamic reference of the type illustrated in Fig. 5.52, then we end up designing in a box. In general, using a dynamic reference voltage is not bad; what is bad is to get into a situation in which one must simultaneously solve two complex, coupled problems. This is always much, much harder than

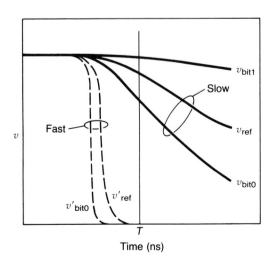

FIGURE 5.52   Waveforms for sense amplifier with dummy line. It is difficult to make a circuit work under both fast and slow process corners.

solving two uncoupled problems. A global design strategy is then to partition a problem in such a way that one need only worry about one thing at a time—*divide and conquer!* ∎

5.6   Level conversion

Having decided on a circuit methodology and valid voltage levels that one will use for the major parts of a design, one is then confronted with the exceptions. There are a variety of places, both internal and external to the chip, that the voltage levels are incompatible with the basic methodology. One must therefore be able to shift the valid voltage levels. One is often required to amplify the signal as well. The most common case is converting a TTL signal to 5 volt MOS logic levels.

The specified valid input voltage levels for TTL are $V_{IL} = 0.8$ volts and $V_{IH} = 2$ volts. There are two problems here for MOS—the low is too high and the high is too low. Let us examine the high level first. In depletion load nMOS, a 2 volt V_{IH} will require a very large β ratio. By itself, this is not a problem. The only tricky point is that since V_{IH} is so low, the worst-case V_{TE0} may be high rather than low. This was discussed in Section 4.4.

The second problem in TTL conversion is that the V_{IL} can be above the enhancement-mode transistor threshold. This means that the high output voltage level out of the high β-ratio inverter will not be at V_{DD}. A second stage is required to fix the V_{IL} level. Thus a simple TTL compatible input consists of two inverters. The first high β-ratio inverter fixes the V_{IH} level and the second inverter fixes the V_{IL} level. If the "standard" process files do not have a specific TTL worst-case file, the worst-case noise file may need to be tweaked for the correct worst-case V_{TE0}.

Another typical problem is lowering a signal centered too close to V_{DD}. For example, in Section 5.3, the output of the charge sharing amplifier is centered fairly high. In this case, a source follower is helpful. The classical source follower is illustrated in Fig. 5.53. While the transistor is in saturation, its current is proportional to $(V_{GS} - V_{TE})^2$. For the drain-to-source current to remain constant, $V_{GS} - V_{TE}$ must be constant. In this regime of operation, we have

$$v_{\text{out}} = v_{\text{in}} - V_{TE0} - \sqrt{\frac{(1+\delta)I}{K'S}}. \tag{5.112}$$

Figure 5.53 also illustrates the transfer function of a simple source follower configuration. A current mirror can be used to bias the bottom transistor. The current mirror transistors should be long to minimize

FIGURE 5.53   Static transfer characteristics of a source follower.

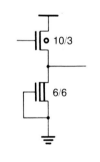

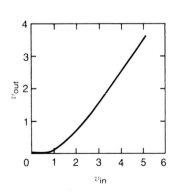

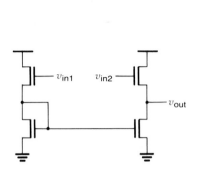

FIGURE 5.54   A differential to single-ended converter.

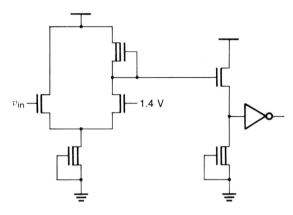

FIGURE 5.55   A TTL input buffer.

processing variations. Figure 5.54 illustrates a technique, common in analog MOS circuit design, for performing double-to-single-ended conversion. The dynamics of the source follower are very good at high frequency because C_{GS} acts to lower the input capacitance and feed the signal forward.

Sense amplifier techniques (see Section 5.5) can also be used to either level shift a differential signal or to compare it to a reference voltage. Figure 5.55 illustrates a simple TTL input buffer using a sense amplifier.

5.7   Input protection

The processing of a VLSI circuit is optimized for the voltages seen internal to the chip, that is, voltages on the order of 5 volts or less. At the interface to the chip, voltages almost three orders of magnitude higher are encountered. Input protection circuitry, often called "lightning arresters," must be designed to protect the chip. These circuits are difficult to design and analyze, and have undergone continual evolution.

Figure 5.56 illustrates a simple electrostatic model of a human being. It consists of a 100 pF capacitor, charged to 2000 volts, in series with a 1.5 kΩ resistor. There are many variations on this model. For instance, some manufacturers test for input protection using a capacitor charged to 300 volts, without a series resistor. Take the case of a finger approaching a MOS chip. When some distance away, the charged finger will arc to the chip. When the voltage is low, say 300 V, the finger will need to approach the chip very closely, and the capacitance will be high, say 300 pF. When the voltage is high, the arc will occur further away and the capacitance will be lower. The discharge will also occur over

FIGURE 5.56   An electrostatic model of a person. (a) Physical configuration; (b) Equivalent circuit.

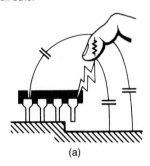

(a)

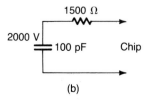

(b)

a larger area of the finger and this will affect the resistance. Our goal is to design a circuit that will enable a person to touch the leads of a packaged chip without ruining the part.[9]

Clamping and filtering the discharge

At 2000 volts, the energy stored in the capacitor ($\frac{1}{2}CV^2$) is enormous: 0.2 mJ. There are several ways in which the static discharge can ruin a MOS device. We will discuss only two. The most obvious mechanism is gate oxide breakdown due to an excessive applied electric field. The critical field for silicon dioxide is about 7×10^6 V/cm. Thus the highest voltage a 300 Å gate can withstand, assuming a perfectly uniform electric field, is only 21 V. The second destructive mechanism is the thermal burnout of p–n junctions or polysilicon links. The heat capacity of silicon is roughly 0.7 J/(g·K). To get a feeling for how much energy is stored in the "human" capacitor, we compute the volume of silicon it can melt. Silicon melts at 1415° C. If we begin with the silicon at 15° C, we are considering a temperature rise of 1400° C. From before, we have an energy of 0.2×10^{-3} J, external to the chip. Assuming half of this energy is dissipated in the silicon, we compute that a volume of 45,000 μm^3 of silicon can be melted[10] by a single static discharge.

The most popular input protection structure consists of a length of n+ active area terminated in the drain of an enhancement-mode transistor, with its source and gate connected to ground. The input n+ line has a typical series resistance of 500 to 1000 Ω, while the transistor has a width between 80 and 100 μm. A lumped equivalent circuit for this structure is illustrated in Fig. 5.57. This structure has been shown to be reasonably effective.

The n+ input resistor is modeled here by a pi structure. It is actually a distributed RC and diode line. This structure has the dual purpose of acting as a low-pass filter to input spikes, and clamping the input to no less than a diode drop more negative than the substrate. The low-pass filter must be designed so it will not impinge on the necessary bandwidth of the input signal. Otherwise, the lower the cutoff frequency the better. The enhancement-mode transistor also has several functions. First, it clamps the internal voltage to no less than a threshold drop more negative than ground. For cases when the chip is running with a negative V_{BB} voltage, this transistor, rather than the p–n diode, can provide the first level of clamping. The second purpose of the transistor is to limit the positive range of the internal voltage. The two mechanisms whereby this can occur are punch-through and avalanche breakdown.

---

[9] "It is difficult to make something foolproof because fools are so ingenious." [Anon.]

[10] This is only a rough calculation. The specific heat is assumed independent of temperature, and heat of transition is neglected. Nevertheless, it should give one a new respect for the problem!

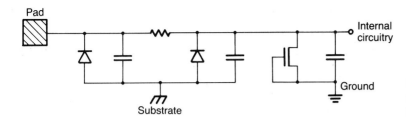

FIGURE 5.57   Circuit model for an nMOS input protection device.

The problem is that the punch-through voltage depends heavily on the effective transistor length L. The junction depth, through two-dimensional effects, also influences V_{PT}. Thus, for short channel devices (those with acceptably low punch-through voltages for protection), the punch-through voltage tends to be too uncontrolled on most processing lines.

Under normal conditions, the avalanche breakdown of a p–n junction occurs at voltages too high to be helpful for input protection (on the order of 40 V). This voltage can be lowered by the presence of the grounded transistor gate. The gate crowds the reverse bias electric field in the p–n junction, causing avalanching to occur at lower voltages. This effect is extremely repeatable, and can be made to cause avalanching around 15 to 17 V. To make the avalanche mechanism dominate over punch-through, the gate of the transistor must be drawn long. The only drawback of this technique is that the breakdown occurs near the surface of the device and hence the protection transistor must be drawn wide to dissipate the required energy without burning up.

When the input protection fails, the part of the p–n junction diode near the input typically burns out. This problem is particularly bad in CMOS processes with an epi layer because the substrate has a very low resistance. There is therefore little to limit the maximum current. Some designers have opted for using polysilicon input protection resistors. This structure does not have the junction burnout problem of the n+ structure. On the other hand, the fact that the poly is suspended over the substrate on a poor thermal conductor causes extremely serious problems. The poly must be drawn very wide to increase its total heat capacity. When this is done, this structure has also proved effective. However, there is not nearly as much experience as with the n+ resistor technique.

Another protection structure is the metal gate field transistor. (These always occur and are normally considered parasitic devices.) Figure 5.58 illustrates such a device schematically. When it is working in the presence of a large positive input transient, it must be operating in its linear region. We have

FIGURE 5.58   Input protection device using a thick field metal gate transistor.

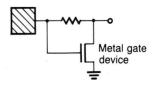

Metal gate device

$$v_{\text{internal}} = \frac{v_{\text{spike}}}{1 + 2K_F' SR(v_{\text{spike}} - V_{TF})}. \tag{5.113}$$

To be effective, the shape factor S must be on the order of 100 or more. K'_F refers to the field device and its gain is severely impacted by fringing fields. These field transistors are not self-aligned. In these structures the breakdown of the CVD oxide can be a problem.

On outputs or clock inputs, the series resistance is not used because one cannot tolerate the performance degradation that would result. For clocks, the normal input enhancement-mode transistor must be drawn over 200 μm wide to provide additional protection. On outputs, the output transistors normally provide protection. Minimum total channel width is typically 500 μm.

Additional considerations for CMOS

For CMOS, the input protection circuit can be an inducer of latchup if suitable guard structures are not used. For example, consider the protection circuit discussed in Section 5.7, which uses for the clamping device an n-channel enhancement transistor with gate and source grounded. Assume further that the process is n-well CMOS.

For a positive-going discharge transient, the transistor enters the avalanche mode. Pairs of holes and electrons will be generated by the avalanche process and a portion of the discharge current will be carried by these carriers. The electric field in the depletion region around the source and drain regions will force most of the electrons toward the drain or source, where they will be collected and flow out to the power lines or signal nodes.

The majority holes will flow through the substrate and out through the contacts that connect the substrate to the substrate bias, which is usually at ground potential for CMOS. This majority current will cause the voltage in the substrate in the vicinity of the protection circuit to rise above ground.

If nearby n+ sources, or drains at ground potential, see a substrate voltage in excess of approximately 400 mV, that source-to-substrate junction will be forward-biased enough to cause forward conduction and a resulting injection of minority electrons into the substrate. A nearby well containing a p-channel device can collect these electrons, and if the pnp bipolar transistor in this well then turns on, a loop formed by the original n+ source (which is part of a lateral npn bipolar transistor) can be excited into latchup.

A negative-going discharge on the input to the protection circuit will be clamped by two mechanisms. Consider first the MOS diode clamping mode. As in the nMOS example, as the drain is pulled negative, it acts as a source with the transistor in a source-follower configuration. This clamps the drain (now the "source") to a threshold drop below ground. Now consider the p–n junction clamping mode. In CMOS, the substrate is usually grounded, rather than biased to a negative potential as with nMOS. Therefore the n+ drain junction will become forward-biased with respect to the substrate. Current flow here will be even larger than for the first mechanism, with the result that the p–n junction clamping mode will usually dominate.

The problem with this second clamping method is that the entire clamping current consists of minority electrons that are injected into the substrate. The diffusion length (or effective lateral diffusion length for an epi process) will be on the order of 100 μm, with the result that these minority electrons can perturb the chip for distances of several hundred microns from the protection circuit.

If there is a potentially latchable pnpn loop within this rather large radius, and these carriers are collected by the well of the pnp transistor in that loop, then that loop can be driven into latchup. (The pnp vertical well transistor would go on first, which would subsequently drive an adjacent lateral npn transistor on.)

Thinking about this example, we see that both positive and negative discharges (either is possible) will cause substrate disturbances that can latch susceptible latch loops even several hundred microns away. The currents that flow during the discharge can be huge—the 1500 V "human" capacitor discharging through approximately 1500 Ω will cause a peak current of 1 A.

Consider a positive-direction discharge, where a portion of the current is avalanche-generated majority current through the substrate to its V_{BB} or ground contact. Assume that a third of the total discharge current appears as avalanche-generated majority substrate current. A typical vertical substrate resistance of 200 Ω (for an epi substrate) that this clamp current flows through (in the positive avalanche case) would result in a 60 V rise in the substrate voltage. Of course, the rise would not be that great since other n-well and n+ source regions tied to V_{DD} would forward-bias to clamp the substrate to V_{DD}. The minority carriers then generated by these forward-biased structures would increase the conductivity of the substrate; unfortunately, these minority carriers could also induce latchup. At any rate, the disturbance would be drastic and could extend over a radius of hundreds of microns.

The solution to this problem is to attenuate the ESD-generated latching disturbances. Two types of guard structures are usually used for attenuating the large disturbing currents that can exist in the substrate during the transient. (Also refer to the latchup discussion in Section 2.3.) One technique is minority carrier attenuation. A stripe of well that totally surrounds the protection cell would collect minority electrons that are emitted by a negative-going n+ emitter. As discussed in Section 2.3, this type of collector is so effective an interceptor when epi is used that the minority currents are attenuated by a factor of four to six orders of magnitude [Troutman 83]. If the process is not an epi process, this well attenuates by only one to two orders of magnitude. The guard

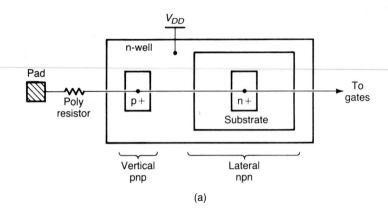

(a)

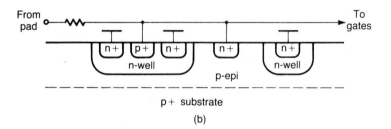

(b)

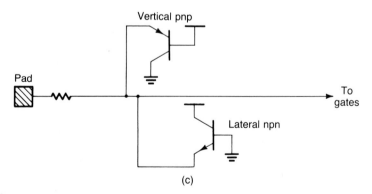

(c)

FIGURE 5.59   CMOS ESD clamp using p+ and n+ diffusions. Clamping is accomplished by p–n junctions to V_{DD} and ground. The two junctions become vertical (pnp) and lateral (npn) bipolar transistors, rather than simple diodes as seen in nMOS. (a) Top view of layout, (b) cross section, (c) equivalent circuit.

well might be made wider to sufficiently stop minority carriers that are passing deeply under the well without being collected (carriers can easily pass under the well since the wafer is about 500 μm thick, whereas the intercepting well is only 3 to 5 μm deep).

The other technique is majority carrier attenuation. With an epi process, the highly doped substrate beneath the epi results in rapid attenuation of the disturbance with distance. For a non-epi process, the substrate surface near the cell should be tied to ground (or V_{BB}) by a contact strip that encircles the cell. This surface contact will prevent the substrate voltage from rising to the point where nearby circuits go into latchup.

A general conclusion about protection circuits with CMOS can be deduced from these scenarios. If the electrostatic discharge occurs while the chip is operating in its system environment, then even though the protection circuit might sufficiently absorb the energy of the discharge, a portion of the surrounding circuit might be left in a latched condition. If the circuit is not destroyed, it can be unlatched only by removing and then reapplying the chip power. To prevent latchup in this situation, additional guard structures are needed to attenuate disturbances that will travel through the substrate to potentially latchable pnpn loops elsewhere on the chip.

CMOS permits more flexibility than nMOS in the design of protection circuits because of the existence of the complementary p-channel device. Another protection circuit sometimes used is shown in Fig. 5.59. A polysilicon resistor absorbs most of the discharge energy. Clamping is done with p–n junction clamps. The presence of p+ diffusion forms a "diode" with the well, which actually becomes a vertical pnp transistor with the well as the base and the substrate as the collector. The n+ to substrate diode becomes a lateral npn transistor once the well collector guard is placed around it to capture minority electrons emitted into the substrate by the discharge. The n+ cathode becomes the npn emitter, the substrate becomes the base, and the well collector guard becomes the npn collector.

5.8   Substrate-bias generation

The parametric generation of voltages that are outside the range provided by external power supplies is an important extra degree of freedom that is often used in MOS circuitry. Voltages can be generated that are above V_{DD} or below ground. By far the most popular application is the generation of a negative V_{BB} voltage on nMOS chips [Cheng 80, Huffman 77, Lee 79, Martino 80, Pashley 76, Puri 78].

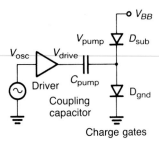

FIGURE 5.60 Idealized model of a V_{BB} generator.

FIGURE 5.61 Ring oscillator (a) and Schmitt trigger oscillator (b).

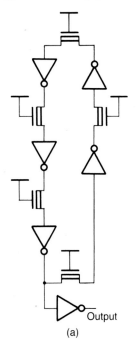

(a)

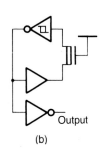

(b)

Other applications include V_{GG} generation and high-voltage drivers for displays. Bootstrapping is a simple example of parametric voltage conversion.

The use of a negative V_{BB} voltage has several advantages over a grounded substrate. It lowers the sensitivity of threshold voltages to the body effect, increases punch-through voltages, lowers the diffusion-to-substrate capacitance without requiring a decrease in substrate doping, and increases the gain of typical inverters. It also lowers subthreshold leakage on clocked depletion transistors and protects the chip against forward-biasing of the substrate due to voltage undershoots at the inputs, which are common with TTL peripherals. If feedback is provided, the V_{BB} voltage can compensate for some device parameter variations.

A negative V_{BB} voltage is generated by pumping electrons out of the ground node and into the substrate. Because the substrate is a giant capacitor (on the order of 1000 pF), this slowly lowers its voltage. A canonical V_{BB} generator configuration is shown in Fig. 5.60. The driver amplifies the ac signal generated by the oscillator and powers the charge pump. The power is coupled through the capacitor C_{pump}. The two diodes D_{gnd} and D_{sub} gate the charge out of the substrate and into the ground node. When the voltage V_{pump} is near its peak value, then D_{gnd} is forward-biased and charge is being pumped into the ground node. The other diode D_{sub} is off. On the other half of the cycle, V_{pump} is near its most negative value and D_{gnd} is off, while D_{sub} drains charge out of the substrate. Of course, charge does not flow across the capacitor.

The oscillator

Two circuit forms are used to implement oscillators for substrate-bias pumps. One is the ring oscillator. For an N stage ring oscillator, where N is an odd number, the oscillator frequency f is approximately $(2N\tau_{\mathrm{delay}})^{-1}$, where τ_{delay} is the delay of a single stage. This relation is only approximate because the loading on all gates is not equal. N should be greater than or equal to 5; otherwise, the voltage swings may be too small. A Schmitt trigger wired in a loop with an RC filter can be used instead of a ring oscillator. Both the ring oscillator and Schmitt trigger configurations are shown in Fig. 5.61. Typical oscillator frequencies range from 5 to 10 MHz. As a practical matter, the oscillator often frequency locks to the system's clock.

The charge pump diodes

The theoretical minimum value of V_{BB} is determined by the peak-to-peak value of V_{pump} and the voltage drops in the two diodes. During the high part of the cycle, V_{pump} must be one diode drop (diode D_{gnd}) above ground to pump charge. On the low side of the cycle, V_{pump} must be a diode drop (diode D_{sub}) below V_{BB} to do any work. Assuming

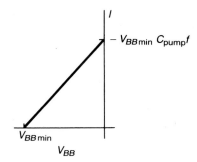

FIGURE 5.62   Ideal I–V characteristics of a V_{BB} generator.

the maximum peak-to-peak voltage of V_{pump} is less than V_{DD}, then $V_{BB\,\text{min}}$ is greater than or equal to $-V_{DD}$ plus two diode drops. Because of leakage currents and parasitic capacitances, such an ideal value is rarely achieved.

Due to the constraints imposed by nMOS technology, D_{gnd} is generally implemented as an enhancement-mode transistor with gate and drain tied together. D_{sub} may be implemented in this fashion, or the p–n diode naturally provided by the technology can be used. If a transistor[11] "diode" is used, the substrate must be connected externally, usually by a drop bond that is inside the chip package and connects the V_{BB} pad to the substrate back connection. When this is done, care must be taken to ensure a good contact between the back of the chip and the drop bond. Note that the p–n diode is always there.

If we had ideal diodes, the current pumped out of the substrate would be

$$I_{\text{avg}} = \Delta V C_{\text{pump}} f, \qquad (5.114)$$

where ΔV is the difference between the substrate voltage V_{BB} and its theoretically optimum value of $V_{BB\,\text{min}}$ (around $2V_T - V_{DD}$). Note that the current is large when ΔV is large—that is, when the pump first starts up. The pump must work at all V_{BB} voltages between zero and its minimum value.[12] I_{avg} increases with the pumping rate f. This ideal response is illustrated in Fig. 5.62.

Equation (5.114) is valid when all of the charge represented by $\Delta V C_{\text{pump}}$ is actually pumped out of the substrate. This is always true for ideal diodes, but for real devices the finite resistance of the forward-biased device severely limits the effectiveness of the charge pump. The principal question is whether the RC time constant of the diodes and pump capacitor is larger or smaller than $1/f$. If the time constant is short (ideal diodes), Eq. (5.114) prevails and the $v_{\text{pump}}(t)$

---

[11] Note that the threshold voltage of this transistor is low because V_{SB} is always zero.

[12] In some nMOS technologies, short channel devices will actually punch-through at voltages below V_{DD} when $V_{BB} = 0$.

waveform is clipped. If the time constant is long, as it will be when V_{BB} approaches $V_{BB\,\min}$, $v_{\mathrm{pump}}(t)$ is roughly sinusoidal.

We will now study this second regime of operation. Focusing our attention on D_{gnd}, we see that current will flow only if $v_{\mathrm{pump}} > V_{TE0}$. In the region around the top half of the cycle, $v_{\mathrm{pump}}(t)$ is well approximated, using a Taylor expansion around the maximum value, as

$$v_{\mathrm{pump}} \approx V_o - \frac{V_o}{2}\frac{\Theta^2}{2}, \tag{5.115}$$

where V_o is the peak-to-peak value of v_{pump} and $\Theta = 2\pi f t$. We define the maximum value of $v_{\mathrm{pump}} - V_{TE0}$ as ΔV. ΔV is zero when V_{BB} reaches its theoretical minimum value. The angles at which v_{pump} exceeds V_{TE0} are between $\pm 2\sqrt{\Delta V/V_o}$, as seen in Fig. 5.63. The current through D_{gnd} is given by

$$i(t) = K'S\left(\Delta V - \frac{V_o\Theta^2}{4}\right)^2 \quad \text{for } -2\sqrt{\frac{\Delta V}{V_o}} \le \Theta \le 2\sqrt{\frac{\Delta V}{V_o}}. \tag{5.116}$$

Integrating to find the charge pumped in one cycle, we have

$$Q = \int i(t)dt = \frac{2K'S}{2\pi f}\int_0^{2\sqrt{\Delta V/V_o}} \left(\frac{1}{4}V_o\Theta^2\right)^2 d\Theta \tag{5.117}$$

or

$$Q = \frac{2}{5\pi f}K'SV_o^2\left(\frac{\Delta V}{V_o}\right)^{5/2}. \tag{5.118}$$

Note that if v_{pump} were a square wave rather then a sine wave, we would then have

$$Q' = \frac{K'SV_o'^2}{2f}\left(\frac{\Delta V}{V_o'}\right)^2. \tag{5.119}$$

FIGURE 5.63   Voltage waveforms. The top of the sinusoidal $v_{\mathrm{pump}}(t)$ is approximated by a parabola.

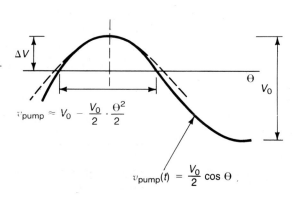

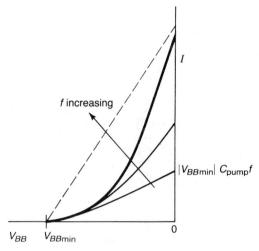

FIGURE 5.64   Characteristics of a nonideal V_{BB} generator.

The average current pumped per cycle, from Eq. (5.118), is

$$I_{\text{avg}} = Qf = \frac{2}{5\pi} K' S V_o^2 \left(\frac{\Delta V}{V_o} \right)^{5/2}. \qquad (\mathbf{5.120})$$

Compare Eq. (5.120) with Eq. (5.114). The average rate of change of the substrate voltage is given by $I_{\text{avg}}/C_{\text{sub}}$, where C_{sub} is the capacitance of the substrate. One can roughly model the effect of I_{avg} on v_{pump} by thinking of it as a load "resistance." We have

$$R_{\text{load}} \approx \frac{\Delta V}{I_{\text{avg}}}. \qquad (\mathbf{5.121})$$

A major difference between this analysis and the one leading to Eq. (5.114) is that in the analysis leading to Eq. (5.114) the waveform was assumed to be clipped, while in this analysis it was not. Indeed, we assumed in this analysis no dependence of $v_{\text{pump}}(t)$ on $i(t)$. This accounts for the different frequency dependences of the two cases. Equations (5.114) and (5.120) account for the limitations of the maximum current out of the charge pump. Figure 5.64 illustrates the general behavior of the charge pump.

The coupling capacitor

The coupling capacitor C_{pump} transfers power from the driver into the charge pump, where it does the work of moving electrons from a higher potential to a lower one. Figure 5.65 illustrates a V_{BB} generator with the important capacitors explicitly denoted. C_{bad} is a parasitic capacitor

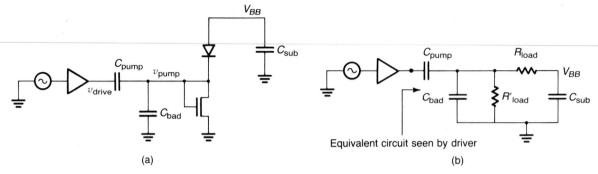

FIGURE 5.65 Models of a V_{BB} generator showing (a) important capacitances and diode implementation and (b) small-signal model.

that acts to lower V_o and hence the minimum value of V_{BB}. Because $C_{\text{sub}} \gg C_{\text{bad}}$ and $2\pi f C_{\text{pump}} \gg G_{\text{load}}$ when $V_{BB} \approx V_{BB\,\text{min}}$, we have

$$v_{\text{pump}} \approx \frac{C_{\text{pump}}}{C_{\text{pump}} + C_{\text{bad}}} v_{\text{drive}} + V_{\text{offset}} . \tag{5.122}$$

If the peak-to-peak voltage of v_{drive} is V_{DD}, we then have

$$V_o \approx \frac{C_{\text{pump}}}{C_{\text{pump}} + C_{\text{bad}}} V_{DD}. \tag{5.123}$$

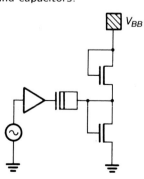

FIGURE 5.66 The use of MOS transistors as diodes and capacitors.

We clearly want to minimize C_{bad} so as to decrease $V_{BB\,\text{min}}$. To make V_o large, we require $C_{\text{pump}} \gg C_{\text{bad}}$ and $C_{\text{pump}} \gg G_{\text{load}}/2\pi f$. To minimize C_{bad} we want to lay out C_{pump} as a large square depletion-mode transistor with source and drain connected together. Ideally, we want the transistor's gate to be connected to the charge pump. This is usually not possible because when $v_{\text{drive}} = V_{DD}$ and $v_{\text{pump}} = V_{TE0}$, the depletion transistor is below threshold. If this occurs, the channel disappears and the value of C_{pump} drops several orders of magnitude. Thus C_{pump} is usually built as illustrated in Fig. 5.66, with the gate connected to the driver. One must also take care that the length of the C_{pump} transistor does not get too long. Because the time constant of this transistor is proportional to its length squared, it is easy to make the geometry of the transistor such that it is too slow to respond to the clock. Note that unless great care is taken, this problem may not be discovered in circuit simulation. A typical size for C_{pump} is 200 μm $\times$ 200 μm. Thus if a 2 μm long transistor has a τ of 0.2 ns, a 200 μm long transistor would have a τ of 2 μs. Putting a couple of contacts internal to the transistor can solve this problem.

The design of the diode shape factor S also involves a trade-off. As S increases, the current in Eq. (5.118) increases, but V_0 decreases because of the increase in the parasitic capacitance C_{bad}.

The driver

A close inspection of Fig. 5.65 shows that, fortunately, the driver does not need to drive an equivalent load of C_{pump}. The driver should be swinging from rail to rail when $\Delta V \approx 0$. Under these conditions, R_{load} is very large and the equivalent load of the driver is C_{pump} in series with C_{bad}, or approximately C_{bad}. When the pump just starts up, $\Delta V \approx V_{DD}/3$ and the minimum differential resistance is approximately $(2K'S\Delta V)^{-1}$. This resistance is in parallel with C_{bad} and in series with C_{pump}. Since this resistance is low at startup, the initial load is approximately C_{pump}. For once, nature is nice to us, and the large load is seen at the beginning when a large swing is not important. The small load is seen later when a large swing is important.

In the steady state, the average current I_{avg} pumped by the bias generator equals the leakage current into the substrate. This leakage is caused primarily by minority carrier injection into the substrate from saturated load devices (hot electron effects). A typical current is 5 μA.

Because the substrate-bias voltage affects several electrical parameters on the chip, such as power dissipation and threshold voltages, feedback can be used to control these parameters via the bias generator. Some feedback is also there naturally because, for instance, the V_{BB} voltage affects the minority carrier injection into the substrate, which in turn affects the V_{BB} voltage. As in all feedback systems, stability can be an issue. Temperature also affects the V_{BB} voltage. At low temperature, the "diodes" are better, leakage currents are less, and the V_{BB} voltage is generally more negative. In general, the natural dependence of V_{BB} on temperature hurts the voltage threshold control but tends to stabilize the power in depletion-load devices. Note that when hot electron effects are dominant, the substrate current can be highest at low temperature.

Layout considerations in the charge pump

The performance of a V_{BB} generator depends on many layout considerations. We have, for instance, already discussed the minimization of $C_{\text{bad}}/C_{\text{pump}}$. Since small potential drops are a problem, a good ground connection to the generator is needed. Substrate-bias generators are usually placed near a ground pad. Metal straps to minimize voltage drops in the diffusion and poly regions might also be required.

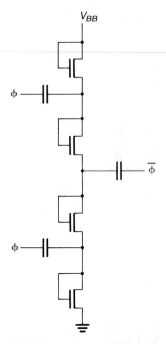

V_{BB}

ϕ

$\overline{\phi}$

ϕ

FIGURE 5.67 A totem pole implementation.

Many issues also affect the choice between a bipolar p–n diode for D_{sub} and a MOS transistor and drop bond. Both types are used on commercial chips. While the p–n diode approach does not require an external bond, it has the potential problem of injecting electrons into the substrate near active circuitry. When this circuitry depends on the storage of dynamic charge, this charge may be annihilated. Diffusion guard bands connected to V_{DD} can help some, but the aspect ratio is a disadvantage. The substrate is about 250 μm thick, while the guard bands are only about 1 μm deep. Besides these electrons there are, after all, the ones that are supposedly pumping down the voltage of the substrate. If we did manage to collect them, we would have neatly defeated ourselves! Another potential problem is the leakage of charge off the node V_{pump}. The thick field transistors composed of poly or metal over field oxide are very leaky devices because they are in reality very short channel transistors (poor scaling). The problem is exacerbated by the fact that the charge pump part of the circuit essentially runs with zero V_{BB}, so the field thresholds in that area of the chip are low. (The body effect on field transistors can be 50 $V^{1/2}$.) If a transistor implementation of D_{sub} is used, the charge pump can be isolated from the rest of the circuitry by using a grounded gate thick field poly over field transistor to isolate the entire region. The poly is connected to the substrate. An external capacitor is sometimes used to stabilize the V_{BB} voltage against noise.

The circuits we have shown can be generalized to provide even lower V_{BB} voltages [Dickson 76]. The general totem pole structure is shown in Fig. 5.67. Keeping the total pump capacitance the same as before, we can compare the old and new designs. Figure 5.68 illustrates the two load lines for these devices. For back gate voltages near –3 volts, we find that the old design tends to be badly resistance limited, while the new design does not. With the new design, the actual V_{BB} voltage is regulated in a feedback loop that starts and stops the oscillator. This is usually a nicer design. The totem pole structure could also be used to provide large positive voltages.

FIGURE 5.68 Comparison of two implementations with equivalent capacitance. At the dotted line the totem pole design is better.

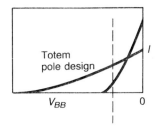

Totem pole design

V_{BB}

I

0

5.9   Layout considerations in circuit design

The principal layout concerns of the circuit designer are the management of capacitive and resistive wiring parasitics on signal lines and the resistance and current-carrying capacity of the power busing. At a global level, the two major techniques one employs to control signal parasitics are block placement and logic modifications. By minimizing wire lengths, one typically minimizes both chip area and wiring parasitics. Floor planning is the exercise of planning the flow of signals, power busing, and logic block placement on a VLSI chip. Floor planning is to layout as functional block diagrams are to logic design.

Floor planning

A VLSI circuit is normally composed of functional blocks. The signals passing between these blocks will represent global wiring when the functional blocks are mapped into hardware. By the correct placement of these blocks, the interconnect wiring can be minimized. In many cases, the geographical assignments of a module on a chip are straightforward. The register file in a microprocessor should be close to the arithmetic logic unit (ALU). The next state logic should be near, or even in, the microcode PLA. In other cases, the placement of a register or functional block is not so clear. Does the instruction register (IR) belong in the data path or in the control section of the chip? Literal instructions must be piped to the data path, while most other instructions influence the control of the machine. The approach in these cases is to count wires. If we place the IR in the data path, how many wires must go to the control section? What if we place the IR in the control section?

Local decoding of signals is often used to save wires. A generalization of this comes about when some function is dependent on information in two different blocks. Consider, for instance, an ALU function that is dependent on both the ALU result and the next microinstruction word. We could take the ALU result, bring it over to the control block, and use it to determine the next microinstrution. That microinstrution would be decoded to control the ALU. This is a straightforward, clean approach. A more efficient technique is to use the ALU result as an input to control logic that actually resides in, or very near, the ALU. The microcode provides an "operate" instruction that controls whether or not that local control is activated. This can save an enormous number of wires and shorten the critical path as well. Figure 5.69 illustrates an example of this technique applied to the selection of a register from the register file. The IR, which in this example is part of the register file, contains a field specifying which register is to be selected. The SELECT

FIGURE 5.69 Register file with decoder. The selection of the register number is made locally.

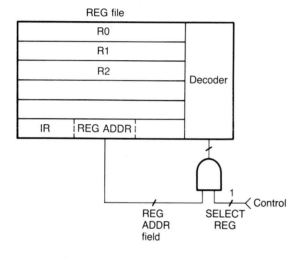

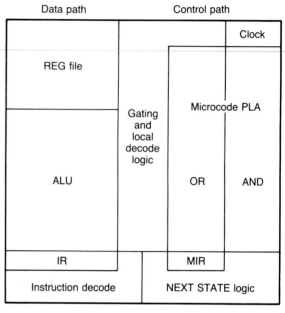

FIGURE 5.70 Floor plan of a simple microprocessor.

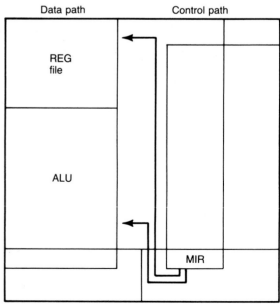

FIGURE 5.71 Routing the data-path control wires.

REG wire from the control section of the chip tells the selector only whether or not to select any register. Local logic determines the register. The microcode never has to know which register is actually selected. If there were thirty-one registers, the number of control wires would be reduced from five to one.

Another technique to save wire is to duplicate logic. Consider, for example, the microcode PLA section of the microprocessor shown in Fig. 5.70. The output of the PLA is latched into the microinstrution register (MIR). The MIR controls many functions spread throughout the data path. Two such control points, together with the wiring, are shown in Fig. 5.71. Note that a wire must traverse the height of the chip. On the other hand, there is already a wire containing essentially the same information. It resides in the OR plane of the PLA. Figure 5.72 illustrates an alternative implementation. The implementation duplicates the MIR, but localizes the intermodule wiring.

The two major interconnect media on a chip are usually run perpendicular to each other. These would be poly and metal in the technologies we have been discussing, or metal-one and metal-two in a more advanced process. Assigning a layer to a particular direction involves many considerations. In many cases, data flows in one direction and control flows in the other. The decision of whether data or control should be run in metal is a compromise among many issues. Control is often accomplished with pass transistors. This implies that data goes in and out of the pass transistor in diffusion and control connects to the gate in poly. That suggests that poly should be used for control and data run

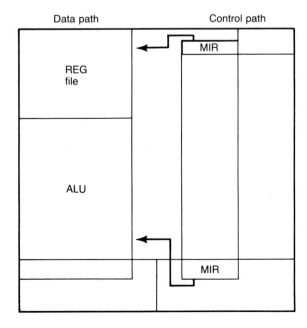

FIGURE 5.72   A better routing of the data-path control wires with a redundant MIR.

in metal, which must drop down to diffusion when it is needed. Figure 5.73 illustrates the major metal signal routing of the microprocessor shown in Fig. 5.70. Signals are not run very far in diffusion because of its high capacitance. This solution is often a good one, but there can be problems. Poly has a high resistance and, with many gates hanging from it, RC time constants can become very bad, causing skew. If the array is very wide, several solutions are possible. One could drive the array from the center, use repeaters, or switch the control function to metal.

With two layers of metal, the situation is a little different, but the philosophy is the same. In both cases, contacts must be dropped to either diffusion or poly to make a transistor. Metal-one and metal-two are not equal. In most technologies with two layers of metal, the upper layer of metal (metal-two) must contact metal-one in order to connect to poly or diffusion. In some technologies, these contacts must be staggered; that is, a metal-two to metal-one via cannot be on a metal-one to poly or diffusion contact. Even if metal-two could contact poly or diffusion directly, there would still be the region under the contact that would not be available to a metal-one wire. In all of these scenarios, a metal-two via is more expensive than a metal-one contact. A reasonable strategy is therefore to minimize the number of metal-two vias. Assume for the moment that in a data path, more contacts are required to data lines

FIGURE 5.73   Plan of major signal metal routing.

Data path   Control path

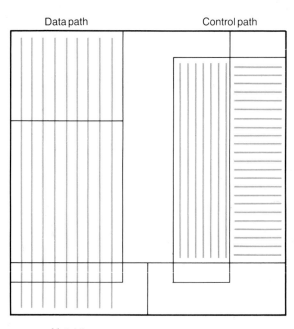

Metal lines

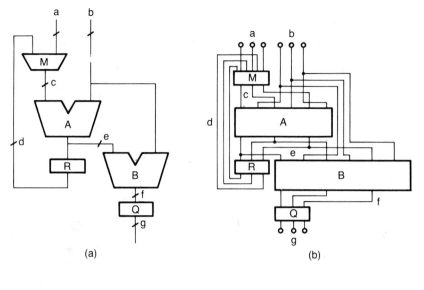

(a)

(b)

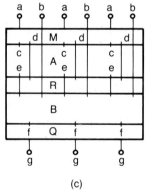

(c)

FIGURE 5.74 Transferring a block diagram into VLSI: (a) the block diagram, (b) transliteration to obtain floor plan, (c) VLSI-style floor plan.

than to control.[13] Then we would choose metal-one for data. This has the advantage that contacts to poly in the control lines can often each be used in several cells.

The translation of functional blocks to VLSI layout is not one of transliteration. Functional block diagrams tend to emphasize logic over wire and not reflect the relative sizes of modules. In Fig. 5.74(a), we illustrate a simple data-path block diagram. In Fig. 5.74(b), we show a transliteration of the function. Note all of the area taken up by wire. Layout is difficult, wire capacitances are large, and the area grows roughly quadratically with the width of the data path because it grows in both length and width. In Fig. 5.74(c), we have a VLSI-style layout. The

---

[13] This has been our experience.

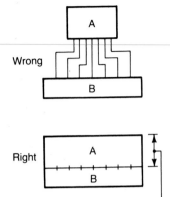

Wrong

Right

Height can probably be reduced, too.

FIGURE 5.75 The usually wrong and usually right ways of performing intermodule wiring.

layout is straightforward and signal paths are short. The area is small and grows only linearly with the width of the data path. In addition, it is easy to see how to bring control in horizontally. The only drawback is one of modularity. Primitive cells can have wires running through them that do not logically "belong."

The layout in Fig. 5.74(c) reflects two major principles. First, cells are connected by abutment rather than by running wires. This improves performance and saves area. To do this, cells are stretched so they match in pitch, as illustrated in Fig. 5.75. The second principle is to avoid end around wiring runs. Figure 5.76 illustrates this principle. It is better to run wires through the cells[14] rather than around. Wires corresponding to the same bit position are interleaved.

In Chapter 1, we saw the concept of leaving spare rows in PLAs for engineering changes. The concept of spares is also appropriate in other layout contexts. For instance, it is good practice to leave spare locations for metal wires in a data-path design. As the design of the data path progresses, uses for these locations will be found. While the uses of these well-structured spares may be fairly random exception handling, bit hacking, and so on, the general floor plan will at least be regular and well planned. Though not all of the spares will be used, the overall efficiency is improved. This technique can be extended to the more random parts of the chip. It pays not to pack those parts too tightly so that changes can be implemented.

In order for two signals to interact they must intersect. From a planning perspective, if one signal wire is to influence another, it is important that they meet somewhere. Note that if one vector of wires is to influence another in some arbitrary manner, then all "cross products" must exist. This results in an array structure. PLAs, ROMs, barrel shifters, and so on are special cases of this simple principle. Arrays are mutable structures that can be changed to fit the layout context in which they are placed. The shape of a ROM is one obvious example. Another more subtle example can be seen in the microprocessor we have been studying. We would really like to take some signals out the side of the microcode PLA OR plane so as to feed them directly into the condition logic. By taking advantage of the sparseness of the array, this can be done as shown in Fig. 5.77. The PLA is no longer quite as regular, but the overall chip may be more regular.

From an overhead standpoint, the larger an array is, the better. This way, the cost of designing such components as high-performance sense amplifiers can be amortized over many transistors. The ability to share common sense amplifiers is, for instance, one of the reasons registers on a chip tend to be located in an array.

---

[14] Care must be taken with regard to coupling noise.

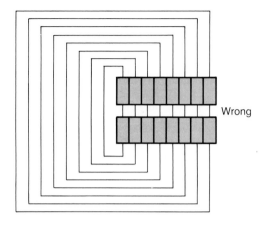

Wrong

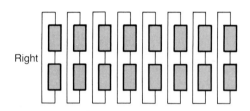

Right

FIGURE 5.76   The usually
wrong and usually right
ways of adding feedback
loops to intermodule wiring.

FIGURE 5.77   Stealing
outputs from the side of a
PLA.

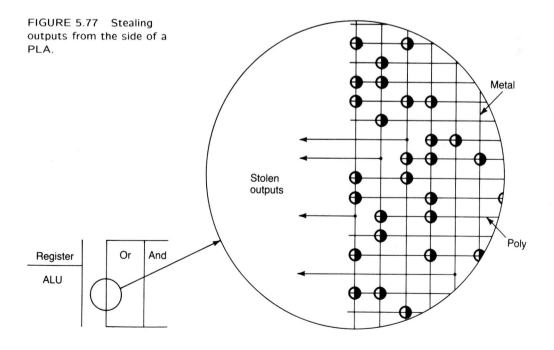

Metal

Poly

Stolen
outputs

Register

ALU

Or   And

Power distribution

One of the most important issues in chip planning is the routing of power. In technologies in which there is only one level of metal, V_{DD} and ground are routed in interdigitated trees. This is illustrated in Fig. 5.78 with the microprocessor from Fig. 5.70. Crossunders are very difficult. When necessary, these are done in low resistance interconnect (poly over buried contact over active area) with a multiplicity of contact cuts. Consider the extreme case of a crossunder that must carry 100 mA. One square of low resistance interconnect might have a maximum resistance of, say, 10 $\Omega/\square$. Thus a square crossunder would drop 1 volt. Over fifty contact 2 μm cuts to the metal on each side would be needed because of metal migration limits. Obviously, 100 mA is an awful lot of current to squeeze through a crossunder. Even 10 mA can be difficult, and 10 mA corresponds only to about twenty nMOS inverters.

Power is usually distributed locally in diffusion since it must get to the sources and drains anyway. For low-power gates, this local power distribution is not too bad, but for high-performance devices, great care must be taken. When two levels of metal are available the general power distribution problem is much easier, though by no means trivial.

Clearly, one of the worst scenarios for power supply noise is when large segments of the chip transition simultaneously. One strategy, therefore, is to distribute power in such a way that parts of the chip that are likely to transition all at once are routed separately. An example of this occurs in the data path. If power is distributed horizontally

FIGURE 5.78   Power and ground routing of a microprocessor. Power and ground are routed as interdigitated trees.

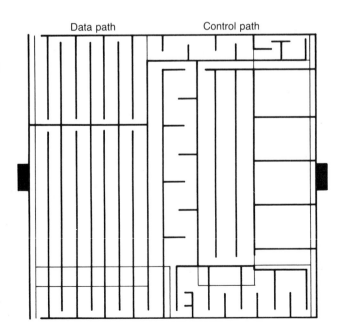

Data path   Control path

across the bit positions, we would expect large surges on the power lines, but if power is distributed vertically along the signal lines, then surge currents should be much smaller because events along the data path are necessarily staggered.

A major problem of high performance chips is bringing power onto the chip. Bonding wires can have anywhere from 0.25 to 2 nH of inductance (about 0.5 to 1 nH/mm). V_{DD} and ground wires are often double-bonded (two wires to the bonding pad) but while this lowers the inductance somewhat, it does not give the expected factor of two unless the wires are kept far apart. This is because there is mutual coupling between the wires. To make matters worse, on a forty-pin DIP the power leads are, by convention, the longest ones in the package. In the package, 10 to 50 nH of inductance is possible [Prokop 78]. One can effortlessly get volts of power supply noise even with a large external bypass capacitor. Often several V_{DD} and ground pads are used. For instance, separate power pins might be used for the output drivers since these drivers cause huge switching transients and can tolerate more power supply noise than the internal circuitry.

Electrical issues in microcell layout

While most microcell layout issues are clearly beyond the scope of this book, several layout issues are appropriate subject material because of their impact on circuit performance.[15] In chapter 4, we discussed noise issues relating to the distribution of power in diffusion and the capacitive coupling of noise, especially to dynamic nodes. The diffusion areas in custom-integrated circuits tend to have a carved-out look to keep the source and drain resistance down. Metal straps are sometimes used to "refresh" ground or V_{DD}. We have seen these used, for instance, in PLAs and ROMs. Leakage through minimum-length pass transistors can be a problem on dynamic nodes and sometimes these transistors are drawn longer than minimum to ease short channel effects.

To expedite the design of the noncritical portions of a chip, standard device sizes are often used. These device sizes become part of the methodology, that is, "When driving a 0.2 pF load, use a size **A** pullup device." These device sizes are obtained by assuming a certain number of levels of logic per clock phase, say five or so, and partitioning the logic gate delay equally among the stages. Thus if a clock phase is 40 ns, then we budget 6 to 8 ns per logic gate. One can also develop guidelines for when to use special circuit forms, that is, "If the load is greater than 0.5 pF, use a superbuffer." These methodological techniques expedite the layout design by increasing the likelihood that cells can be reused and

---

[15] The serious student of layout techniques is reminded that the best "books" on layout are in silicon. A good microscope is an important tool for the study of both layout and circuit techniques.

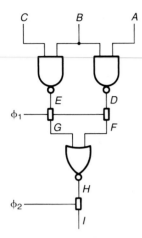

FIGURE 5.79 A collection of logic gates.

by helping the layout artist to learn quickly the most effective layout idioms for a given chip.

On occasion, chips will be designed so that the PLAs can be reprogrammed by changing only one mask (usually active area or contact). A speed penalty often occurs when this degree of freedom is required, and one is counting on PLA or ROM changes being the only necessary chip changes. In some contexts this makes sense, but not as a general rule.

In nMOS technology with buried contacts, the standard layout idiom is to place the transistors under the metal. Various nMOS layouts are illustrated throughout the text.

In CMOS technology, the primary concern is the management of the well areas. There are two major layout choices. The data can be run either parallel or perpendicular to the flow of metal. Figure 5.79 illustrates a simple logic structure. In Fig. 5.80, the transistor schematic is drawn to suggest the flow of information parallel to the

FIGURE 5.80 One topological strategy for implementing the logic in Fig. 5.79.

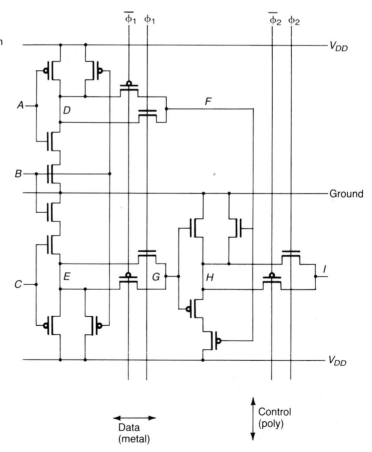

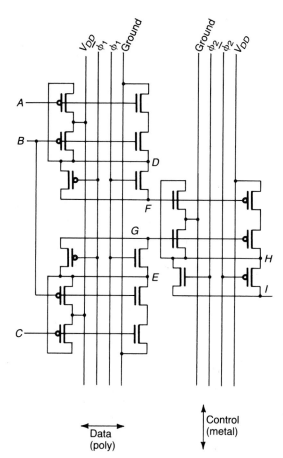

FIGURE 5.81   Another topological strategy for implementing the logic in Fig. 5.79. Clocks are run in metal, and the well edges are well used. Plugs are easily placed near the well edge.

shared metal power lines. In Fig. 5.81, this schematic is redrawn in a configuration suggesting data flow perpendicular to the metal power lines. The second layout idiom has several advantages. Wells are easily plugged on the periphery rather than the center. This is better from a latchup perspective. Note also how much more regular and compact the schematic is. This will be reflected in the final layout. Clocks are also run in metal.

We have discussed poly propagation problems in Chapter 2 and elsewhere. We should point out here that making the poly nonminimum width does help the RC time constant because of fringing fields and the additional parasitics due to gates. As with any other parasitic circuit effect, there are ways of using even poly propagation delay to advantage. For instance, by gating output drivers with a poly line, one can stagger the current surges and hence lower noise spikes.

The detailed layout of very wide transistors can have a strong impact on the performance and area of a VLSI chip. Large transistors

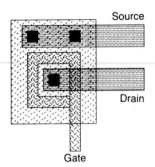

FIGURE 5.82   A doughnut-shaped pulldown transistor.

often occur in the final stage of pad and clock drivers. There are two important issues. The first concerns parasitic capacitances and the second is our constant nemesis, poly resistance. We have seen that the edge capacitance of diffused lines provides a major contribution to the capacitive load on a logic gate. This is because of the field implant. While it is true that for convenience we count the line where diffusion meets poly at the edge of the transistor as diffusion edge capacitance, this region actually has a different and usually smaller value of capacitance. Even when it does not, it is still, of course, helpful to minimize the diffusion area. The "closed" or "doughnut" transistor pulldown illustrated in Fig. 5.82 is a good structure from both of these points of view. There are some corresponding disadvantages, however. The corners contribute almost no extra current-carrying capability to the structure, but they do contribute to the gate input capacitance. On some processes with a high ratio of power supply voltage to channel length, the high electric fields in these corners may even cause the device to break down. In this structure, the series resistance of the outer ring of diffusion must also be considered.

The pad drive transistors of a high-performance VLSI circuit can easily be several millimeters wide.[16] In this case, the resistance of the diffusion and poly lines is extremely critical. Figure 5.83 illustrates two structures that are used to build very wide devices. Poly resistance is minimized by paralleling short gate segments, and the diffusion resistance is kept low by strapping it with metal. Figure 5.83 illustrates single transistor layouts. Multiple devices can be placed side by side or interleaved as appropriate to the actual technology at hand.

One occasionally needs to match circuit elements in digital MOS design. The key is to match the components in such a way that they

---

[16] Emitter followers are sometimes used for output drive pullups in CMOS technology.

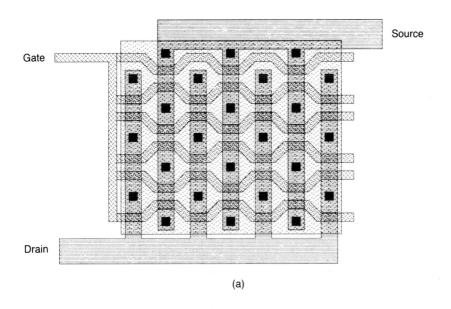

(a)

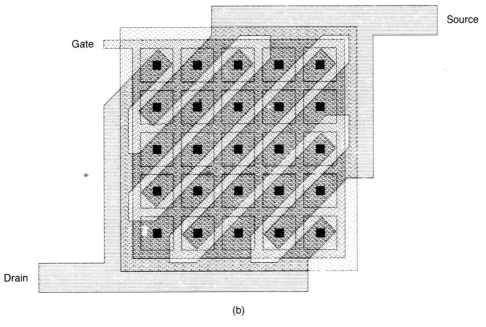

(b)

FIGURE 5.83   Very wide transistors that effectively deal with *RC* time
constant problems. By shorting together equipotential nodes in (a) one
obtains the waffle-iron design in (b).

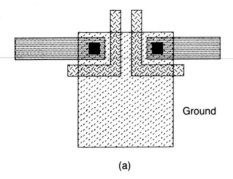

Ground

(a)

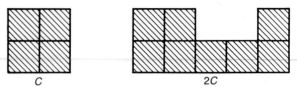

FIGURE 5.85 Capacitors matched in area and edge dimensions.

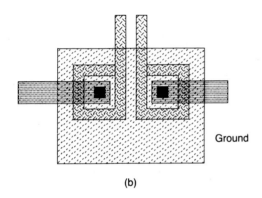

Ground

(b)

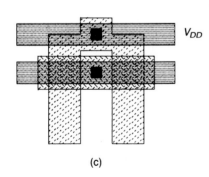

V_{DD}

(c)

FIGURE 5.84 Transistor pairs: (a) drive not symmetric under misalignment; (b) and (c) drive symmetric under misalignment.

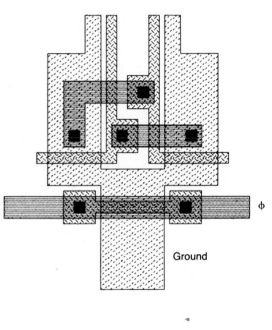

ϕ

Ground

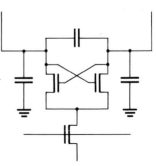

FIGURE 5.86 Drive and loading symmetric under misalignment.

remain matched even under misalignment. Figure 5.84 illustrates some matched and unmatched transistors. In many cases, it is not only important to match transistor drives but also capacitances or resistances. Figure 5.85 illustrates two capacitors, one with double the capacitance of the other. It is important that both capacitors are scaled both in area and edge.[17] Note that scaling the edge length, to first-order, means taking care of uniform shrinks or bloats as well as fringing. A circuit that needs to be matched in resistance, capacitance, and drive is the clocked sense amplifier from Section 5.5. Figure 5.86 illustrates a noncompact but balanced implementation of the cross-coupled structure.

5.10   Perspective

Several sections in this chapter dealt with circuit optimization. Given a critical path, there are six general circuit optimization techniques.

1. Change the widths and lengths of various transistors in the schematic. This is the most straightforward technique and has the minimum impact on the layout. (Section 5.1)

2. Add buffer stages between a high-impedance source and a low-impedance load. There is an optimum number of buffer stages. (Section 5.2)

3. Decouple the loads on a high fan-out node and optimize each of the resulting paths separately.

4. Use clocked buffers to take advantage of the low-impedance characteristics of a strong system clock. This technique can be applied only in cases where an appropriate clock edge exists. The usefulness of clocked circuit techniques is one of the motivations for a multiplicity of clock edges in the architectural specification of a chip. (Section 5.2)

5. Precharge heavily capacitive nodes to achieve both better speed performance and, in nMOS, lower static power dissipation. Precharging improves the speed performance of a MOS circuit when the delay is significantly different between a rising and falling output transition. As with technique (4), precharging requires appropriate clock edges, though they can sometimes be generated with timing chains if the system architecture allows.

---

[17] These are only the most elementary considerations. In the design of analog MOS circuits, one would also make sure the number of corners matched and that the $2C$ capacitor was placed half on one side of the smaller capacitor and half on the other. This is to compensate for linear gradients in the oxide thickness.

6. Use specialized circuits. These circuits are particularly useful for high fan-in or fan-out situations. Common techniques are differential and single-ended sensing and superbuffers. (Sections 5.2 through 5.4)

In addition to these circuit domain techniques, optimizations that can be performed in other domains of abstraction include the domains of machine organization, logic, and layout.

Special situations arise in a chip design that require special circuits. These often appear at the inputs (input protection, Schmitt triggers, level conversion) and outputs (output drivers, parametric voltage generation) of a chip. The effective use of large regular structures often requires special interface circuitry.

Problems

5.1 Analyze the sense amplifier illustrated in Figure 5.87. What are its advantages and critical layout issues? Discuss the sensitivity of the output to device threshold matching [Furuyama 81].

FIGURE 5.87

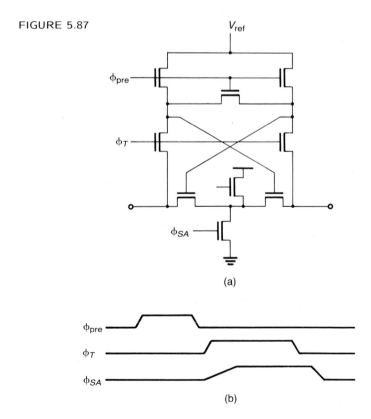

(a)

(b)

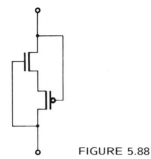

FIGURE 5.88

5.2 What is the peak current through the two-terminal element, called a "lambda diode," illustrated in Fig. 5.88?

5.3 Design the circuitry inside the black square in Fig. 5.89 to condition the bit lines and sense the read as quickly as possible. Use the 2 μm nMOS

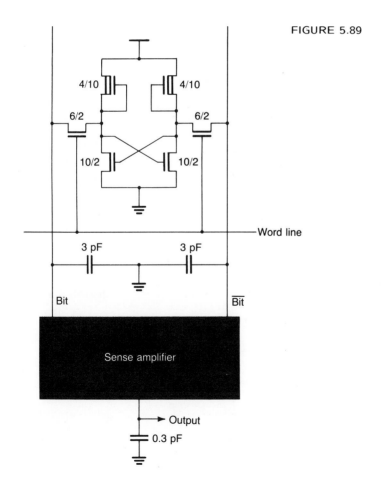

FIGURE 5.89

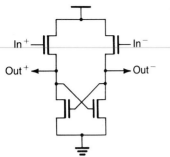

FIGURE 5.90

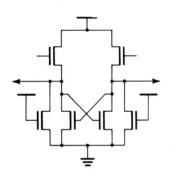

FIGURE 5.91

technology. Do not use more than 1 mW of worst-case power. Invent as many clocks as you need. Run SS simulations to show access time, FF to show power, and FS and SF to prove functionality. Assume that clocks do not vary as a function of processing. To prove your design, read a one, then a zero, then a one. Do a rough layout to extract parasitics. △

5.4 The circuit in Fig. 5.90 can be used to level shift a differential input signal. Investigate the circuit analytically and with SPICE and report your findings. What role does the differential gain of the inverters play? Now look at the circuit in Fig. 5.91. This circuit has better behavior. Why? △

5.5 Show that the optimum ratio of pullup to pulldown device width (assuming identical effective channel lengths) for an infinite string of identical CMOS inverters is given by

$$\frac{W_{pu}}{W_{pd}} = \sqrt{\frac{\mu_n}{\mu_p}}$$

to achieve minimum pair delay. Ignore interconnect capacitance.

5.6 Design a 14.5 MHz crystal oscillator using either of the example processes. An equivalent circuit for the crystal is given in Fig. 5.92. Use $C_o = 5.2$ pF, $C = 0.027$ pF, $L = 4.4$ mH, $10 \leq R_1 \leq 25$, and $50 \leq R_2 \leq 100$. Make sure that the oscillator starts up in the fundamental mode. A suggested schematic is shown in Fig. 5.93. △

5.7 Write a computer program to design a CMOS Schmitt trigger, given an input specification of the trigger points, the load capacitance, and the worst-case rise and fall times. Such a program represents procedural circuit design. Compare your predictions with SPICE simulations. How technology-independent do you think such a procedure can be made? △

5.8 It has been claimed that the precharged NOR gate in Fig. 5.94 is less sensitive to short channel effects than the one in Fig. 5.95 [White 80]. If this claim is true, why is it true?

5.9 Derive a simplified expression for the I–V characteristics of the circuit shown in Fig. 5.96.

FIGURE 5.92

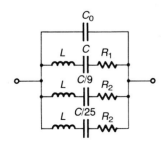

FIGURE 5.93

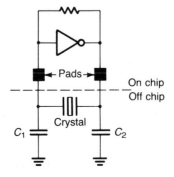

FIGURE 5.94

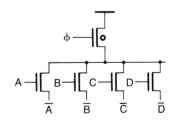

5.10 Bootstrapped buffers depend on the bootstrap capacitor being well charged before the pulldown is released. In the buffers we have seen so far, there is little correlation between the capacitor charging time and the delay. Design a bootstrapped buffer circuit in which these times are better correlated. Simulate your design over all process corners.

5.11 In the bootstrapped buffer of Figure 5.22, we have the potential problem that charge will leak off M_4 when v_2 is supposed to be above V_{DD}. Assume the presence of a clock waveform ϕ. Design a charge pump circuit to keep v_2 high when it is supposed to be high.\

5.12 One of the limitations to the circuit in Fig. 5.22 is the resistance of M_2. Design and simulate a buffer that uses an extra stage to gate M_2 on and off (allowing it to be wider). Assume a load of 100 pF.

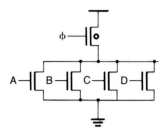

FIGURE 5.95

5.13 A bus for connecting MOS chips has a 100 Ω external pullup to a 5 V power supply. Each chip is connected to the bus by the drain of a single on-chip enhancement-mode transistor with its source grounded. We want a minimum voltage swing of 1 V (pull the bus down to 4 V). Swings larger than this dissipate excess power and contribute to the noise external to the chip. Design a circuit that will limit the gate voltage on the output transistor in such a way that it controls the output swing, making it reasonably process-insensitive. Be careful that your circuit does not limit the worst-case speed of the driver.

5.14 Explain how to use charge pumping to perform level shifting. This capability is critical for input voltages near or below V_{TE}.

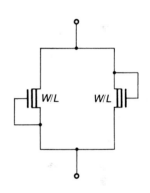

FIGURE 5.96

References

[Arai 78] E. Arai and N. Ieda, "A 64-kbit Dynamic MOS RAM," *IEEE J. Solid-State Circuits* **SC-13**: 333–338, 1978.

[Bandler 79] J. W. Bandler and M. R. M. Rizk, "Optimization of Electrical Circuits," *Engineering Optimization*, Mathematical Programming Study 11, North-Holland, Amsterdam, 1979.

[Barnes 80] J. J. Barnes and J. Y. Chan, "A High Performance Sense Amplifier for a 5 V Dynamic RAM," *IEEE J. Solid-State Circuits* **SC-15**: 831–838, 1980.

[Blaser 78] E. M. Blaser and D. A. Conrad, "FET Logic Configuration," *IEEE International Solid-State Circuits Conf.*, 14–15, San Francisco, Calif., 1978. (Cascode-based logic gates)

[Brayton 80] R. K. Brayton and R. Spence, *Sensitivity and Optimization*, Computer-Aided Design of Electronic Circuits, **2**, Elsevier, New York, 1980.

[Carr 72] W. N. Carr and J. P. Mize, *MOS/LSI Design and Applications*, McGraw-Hill, New York, 1972.

[Chan 80] J. Y. Chan, J. J. Barnes, C. Y. Wang, J. M. DeBlasi, and M. R. Guidry, "A 100 ns 5V Only 64K × 1 MOS Dynamic RAM," *IEEE J. Solid-State Circuits* **SC-15**: 839–846, 1980.

[Cheng 80] E. K. Cheng and S. Domenik, "Substrate Bias Generation Design," *Lambda*, **1**: 74–75, 1980.

[Cooper 77] J. A. Cooper, J. A. Copeland, and R. H. Krambeck, "A

CMOS Microprocessor for Telecommunications Applications," *IEEE International Solid-State Circuits Conf.*: 138–139, Philadelphia, Pa., 1977. (pseudo-nMOS)

[Crawford 67] R. H. Crawford, *MOSFET in Circuit Design*, McGraw-Hill, New York, 1967.

[Dickson 76] J. F. Dickson, "On-Chip High-Voltage Generation in NMOS Integrated Circuits Using an Improved Voltage Multiplier Technique," *IEEE J. Solid-State Circuits* **SC-11**: 374–378, 1976.

[Dingwall 77] A. G. F. Dingwall, R. E. Stricker, and J. O. Sinniger, "A High-Speed Bulk C^2L Microprocessor," *IEEE International Solid-State Circuits Conf.*: 136–137, Philadelphia, Pa., 1977.

[Elmasry 81] M. I. Elmasry, *Digital MOS Integrated Circuits*, IEEE Press, New York, 1981.

[Furuyama 81] T. Furuyama, S. Saito, and S. Fujii, "A New Sense Amplifier Technique of VLSI Dynamic RAMs," *IEEE International Electron Device Meeting*: 44–47, Washington, D.C., 1981.

[Gallace 77] L. J. Gallace and H. J. Pujol, "The Evolution of CMOS Static-Charge Protection Networks and Failure Mechanisms Associated with Overstress Conditions as Related to Device Life," *Proc. IEEE International Reliability Physics Symposium*: 149–157, Las Vegas, Nev., 1977.

[Glasser 84] L. A. Glasser and L. P. J. Hoyte, "Delay and Power Optimization in VLSI Circuits," *21st Design Automation Conf.*: 529–535, Albuquerque, N. M., 1984.

[Heller 76] L. G. Heller, D. P. Spampinato, and Y. L. Yao, "High-Sensitivity Charge-Transfer Sense Amplifier," *IEEE J. Solid-State Circuits* **SC-11**: 596–601, 1976.

[Heller 79] L. G. Heller, "Cross-Coupled Charge-Transfer Sense Amplifier," *IEEE International Solid-State Circuits Conf.*: 20–21, New York, 1979.

[Huffman 77] D. Huffman, B. Green, and D. Segers, "Minimizing Threshold Voltage Temperature Degradation with a Substrate Bias Generator," *MOSTEK Applications Information*, 1977.

[Ipri 77] A. C. Ipri, "Lambda Diodes Utilizing an Enhancement Depletion CMOS/SOS Process," *IEEE Trans. Electron Devices* **ED-24**: 751–756, 1977.

[Kang 81] S. Kang, "A Design of CMOS Polycells for LSI Circuits," *IEEE Trans. Circuits and Systems* **CAS-28**: 838–943, 1981.

[Knepper 78] R. W. Knepper, "Dynamic Depletion-mode: An E/D MOSFET Circuit Method for Improved Performance," *IEEE J. Solid-State Circuits* **SC-13**: 542–548, 1978.

[Koomen 72] J. Koomen and J. van de Akker, "A MOST Inverter with Improved Switching Speed," *IEEE J. Solid-State Circuits* **SC-7**: 231–237, 1972. (Bootstrapped superbuffer)

[Lee 79] J. M. Lee, J. R. Breivogel, R. Kunita, and C. Webb, "A 80ns 5V-Only Dynamic RAM," *IEEE International Solid-State Circuits Conf.*: 142–143, New York, 1979. (V_{BB} generation)

[Lin 75] H. C. Lin and L. W. Linholm, "An Optimized Output Stage for MOS Integrated Circuits," *IEEE J. Solid-State Circuits* **SC-10**: 106–109, 1975.

[Lynch 74] W. T. Lynch and H. J. Boll, "Optimization of the Latching Pulse for Dynamic Flip-Flop Sensors," *IEEE J. Solid-State Circuits* **SC-9**: 49–55, 1974.

[Martino 80] W. L. Martino, Jr., J. D. Moench, A. R. Borman, and R. C. Tesch, "An On-Chip Back-Bias Generator for MOS Dynamic Memory," *IEEE J. Solid-State Circuits* **SC-15**: 820–825, 1980.

[Millman 79] J. Millman, *Microelectronics*, McGraw-Hill, New York, 1979.

[Matson 83] M. Matson, *Circuit Level Optimization of Digital MOS VLSI Designs*, Ph.D. Thesis Proposal, Massachusetts Institute of Technology, Cambridge, Mass., June 1982.

[Mohsen 79] A. M. Mohsen and C. A. Mead, "Delay-Time Optimization for Driving and Sensing of Signals on High-Capacitance Paths of VLSI Systems," *IEEE Trans. Electron Devices* **ED-26**: 540–548, 1979.

[Pashley 76] R. D. Pashley and G. A. McCormick, "A 70 ns 1K MOS RAM," *IEEE International Solid-State Circuits Conf.*: 138–139, 238, 1976.

[Penney 72] W. M. Penney and L. Lau, *MOS Integrated Circuits*, Van Nostrand, Philadelphia, Pa., 1972. Reprinted in 1979 by Robert E. Krieger, New York.

[Proebsting 75] R. J. Proebsting, "Low Power, High Speed, High Output Voltage FET Delay-Inverter Stage," U.S. Patent 3.898.479, August 1975.

[Prokop 78] J. S. Prokop and D. W. Williams, "Chip Carriers as a Means for High-Density Packaging," *IEEE Trans. Components, Hybrids, and Manufacturing Technology* **CHMT-1**: 297–304, 1978.

[Puri 78] Y. Puri, "Substrate Voltage Bounce in NMOS Self-Biased Substrates," *IEEE J. Solid-State Circuits* **SC-13**: 515–519, 1978.

[Schmidt 65] J. D. Schmidt, "Integrated MOS Random-Access Memory," *Solid-State Design*: 21–25, 1965. (sRAM)

[Schroeder 77] P. R. Schroeder and R. J. Proebsting, "Clock Generator and Delay Stage," U.S. Patent 4.061.933, December, 1977.

[Stein 72] K.-U. Stein, A. Sihling, and E. Doering, "Storage and Sense/Refresh Circuit for Single-Transistor Memory Cells," *IEEE J. Solid-State Circuits* **SC-7**: 336–340, 1972.

[Tokuda 83] T. Tokuda, K. Okazaki, K. Sakashita, I. Ohkura, and T. Enomoto, "Delay-Time Modeling for ED MOS Logic LSI," *IEEE Trans. Computer-Aided Design* **CAD-2**: 129–134, 1983.

[Troutman 83] R. R. Troutman, "Epitaxial Layer Enhancement of n-Well Guard Rings for CMOS Circuits," *IEEE Trans. Electron Devices Lett.* **EDL-4**: 438–440, 1983.

[Vergnieres 80] B. Vergnieres, "Macro Generation Algorithms for LSI Custom Chip Design," *IBM J. Res. Develop.* **24**: 612–620, 1980.

[White 80] L. S. White, Jr., N. H. Hong, and D. J. Redwine, "A 5V-Only 64K Dynamic RAM," *IEEE International Solid-State Circuits Conf.*: 230–231, 1980.

[Wooley 80] B. A. Wooley, "Course notes for EECS 243," University of California, Berkeley, 1980, unpublished.

[Yee 78] Y. S. Yee, L. M. Terman, and L. G. Heller, "A 1 mV MOS Comparator," *IEEE J. Solid-State Circuits* **SC-13**: 294–297, 1978.

CLOCKS AND COMMUNICATION

6

For one signal to influence another they must intersect, not only in space, but in time. Choreographing the thousands of signals in a VLSI system so that each part of the data arrives where it is supposed to, and when it is supposed to, is an extremely complex task. This is particularly true if the design is intended for high performance.

One way to cope with this complexity is to employ a clocking methodology. Its role is to synchronize the orderly and controlled flow of information in the digital system, as well as to reduce the complexity of the circuit design by restricting the valid logic forms. That is, as we did with the logic gate methodology in Chapter 1, we simplify the design process by restricting the forms the circuits may take.

In a digital circuit, we must restore both logical values and time. This means we must make sure that we always know when it is valid to interpret an analog voltage waveform as representing a logical value. As signals propagate through the network, they undergo delays that cannot be precisely predicted. (For instance, if we want to run a computer with a 200 ns cycle time for a year and not use a clock or other method of resynchronization, we would need to know all of the delays in the computer to an accuracy better than 6×10^{-22} s.) Since we need to be able to bring signals together in time, we must develop a technique for dealing with delay uncertainties. One method is to equalize the delays by periodically holding up all signals. This method, called "clocking," equalizes all of the delays by making them all equally bad. Another method is to hold up all signals only until the slowest arrives. This is called "self-timed signaling." The more popular method for restoring

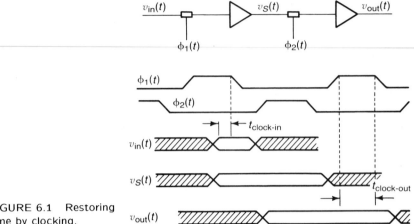

FIGURE 6.1 Restoring time by clocking.

signals in time is to use clocks. Figure 6.1 shows the effect of propagating a signal $v_{in}(t)$ through a set of dynamic registers that are gated by a pair of nonoverlapping clocks. As the signal progresses from v_{in} to v_S to v_{out}, it gets both delayed and restored. It is clearly critical that the time t_{clock} during which both $\phi_1(t)$ and $v_{in}(t)$ are valid should not go to zero (we are assuming, for simplicity, that the registers respond instantaneously). Whereas we apply power to a logic gate to restore voltages, the only way to restore a signal in time is to delay it.

6.1 Single-phase clocking

FIGURE 6.2 Single phase clocking, $T_H + T_L = T_P$. The combinational logic must be neither too fast nor too slow.

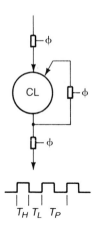

Clocks are used in a digital system to hold up a signal until it is time for it to begin to move through the next stages of logic. Registers are used in conjunction with the clocks so that a signal can be stored at a location until it is needed.

One common way of composing a collection of registers and clocks in TTL is to use edge triggered flip-flops. While this is sometimes done in MOS, it is much more common to use pass transistor registers. A simple example of such a system is illustrated by the single-phase finite state machine illustrated in Fig. 6.2. Single-phase clocking, while the most transistor-efficient of all clocking methodologies, is insidiously complex. This is due to a two-sided constraint on the speed of the logic elements in the block of combinational logic. Let $T_L + T_H = T_P$, where T_P is the period of the clock, T_H is the time the clock is high, and T_L is the time the clock is low. Signals propagate through the combinational logic at different speeds depending on their logical value. We are concerned with two extremes in time. First, we are concerned with the time that it takes the slowest valid logic signal to propagate through the logic block. If we want the clocking to be data-independent, then this maximum propagation delay must be less than T_P. Second, we are concerned with the shortest time it might take a signal (even an invalid signal such as a glitch generated by an internal hazard) to reach the output of the logic block. This time must be greater than T_H. Let τ_{CL} represent the range of combinational logic delays, from the shortest invalid delay to the longest valid delay. We have

$$T_H < \tau_{CL} < T_P, \tag{6.1}$$

which means that every delay through the combinational logic block must be greater than T_H and less than T_P. It is the two-sided nature of this constraint that makes single-phase clocking so difficult to implement. Not only do we need to worry about the critical slow path, but we must also find the critical fast path. And we must design our circuits so that not only is the slow case fast enough, but the fast case is slow enough! We can be free of the two-sided constraint by using multiphase clocking.

6.2   Two-phase clocking

Figure 6.3 illustrates a simple change to the circuit in Fig. 6.2, abolishing the two-sided constraint. Two-phase nonoverlapping clocks are used. T_1 is the time that ϕ_1 is high, T_3 is the time that ϕ_2 is high, T_2 is the low time between ϕ_1 and ϕ_2, and T_4 is the low time between ϕ_2 and ϕ_1. These clocks are also shown in Fig. 6.4. Using the same analysis

FIGURE 6.3   Two-phase clocking circuit solves the two-sided constraint problem.

FIGURE 6.4   Nonoverlapping waveforms for two-phase clocking.

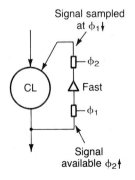

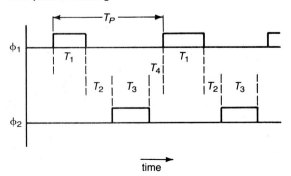

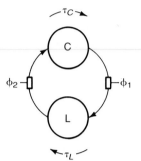

FIGURE 6.5   The combinational logic can be broken apart. There is considerable freedom in how the total delay is partitioned between the two modules.

methods as before, we find

$$\tau_{CL} < T_1 + T_3 + T_4 \tag{6.2}$$

or

$$\tau_{CL} < T_P - T_2, \tag{6.3}$$

where

$$T_P \equiv T_1 + T_2 + T_3 + T_4. \tag{6.4}$$

A signal starts to propagate through the combinational logic block at the time that ϕ_2 goes high, and must finish before ϕ_1 goes low. Note that we reduced the constraint, which is two-sided in Eq. (6.1), to a single-sided constraint in Eq. (6.3). Signals—valid or invalid—can propagate through the combinational logic as fast as they please and this will not cause a race condition. The price we paid for this was threefold: an increase in the latch[1] circuitry from one pass gate to two pass gates and a buffer, a more complicated clock to generate, and the requirement to bus many more clock wires around the chip. Since the medium is VLSI, we are more concerned with the added clock wires than with the added transistors, but even doubling the number of clock wires is a small price to pay for making the design problem merely hard. Conventional circuits such as microprocessors have used as many as eight clocks.

The existence of two clock phases gives an extra degree of freedom. We note from Eq. (6.3) that the time period T_2 is wasted in every cycle. In certain circumstances, this time can be gained back. Assume that we can break the combinational logic block into two sections as shown in Fig. 6.5. We now have a two-dimensional constraint on the delays through the blocks. We have

$$\tau_C < T_1 + T_3 + T_4, \tag{6.5}$$

$$\tau_L < T_1 + T_2 + T_3, \tag{6.6}$$

---

[1] We actually turned the latch into a flip-flop.

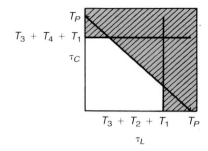

FIGURE 6.6 Constraint space for two-phase nonoverlapping clocks. One can trade delay in C for delay in L.

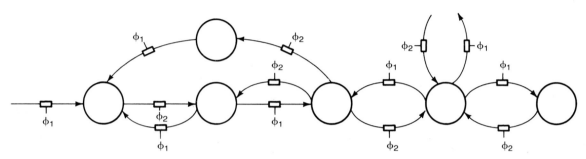

FIGURE 6.7 A two-phase subsystem. Either the inputs or the outputs are clocked on ϕ_1.

and

$$\tau_L + \tau_C < T_P. \tag{6.7}$$

Any combination of delays τ_L and τ_C that satisfies all of these constraints is permitted. The solution space is shown in Fig. 6.6. Note that if τ_L is large, τ_C must be made small, and vice versa. This freedom to move the delay specifications back and forth between modules without changing the clock waveform is characteristic of systems that are not edge triggered. It is sometimes exploited despite the design complexity it adds. Note also that the total delay $\tau_{CL} = \tau_C + \tau_L$ can approach T_P.

Multiple combinational logic blocks are shown in Fig. 6.7. Every feedback path must be cut by both a ϕ_1 and a ϕ_2 pass gate. In the simple methodology illustrated, every combinational logic block has all inputs cut on one phase and all outputs cut on the opposite phase. While the constraints given by Eqs. (6.5) and (6.6) are still valid for each block, the constraint in Eq. (6.7) is now quite complex. A methodology restriction, such as $\tau_C < T_1$ and $\tau_L < T_3$, is often employed to simplify the design. Local exceptions are made where needed for performance.

Precharging

We know that precharging is an extremely useful circuit technique. Figure 6.8 illustrates a simple system consisting of two combinational logic blocks and a precharged bus. The bus is precharged on ϕ_2. All

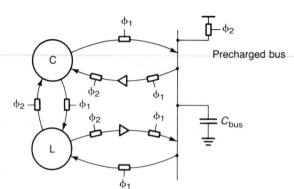

FIGURE 6.8 Interfacing with a precharged bus. Some signals must be held up, and all inputs to the bus must be glitch free.

bus inputs and outputs are taken during ϕ_1. Logic, inside the blocks or from some external control path, makes sure that only one subsystem is trying to drive the bus on any one cycle. Inputs to the bus must be stable during ϕ_1; if there are glitches on the bus input while ϕ_1 is high, the bus could be accidentally discharged. Looking at the two networks C and L, we see that the inputs to C occur during ϕ_2 and the inputs to L occur during ϕ_1. Thus C has no possibility of being stable during ϕ_2, and L has no possibility of being stable during ϕ_1. On the other hand, while L has a possibility of being stable during ϕ_2, (and C during ϕ_1), there is, as yet, no guarantee. The two-phase clocking methodology needs to be changed to accommodate precharged circuits. It is clear that the output of the L network, which cannot be stable on ϕ_1, may not be directly connected to the precharged bus. The output must be delayed by a phase until it can be safely used. The same is true, in reverse, for the input from the bus to network C.

The network in Fig. 6.8 can be redrawn to look similar to the network in Fig. 6.7. The result is Fig. 6.9. Note that for all but one of the combinational logic blocks, the combinational logic blocks have data coming in on one phase and going out on the other. The exception

FIGURE 6.9 Model of the precharged bus circuit in Fig. 6.8.

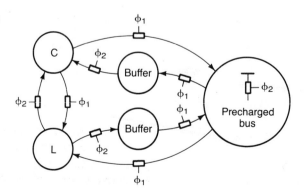

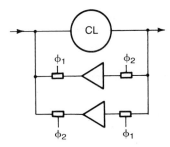

FIGURE 6.10   A circuit that uses the logic twice per cycle.

is the precharged bus, which has a ϕ_1 latch on both the input and output. Generalized to multiphase clocking, we find two different types of modules. In the simple case, the input signals are gated by a ϕ_j clock and the outputs are gated by ϕ_i, where $i \neq j$. For the precharged modules, the inputs and outputs are gated by the same phase clock. The delay through a precharged module must be less than the constraint on delay through a simple module. Thus we find a trade-off. By precharging, we can make a module faster, but we also tighten the timing constraints. That is, a precharged module is a circuit that is used on two phases rather than one. Since we do not want these two signals to interfere with each other, the timing requirements are more severe than when a module is used only once per cycle. Figure 6.10 illustrates another circuit that uses a module more than once per cycle.

The interaction between a precharged and simple module can be studied by examining the circuit shown in Fig. 6.11. The delay τ_C through the precharged module is constrained by

$$\tau_C < T_1. \tag{6.8}$$

Equations (6.6) and (6.7) provide the other two constraints. The constraint space is illustrated in Fig. 6.12. When two precharged modules

FIGURE 6.11   A circuit with one precharged module. The precharged module is clocked on both the input and output on ϕ_1.

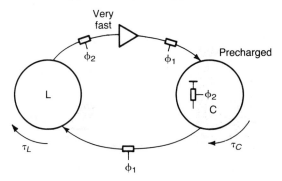

FIGURE 6.12   Constraint space for the circuit in Fig. 6.11.

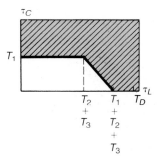

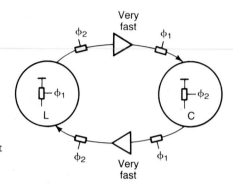

FIGURE 6.13 A circuit
with two precharged
modules.

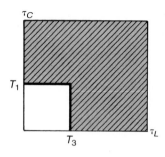

FIGURE 6.14 Constraint
space for the circuit in
Fig. 6.13. The constraint
space has shrunk due
to additional constraints
generated by precharging.

are connected together, the results are as shown in Fig. 6.13. Yet another constraint is added. We have

$$\tau_L < T_3. \tag{6.9}$$

The constraints from Eqs. (6.5) through (6.7) are now inactive. Note how much less freedom we have with the precharged modules illustrated in Fig. 6.14. Moreover, we have

$$\tau_L + \tau_C < T_P - (T_2 + T_4), \tag{6.10}$$

where $T_2 + T_4$ represents wasted time. Since, despite the tighter timing constraints, we need to do precharging to lower power dissipation in most nMOS systems, we try to minimize the nonoverlapping time (T_2 and T_4) in the design of the clock waveforms.

Clocked logic is a generalization of precharging and shift register techniques. Fig. 6.15 illustrates a collection of precharged logic gates.

FIGURE 6.15 A pipelined,
precharged circuit representing an old style of design.

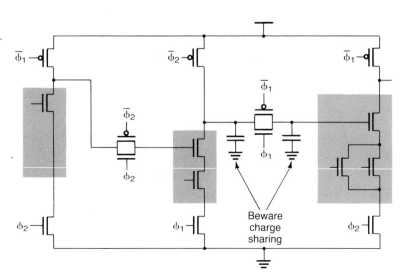

The signal propagates through only one level of logic per clock phase. Systems of this sort are the ultimate in pipelined machines. Four-phase clocks are often used. During one phase the circuit is precharged, during the next phase it is evaluated, and during the third phase it is held stable for the next stage. The fourth phase is required so signals can skip a level of logic and thus get both a signal and its complement to the inputs of the same stage of logic at the same time. The LSI-11 is an example of this older style of design. Charge sharing is a major issue in this methodology, which was invented to lower power dissipation in pMOS circuits at a time when nMOS was not manufacturable.

■■■■■■■■ EXAMPLE 6.1

The design of nMOS buffers for the circuit in Fig. 6.13

This is a perfect example of where clocked buffers can be used. For instance, we can use the buffer shown in Fig. 5.12(d).

Figure 6.16 is Fig. 6.13 redrawn.   ■

Domino and rippling logic

We have seen precharging where only a single level of logic is precharged and evaluated each cycle. It is also possible to precharge logic gates so that multiple levels of logic are evaluated during each phase. The most straightforward way to do this is to precharge only the first stage of logic and use static gates for the remaining stages. We can, however, go beyond this.

If we had a logic family in which single transitions on the input lines could result in no more than a single transition on the output, then we could combine these gates to form larger logic blocks that would have

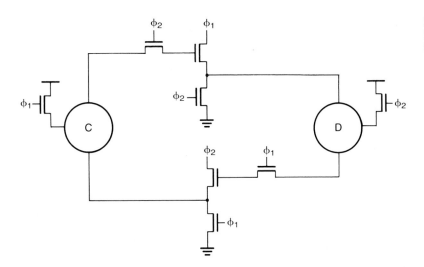

FIGURE 6.16 nMOS implementation of the circuit in Fig. 6.13.

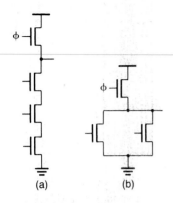

FIGURE 6.17 Two precharged logic gates, (a) NAND and (b) NOR.

the same property, providing there were no cycles. This is the property needed for precharged circuits. If the gates are to be precharged, only one transition need be considered. Examine the precharged NAND gate in Fig. 6.17(a). A consistent state exists when all inputs are precharged low and the output is precharged high. Providing each input can make no more than one transition (from low to high), then the output can also make no more than one transition (from high to low). The same is true for the NOR gate shown in Fig. 6.17(b). In fact, any precharged logic gate has this property if the n-channel pulldown network is composed of the parallel/series combination of enhancement-mode transistors. If we followed this gate by an inverter, the output of the inverter would make no more than one transition (from low to high). Thus one of these precharged complex gates followed by an inverter produces just the sort of input for the next stage that we postulated we would need.

We now have a consistent (though not universal) logic family consisting of AND, OR, and complex noninverting gates. This is domino logic [Krambeck 82], so called because the outputs fall sequentially, like a row of dominos. Because each gate makes no more than one transition, dynamic power is minimized. Fig. 6.18 illustrates a domino logic circuit. The PLA illustrated in Fig. 1.70 is also a member of the domino logic family.

Rippling can also occur in pass structures. In Chapter 8 we will examine a technique called the "Manchester carry chain." The precharged version of that circuit also exhibits a rippling behavior.

FIGURE 6.18 Domino CMOS.

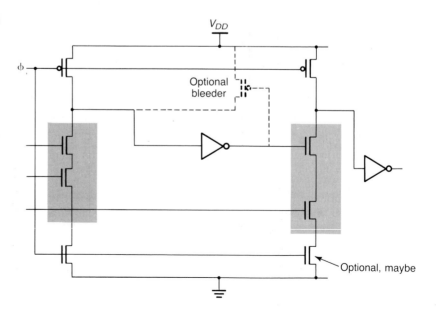

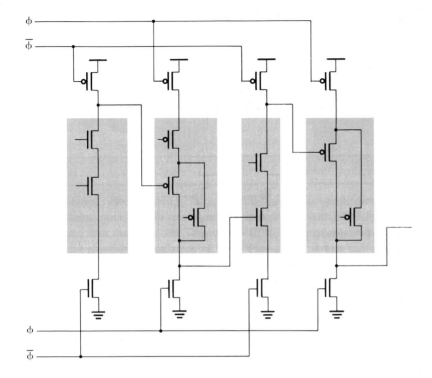

FIGURE 6.19   NORA form rippling logic.

We can use duality to observe that if we switched all the p-type devices to n-type, and all of the highs to lows, we could build a mostly p-type domino logic gate. Moreover, we can alternate precharged n-type and precharged p-type gates, as shown in Fig. 6.19 [Goncalves 83].

Domino logic is not free of charge sharing problems. Nodes internal to a NAND-type totem pole may not be precharged high because the inputs are low and hence these nodes are isolated. Charge sharing can occur if, for instance, the topmost transistor in the pulldown network turns on. This problem is not too serious, however, because the layout of the gate typically ensures that the output node has much more capacitance than the internal nodes. Moreover, charge pumped into the stack by the high-going transient on the gate is in the correct direction to help. For very long stacks, internal nodes can be precharged.

Noise is a serious issue for the second type of rippling logic that we saw in Fig. 6.19. High-impedance precharged nodes coupled with a low-noise robustness make these gates extremely susceptible to capacitive coupling noise—especially when thresholds are placed low for performance reasons or because of drain-induced barrier lowering. (See the discussions on noise in Chapter 4.)

■■■■■■■■■ EXAMPLE 6.2

Redrawing Fig. 6.8 so both C and D can be precharged
This involves adding more delay registers, as shown in Fig. 6.20. The one subtlety involves the two parts of the circuit that are precharged on the same phase. These can be directly connected only if a domino or rippling circuit can be devised. Otherwise, the global timing will need to be changed.   ■

Dynamic considerations

Using pass devices as registers has one obvious implication—because the charge storage mechanism is dynamic, the clock cannot be stopped indefinitely. On the other hand, it is nice to design machines that can be single-stepped for debugging purposes. Several circuits for building static latches are illustrated in Fig. 6.21.

There are other places where these circuits are useful. When the clock is gated by a logic signal, dynamic nodes can exist that are seldom

FIGURE 6.20   Implementation of the precharged bus circuit of Fig. 6.8 with extensive precharging.

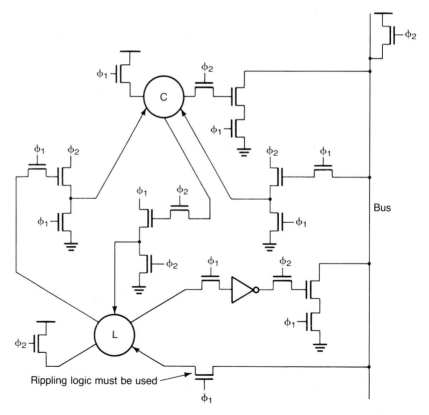

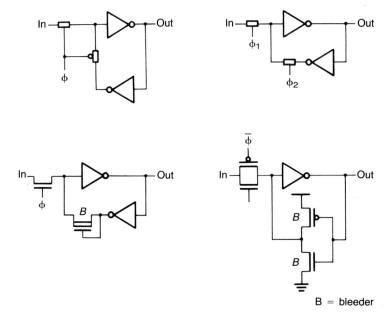

FIGURE 6.21   Static and semi-static latches. "B" stands for bleeder in this figure.

B = bleeder

refreshed. Figure 6.22 illustrates such a case. Another bug of this sort is illustrated in Fig. 6.23. Here, if A and ϕ are true at mutually exclusive times, charge can be pumped onto or off the dynamic node. Both of these bugs can be avoided by either using a clocking methodology in which all dynamic nodes are refreshed every cycle or by using static latches (in which case the node is continually refreshed).

In terms of area and clocks, dynamic pass gates are much more expensive in CMOS circuits than in nMOS circuits. Static registers can thus have a smaller incremental cost in CMOS than in nMOS. One key to using static registers effectively is to use ratioed CMOS latch designs. One technique, which uses the cross-coupled complementary

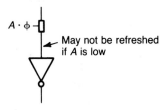

FIGURE 6.22   The inverter input may not be refreshed for a very long time.

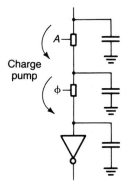

FIGURE 6.23   Charge can be unintentionally pumped onto or off the inverter input.

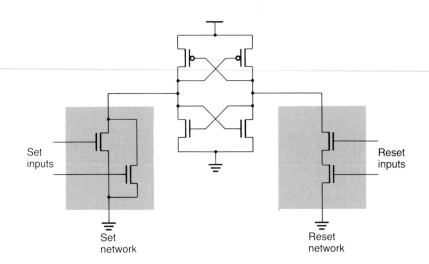

FIGURE 6.24 A ratioed CMOS network. The set and reset networks can often share transistors.

CMOS inverter as a building block, is illustrated in Fig. 6.24. This circuit dissipates only dynamic power, provided that the set and reset networks are never activated simultaneously. Figure 6.25 illustrates an example of this circuit style. The set and reset networks can often share transistors

FIGURE 6.25 Some logic and its implementation using the technique of Fig. 6.24. For proper operation $XY(A+B)\phi = 0$.

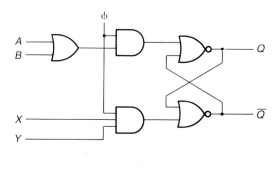

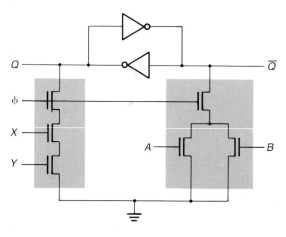

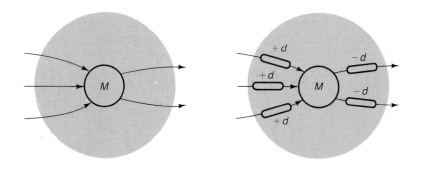

FIGURE 6.26 Retiming a circuit module by adding delay to the input and by removing it from the output.

[Heller 84]. To improve the speed of the ratioed pullup, precharging can be used.

Precharging of arrays can cause very large current spikes, which are a source of noise. Staggered precharge clocks can be used to limit the maximum current. Precharging to $V_{DD}/2$ is also a useful technique in some situations.

Clock skew

Controlling clock skew in large systems is always a problem, and VLSI systems are no exception. In fact, the problem in a VLSI system is slightly worse because the clock transition time is fairly long compared to the propagation delay through a fast MOS gate. Uncertainties in our ability to exactly predict clock timing lowers the performance of a VLSI system. In extreme cases it can even add new design constraints to the problem.

To examine the effects of uncontrolled delay on the system timing constraints, we look at a simple delay transformation that preserves the overall timing of a module. In Fig. 6.26, we show a module with an arbitrary number of inputs and outputs. If we add a delay d (which can be positive or negative) to every input and subtract the same delay from every output, the module timing remains unchanged. The transformed module is also shown in the figure.

This transformation allows us a simple way to look at the influences of skew on our canonical two-phase nonoverlapping clock system. If one of the clock phases has a delay uncertainty d, this delay uncertainty can be transformed to appear in the loop, as shown in Fig. 6.27. Thus a delay

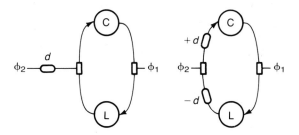

FIGURE 6.27 Retiming a circuit with clock skew. A circuit with known delays through the combinational logic and uncertain clock skew can be transformed into a system with uncertain delays through the combinational logic and no clock skew.

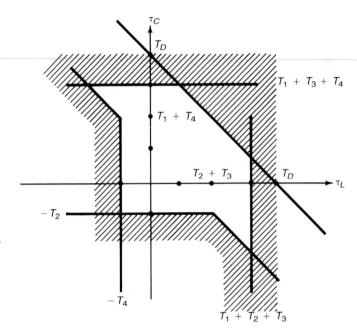

FIGURE 6.28 Full constraint space for two-phase nonoverlapping clocks.

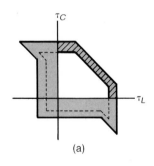

(a)

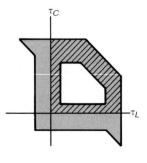

(b)

FIGURE 6.29 Shrinking of the constraint space due to skew: (a) moderate skew, (b) very bad skew, which introduces two-sided constraints.

d is effectively added to one of the pieces of combinational logic, while the same delay is subtracted from the other piece of combinational logic. In terms of our earlier picture of delay constraints (Fig. 6.6), this delay shifts the τ_C- and τ_D-axes in the $(-1, 1)$ direction. However, we have not characterized the constraint space with respect to negative delays. While these negative delays are not physical in an ideal system, in the transformed system they do have meaning. Figure 6.28 illustrates the complete constraint space for a two-phase nonoverlapping clock system. Figure 6.29 illustrates how the constraint space shrinks as the clock uncertainty grows. Note that in extreme cases the clocks can become overlapping, putting us back to two-sided constraints.

Several practical issues come up in relation to performance and clock skew. As we have seen, clock uncertainty can impact the maximum delay constraint of a module. This might be important when the process parameters come out "slow." It is important to realize that clock skew can almost be an orthogonal issue from logic circuit delay. Poly resistance can, for instance, affect the clock waveforms much more than the waveforms of the logic.[2] Also, the clock generation circuits are typically special circuits that have a different worst-case corner than the rest of the circuit. Indeed, since there is a large premium on making the dead time between clocks very short when precharging is used extensively, keeping the clock nonoverlapping under "fast" process conditions can be a real challenge. Note also that the usual bootstrapped clock buffers

---

[2] There is always a great layout temptation to distribute clocks with poly.

common in nMOS design have their own worst-case, which is different from that of the rest of the circuitry.

To account for all of this, timing margins can be used.[3] Typical timing margins are on the order of 15%. Some timing margin is built into the conservative estimation of layout parasitics during the initial phase of the circuit design.

In CMOS systems, generating $\overline{\phi}_1$ and $\overline{\phi}_2$ (as well as ϕ_1 and ϕ_2) is an additional burden.[4] The basic choice involves whether $\overline{\phi}_1$ and $\overline{\phi}_2$ are generated locally or globally. There is no general rule. A global clock can better control the generated clock skew, while local inverters require fewer wires. The situation is further complicated by the fact that CMOS circuits are more static than nMOS circuits. This enables us to effectively use the T_2 and T_4 times, which now need to be longer to ensure "nonoverlappingness."

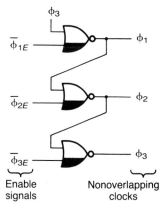

FIGURE 6.30 An example of a two-phase circuit.

■■■■■■■■■■■■■■■ EXAMPLE 6.3

A precharged circuit
The circuit in Fig. 6.30 involves the use of a transparent latch and precharged NOR gate. Redraw the circuit to look like one of the clocked systems described in this section. A redrawn version of this schematic is shown in Fig. 6.31. ■

■■■■■■■■■■■■■■■ EXAMPLE 6.4

Redrawing the circuit in Fig. 6.32, as in Example 6.3

See Fig. 6.33. ■

Generators for nonoverlapping clocks

In Section 5.2, we examined clock driver circuits. In this section, we examine how those circuits are combined with logic to build clock generators.

Fig. 6.34 illustrates clock generators useful for generating nonoverlapping clocks. Feedback assures that the clock phases do not overlap. The buffers in these generators can be built to take advantage of the fact that more than one clock phase is available. For instance, Fig. 6.35(a) illustrates a clock driver for use in a chip with two-phase nonoverlapping clocks. As shown in Fig. 6.35(b), ϕ_1 holds the output low with transistor

---

[3] If timing margins are not explicitly used, then they come implicitly out of the worst-case delay models, which, we remember, are designed to encompass a range of process variations. The model files end up encompassing model inaccuracies as well.

[4] There have been several approaches to try to avoid the necessity of generating these complementary clocks. One is to let $\phi_1 = \overline{\phi}_2$ and another is to use only nMOS pass devices. Each has severe limitations, though they have been used successfully.

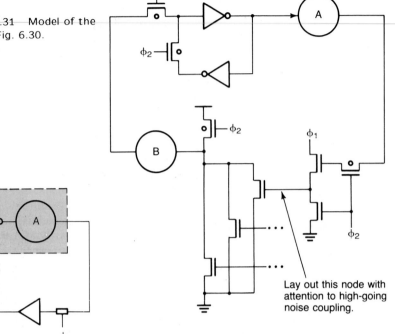

FIGURE 6.31 Model of the circuit in Fig. 6.30.

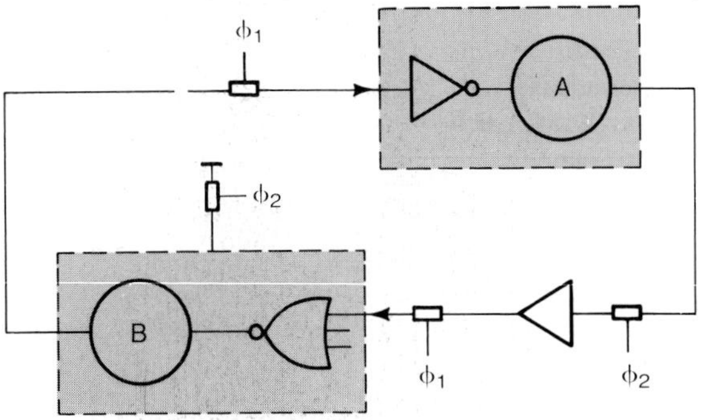

Lay out this node with attention to high-going noise coupling.

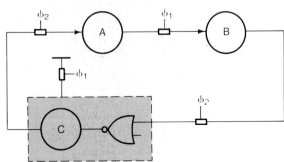

FIGURE 6.32 Another example circuit.

FIGURE 6.33 Model of the circuit in Fig. 6.32.

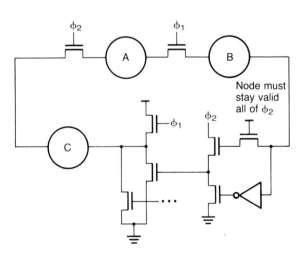

Node must stay valid all of ϕ_2

FIGURE 6.34 Generalized clock generation for nonoverlapping clocks.

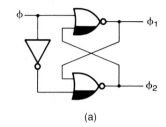

(a)

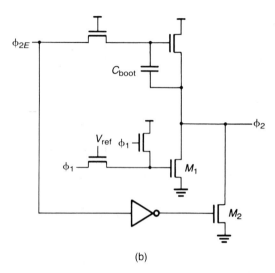

(b)

FIGURE 6.35   (a) Logic for a two-phase nonoverlapping clock generator. (b) Circuit for the ϕ_2 clock driver. The ϕ_2 driver is held low until ϕ_1 drops very low ($V_{\text{ref}} - V_{TE}$).

M_1 until ϕ_1 falls below $V_{\text{ref}} - V_{TE}$. This allows C_{boot} to become fully charged. The delay between the time that the enable signal ϕ_{2E} goes high until ϕ_1 drops to below $V_{\text{ref}} - V_{TE}$ implicitly provides the delay that we needed to provide explicitly in Chapter 5. V_{ref} might be set to about 2 volts in a 5-volt process.

When bootstrapping is not used on the output drivers, there is no need to tolerate the surge currents required to hold the boot node low while the boot capacitor is being charged. Since there will typically be several stages of buffering, one can tailor the amount of nonoverlap time by selecting the stage at which one picks off the clock control signals. Figure 6.36 illustrates the phase two driver for a three-phase nonoverlapping clock generator. Each pipelined stage controls its corresponding stage of the phase three driver and is controlled by the corresponding stage of the phase one driver.

As we saw early in Section 6.2, the length of time during which both clocks are low can severely impact the overall performance of the chip—especially when precharging is used. Maintaining a short, but guaranteed, nonoverlap interval is one of the major challenges in the generation of nonoverlapping clocks.

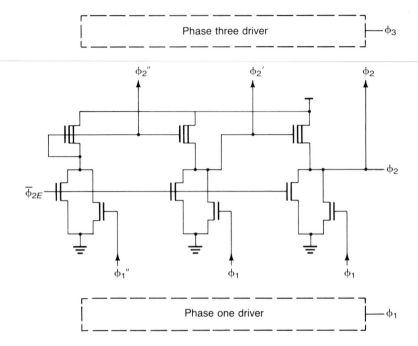

FIGURE 6.36 A pipelined ϕ_2 driver for minimizing the nonoverlap time.

6.3 Overlapping clocks

In all digital systems, the clock should correspond to some naturally occurring logical interval in the machine. For instance, in a microprocessor the natural cycle is the machine's microcycle. A two-phase methodology would further divide the microcycle into two phases. There are many cases, however, when the logic is more naturally subdivided into three or more phases.

Consider the microprocessor organization illustrated in Fig. 6.37. The data path consists of a register file and an ALU. The control path consists of a PLA-driven finite state machine. A natural clocking methodology and phase assignment for this machine is given in Table 6.1.

TABLE 6.1 A natural clocking methodology and phase assignment for the microprocessor illustrated in Fig. 6.37

Phase	Data path	Control path
ϕ_1	Read reg file	Update μaddr latch
ϕ_2	ALU operation	Access PLA AND Plane
ϕ_3	Write reg file	Access PLA OR Plane, update MIR

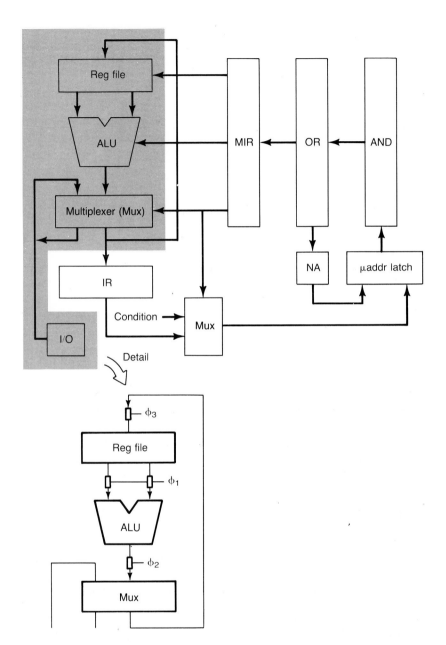

FIGURE 6.37   Three-phase overlapping clocks applied to a simple
microprocessor. Clocked pass devices often appear in the block diagrams,
because nothing models as a pass transistor like a pass transistor.

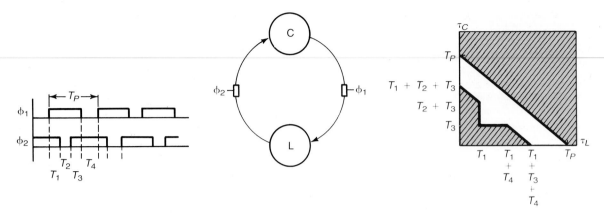

FIGURE 6.38   Waveforms and constraint space of two-phase overlapping clocking, which combines the worst of several techniques.

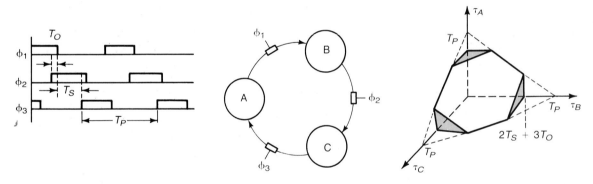

FIGURE 6.39   Waveforms and constraint space of three-phase overlapping clocking.

FIGURE 6.40   Precharging a three-phase system is just like precharging a nonoverlapping clock system.

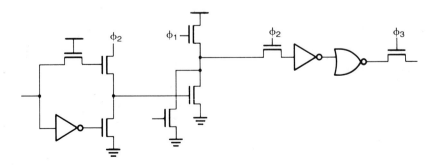

A three-phase clocking methodology is well suited to this simple machine. A major advantage of having three or more phases in a machine is that this can relieve the requirement for generating nonoverlapping clocks, which can be quite difficult to engineer in large high-performance systems. Figure 6.38 illustrates the constraint space for a two-phase overlapping clock methodology. This technique obviously combines the worst of several worlds. The simplest manageable overlapping clocking methodology involves the use of three-phases.

Figure 6.39 illustrates the canonical three-phase system with the constraint space. The constraint space is now three dimensional. All constraints are single-sided, provided that all three phases do not simultaneously overlap. This is a fairly easy constraint to guarantee, even in the presence of severe clock skew. (Clock skew still negatively impacts performance, of course.)

In an overlapping clock system of the sort described here, all cycles must be cut by at least three phases. Precharging is done in basically the same manner as with nonoverlapping clocks. The difference is that there is some short time during which both the pullup and pulldown are on. As before, however, if a module is precharged on ϕ_{i-1} in a three-phase system, it has pass transistors gated by ϕ_i before and behind it, as illustrated in Fig. 6.40.

Generators for overlapping clocks

The generation of overlapping clocks is much easier than the generation of nonoverlapping clocks because uncertainties in the timing of adjacent clock edges are not quite as important to the functionality of the circuit. Edge skew, particularly in precharged circuits, is still important to performance, but this is a somewhat milder constraint because it is important only for slow process conditions. In nonoverlapping clock methodologies, the clocks must never overlap. In the typical overlapping clock methodology, the clocks can be nonoverlapping without ruining the functionality of the circuit. On the other hand, the minimum number of overlapping clock phases is three, whereas it is two for nonoverlapping clocks.

The most effective way to time the edges in an overlapping clock generator is for the ϕ_{i+1} output to kill the ϕ_i output. This is shown in Fig. 6.41. While logically the inputs are able to cause the output to go low, they do not have the drive to do this effectively. A typical phase driver is shown in Fig. 6.42. A Johnson counter can be used to generate the enable signals. The Johnson counter is illustrated in Fig. 6.43. The D-type flip-flops should be static.

FIGURE 6.41   Clock generator of a three-phase overlapping clock. Each phase kills the one before.

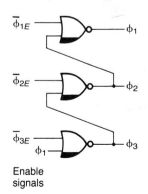

Enable signals

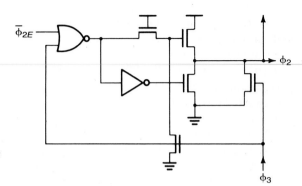

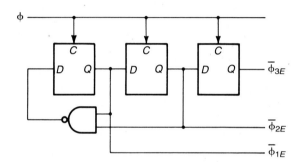

FIGURE 6.42 Detail of the phase two clock buffer.

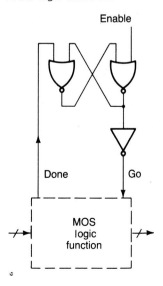

FIGURE 6.43 Johnson counter used to generate enable signals. Latches should be static.

6.4 Self-timed MOS circuits

FIGURE 6.44 A self-timed MOS logic function.

Self-timed circuitry is often used in MOS logic because of its performance and component efficiency. The essence of the control mechanism for a self-timed system consists of two signals and one state variable. The first signal is an input to the MOS logic function and acts to initiate the computation ("GO"). The second signal is an output of the logic function and signals the completion of the computation ("DONE"). This signal may act as the "GO" signal to the next logic block. A latch stores the information on whether the logic is inactive or active. A block diagram of a self-timed logic function is shown in Fig. 6.44. The GO–Compute–DONE cycle is sometimes modified to include a reset function, that is, GO–Compute–DONE–RESET–GO–.... The completion signal can be generated in two ways:

1. By using a signaling scheme that combinationally detects a completion.

2. By using an analog delay that mimics the logic operation to be performed.

In the first technique, the total system is usually asynchronous and an enormous area overhead is paid to realize the additional logic associated with the signaling scheme [Mead 80, Chapt 7]. Since the "DONE" signal is produced in a data-dependent manner, there are potential performance advantages to the technique because it can exploit the statistical nature of a particular problem. Nevertheless, there are few LSI systems that use this technique.

The second option is almost always chosen in MOS systems because of its component efficiency. For example, in an array structure, the completion signal can be generated by using a dummy element that is activated on every cycle. The major disadvantage of this scheme is that the dummy element must be designed to be marginally slower than the slowest path through the logic. If the completion signal is used to trigger a subsequent stage, the delay will be greater than necessary for paths that are not worst case. Note, however, that there is not much variation among delay paths for an array structure.

The key to this second form of self-timed logic is our ability to model the delay through a large block of logic with a much smaller circuit. This works only because process parameters track on an LSI chip. These self-timed techniques are often used in synchronous systems. They may be thought of as clever ways of generating the extra clocks needed in a portion of the chip. For instance, we might want both to precharge and to evaluate a logic function during ϕ_1, but we do not have the necessary extra clock. It could be generated in a self-timed manner by detecting when the precharge completes. The system will remain synchronous if we guarantee, through worst-case circuit simulation, that the combined precharge and evaluation are done before ϕ_2 starts. Dynamic RAMs use dozens of self-timed clocks.

■■■■■■■■■■■■■■■■■■■■■ EXAMPLE 6.5

Self-timed data path

The data path of an nMOS microprocessor provides many opportunities to use self-timed logic. For the three-phase data path we described earlier, a write to the data path on ϕ_3 is followed by a read on ϕ_1. Assume a dual rail bus and a standard six-transistor memory cell, as illustrated in Fig. 6.45.

The write amplifier drives the bus differentially during ϕ_3, and the memory cell, which had been selected during ϕ_3, is written. We want to be able to access a row at the beginning of ϕ_1. For proper operation, the following sequence must take place.

- Deselect old row
- Precharge and balance data bus

- Select new row
- Enable sense amplifier

All four operations can be accomplished in a self-timed fashion during the single ϕ_1 clock time. First we must construct the self-timed precharge circuit. If we assume that a cell is always selected during each machine cycle, then the bus will always be in a $1:0$ or a $0:1$ condition. A simple NAND gate on A and $\bar{A}$ can be used to detect completion of the precharge. Assuming that all columns are identical,

FIGURE 6.45 A register file that must perform several operations on ϕ_1.

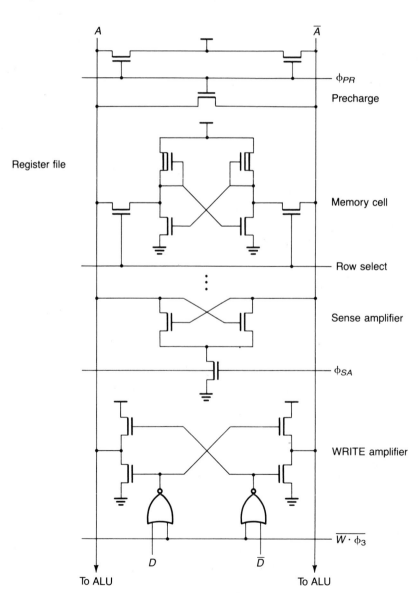

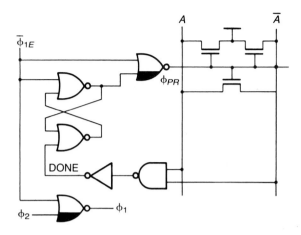

FIGURE 6.46   Self-timed precharge logic.

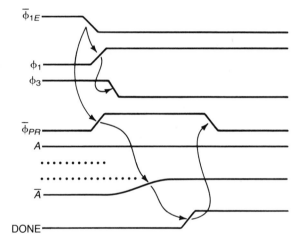

FIGURE 6.47   Timing of precharge logic.

the detector need operate on only one pair of lines. If we assume a clock-generation technique similar to the three-phase overlapping scheme previously developed, then an appropriate generator for the precharge clock is as shown in Fig. 6.46. The timing diagram is illustrated in Fig. 6.47.

Next we consider the interaction of the row selects with the precharge. Recall that during ϕ_3 we are conditionally writing a previously selected cell and that during ϕ_1 we want to read a new cell. During precharge we want to deselect all cells, therefore precharge time can also be used for propagating new row select data into the row drivers. The logic in Fig. 6.48 will accomplish this task. The timing diagram in Fig. 6.49 illustrates the waveforms, where row N was selected on the previous cycle, and row $N+1$ on the new cycle.

All that remains is to trigger the sense amplifiers at the appropriate moment. The sense amp clock ϕ_{SA} must start to rise only after the row

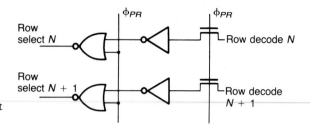

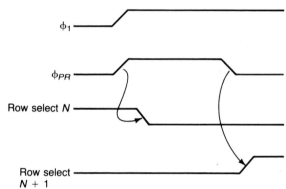

FIGURE 6.48   Row select
logic.

FIGURE 6.49   Timing of
row select logic.

FIGURE 6.50   Logic to
trigger the sense amplifier.

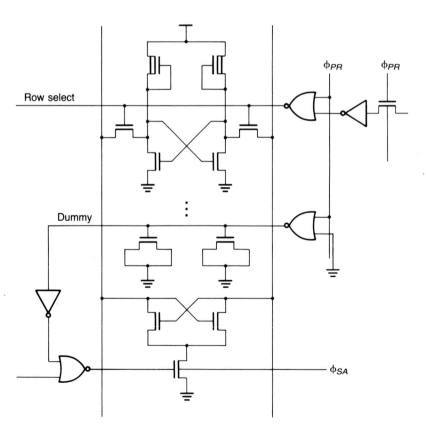

select has been reasserted and the selected cell has tipped the balance on the bus rails by an appropriate noise margin. There are many ways to accomplish this. The simplest is to add a dummy row driver and either model the capacitive load or actually place a complete dummy row in the array. The second solution is illustrated in Fig. 6.50. The complete dummy row is especially appropriate if the row lines are poly, and the poly delay must be accounted for in the triggering of the sense amps. Figure 6.51 illustrates the final timing diagram.

A and $\overline{A}$ must presumably be stable before ϕ_2 starts. Circuits such as the one we have described are fairly tricky to design because their worst-case is not easily identifiable. They are in fact good candidates for statistical simulation approaches. It is intriguing that the success of the self-timed circuits we have discussed depends more on correct hardware models (the on-chip delay lines) than on software models (used for circuit simulation). ∎

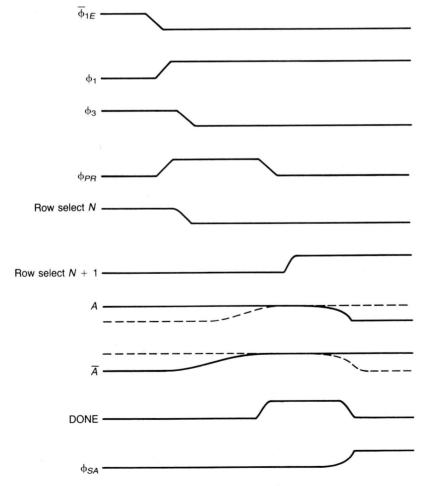

FIGURE 6.51   The complete timing of the self-timed register file.

6.5   Synchronizers and arbiters

Most synchronous digital systems have at least one asynchronous input (an input that can come at any time in relation to a clock edge). These inputs must be synchronized. As an example of two asynchronous events, consider a computer that polls someone's terminal to see if a key has been typed. And consider the typing of the key. There is no reason for the person's typing of the key to have any timing relationship to the computer's internal clock. The computer is most likely a synchronous machine, which is to say it has an internal clock that mandates when certain events must happen. It has a sense of time. When synchronous systems must deal with asynchronous signals, there is the possibility of big trouble. Let us examine the terminal scenario again. There is the possibility that just as the computer checked the terminal, the key was typed. It probably does not matter to anyone whether the computer decides that no, the key was not typed in time and it will be recorded next time the terminal is polled, or yes, the key was typed in time. The critical issue is that the computer must decide one way or the other, and the problem is that it might take the computer arbitrarily long to make the decision. If the computer has not decided, then there will be a signal inside the computer that is somewhere between a high and a low. That is not in itself a problem. The problem comes when two different gates interpret that signal differently. This will cause an inconsistent and probably faulty condition to be set up. (For instance, read register R0 onto the bus vs. read register R2 onto the bus; error: both get ORed onto the bus; nonsense.)[5] The probability of such an event is small, but becomes greater the more times the key is pressed.

In the general case, a circuit is needed to "decide" whether one of two events has occurred. This circuit is called a synchronizer or arbiter [Couranz 75, Kinniment 76, Kung 82, Veendrick 80]. An arbiter is a piece of logic that must decide which of two events has occurred first. With a synchronizer, which is a special case of an arbiter, the decision must be made by a certain time. An arbiter in which one of the inputs is a clock is a synchronizer. The arbiter is classically realized with a cross-coupled latch. We will now look at the issues involved in the design of a good arbiter. We first must decide on a definition of "good" if, as we contend, there is a finite chance that any arbiter could take, say, three seconds to decide. The answer comes in probabilities. From quantum mechanics and the uncertainty principle, we know there is a finite chance that almost anything can happen. The atoms in the words on this page could, for instance, spontaneously rearrange themselves so

---

[5] The classic arbiter example is that of that is two people walking toward each other in the middle of a hallway. The complex (time-consuming) dances that occur as they try to get out of each other's way are a manifestation of the arbiter problem.

that some words were spelled drawkcab. But we really do not worry about this happening. It is sufficiently improbable. Our approach, then, is to make a computer crash due to an asynchronous event sufficiently improbable that we need not worry about it. This is straightforward to do, but expensive. As we will see, we can make the probability of an error due to an asynchronous event exponentially small just by waiting. For reliable system operation, we must unfortunately wait a long time— tens of nanoseconds for our example 2 μm processes. It usually works out that the decision on an asynchronous event must be delayed for a phase in a two- or three-phase system. This clearly would not matter for the terminal application we discussed, but it could be very important in, say, the data communication between two high-speed coprocessors. For this reason, most modern system buses are made to be synchronous where possible.

We will now look at two different issues. First, we will perform an analysis of a simple noiseless synchronizer circuit, and second, we will examine the effects of noise on the behavior of the synchronizer.

A synchronizer circuit

Figure 6.52 illustrates a simple arbiter circuit. The key to this circuit is the cross-coupled latch. In this circuit a signal is brought into the latch. When ϕ_1 goes low, we sample the input signal. It will likely be either a high or a low, but it is possible for it to be somewhere in between. If it is above the metastable point of the latch, the latch will eventually flip to the same side as if the input were a solid high, and vice versa for an input below the metastable voltage. Let us assume that the input voltage has been characterized as moving through the region around the metastable point at a rate of dv_{in}/dt. If $v_{in}(t)$ has no timing relationship to $\phi_1(t)$, then $v_{in}(t)$ can be at any voltage. Let us expand the input voltage about the metastable voltage V_{MS}, as shown in Fig. 6.53. Also

FIGURE 6.52   A synchronizer.

FIGURE 6.53   Synchronizer waveforms showing possible output trajectories after the input is latched.

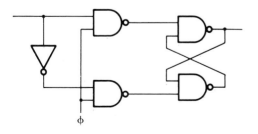

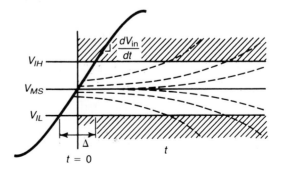

labeled in the figure are the two voltages V_{IH} and V_{IL}. The time Δ it takes the input to cross from V_{IL} to V_{IH} is

$$\Delta = (V_{IH} - V_{IL})\left(\frac{dv_{in}}{dt}\right)^{-1}. \tag{6.11}$$

What is the probability of a fault if we wait for a time T? First looking at $T = 0$, we see that an error occurs if the input signal is anywhere between V_{IL} and V_{IH}. Let us assume that transitions that must be synchronized happen at a rate of f_{in}. In that case, the average time between events is f_{in}^{-1}. This means that the probability P_E that any one event will be incorrectly synchronized if we do not wait at all is

$$P_E = f_{in}\Delta. \tag{6.12}$$

Since we assume this probability is very small, the probability P_S that an error will occur in one second is

$$P_S = f_\phi P_E = f_\phi f_{in}\Delta \tag{6.13}$$

since there are f_ϕ chances for an error each second, where f_ϕ is the sampling frequency. The error rate is thus P_S. The mean time to failure is P_S^{-1} or

$$\text{MTF} = \frac{1}{f_\phi f_{in}\Delta}. \tag{6.14}$$

This analysis is valid[6] for $T = 0$. When $T > 0$ the error rate goes down substantially.

Voltages in the linear region of the synchronizer (assuming a dominant single pole) obey the equation

$$v(T) - V_{MS} = (v_{in}(0) - V_{MS})e^{T/\tau_{switch}}, \tag{6.15}$$

where τ_{switch} is the time constant of the system, $v_{in}(0)$ is the initial condition when ϕ goes low, and $v(t)$ is the time evolution of the latch voltage.

We can say that if a voltage is between V_{IL} and V_{IH} at time $t = T$, then at $t = 0$ it must have been between

$$V_{MS} + (V_{IH} - V_{MS})e^{-T/\tau_{switch}} \tag{6.16}$$

and

$$V_{MS} - (V_{MS} - V_{IL})e^{-T/\tau_{switch}}. \tag{6.17}$$

---

[6] For now we are neglecting the propagation time through the latch.

The difference between Eqs. (6.16) and (6.17) defines a new effective value of Δ. The error rate $P_S(T)$ becomes

$$P_S(T) = f_\phi f_{\text{in}} \Delta e^{-T/\tau_{\text{switch}}}, \qquad (6.18)$$

with a MTF of

$$\text{MTF} = \frac{e^{T/\tau_{\text{switch}}}}{f_\phi f_{\text{in}} \Delta}. \qquad (6.19)$$

Note how the reliability of the system increases exponentially with time. Given a required MTF, we can find T. We have

$$T = \tau_{\text{switch}} \ln\left(\text{MTF} f_\phi f_{\text{in}} \Delta \right). \qquad (6.20)$$

Let us take an example of MTF $= 10$ yr, $f_{\text{in}} = 1$ MHz, $f_\phi = 10$ MHz, $\Delta = 10$ ns, and $\tau_{\text{switch}} = 1$ ns. Ten years is 3×10^8 seconds. We have $T = 31$ ns, which is about 30 time constants!

The values f_ϕ, f_{in}, and Δ are all fairly straightforward to determine. What about τ_{switch}? It can be found analytically (see Section 5.4). It is a function of the gain bandwidth product of the circuit. Another, probably more accurate, approach is to find τ_{switch} by simulation. If we arrange the initial conditions for the simulator such that the latch voltages are very close to the metastable point, then the time constant of the decay we see is exactly τ_{switch}. We have

$$\tau_{\text{switch}} = (t_2 - t_1) \ln^{-1}\left(\frac{v(t_2) - V_{MS}}{v(t_1) - V_{MS}} \right). \qquad (6.21)$$

The extracted value of τ_{switch} is used to calculate the necessary value of T. Note that errors in τ_{switch} are very important since the MTF is exponentially dependent on τ_{switch}. So including some margin for error in T is prudent. Also included in T should be the delay through the arbiter circuit. This is typically a few τ_{switch}.

Synchronizer bugs are extremely difficult to trace. For this reason the number of synchronizers in a system should be minimized (that is, zero or one).

To avoid the synchronizer problem, systems can sometimes be phase-locked [Gardner 79] so that they no longer have an arbitrary timing relationship. Phase-locking is often used in large systems where it is easy to distribute accurate frequency but not phase information. When two systems have independent but well-defined frequencies, it is possible to avoid much of the synchronizer timing penalty by predicting future synchronization events and resolving them ahead of time [Ward 83]. In other words, the speed with which an unpredictable event can occur is related to its bandwidth rather than its frequency. The purpose

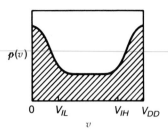

FIGURE 6.54 Initial probability distribution.

of a synchronizer is to resolve unpredictable events. Therefore the performance of the synchronizer is fundamentally limited not by the frequency of incoming signals, but by their bandwidths.

The influence of noise on synchronizer performance

Noise of low amplitude does not affect the performance of the arbiter circuit. Since this is somewhat counterintuitive, we will demonstrate why this is so. To do this we must show that any noise added to the system does not change the equations earlier in this section. That is, we want to show that noise is a completely orthogonal issue.

Divide the voltage interval around the metastable point into a logarithmic scale. For instance, 1 to 10 μV, 10 to 100 μV, and so on, as shown in Table 6.2.

TABLE 6.2 Initial probability distribution

Voltage range	Probability
1–10 μV	P_0
10–100 μV	$10P_0$
100–1000 μV	$100P_0$
1000–10,000 μV	$1000P_0$

Given that the initial voltage can be at any voltage, if P_0 is the probability that the voltage is in the interval from 1–10 μV, then it has a probability of $10P_0$ of being in the interval from 10–100 μV since this interval is ten times as wide. Figure 6.54 illustrates the initial voltage probability distribution. The important question concerns what happens as a function of time. The voltages in the system obey Eq. (6.15). Thus it would take a voltage offset at 1 μV exactly as long to move to 10 μV as it would take a voltage at 10 μV to move to 100 μV, and so on. Thus with the initial probabilities given in Table 6.2, at $T = \tau_{\text{switch}} \ln 10$ the probabilities would be those shown in Table 6.3.

We have shown that in the linear region about V_{MS}, the probability density decays uniformly. The overall effect is shown in Fig. 6.55. If the initial voltage is just as likely to be at V_{MS} as at $V_{MS} + 1$ mV, then two

FIGURE 6.55 Later probability distribution.

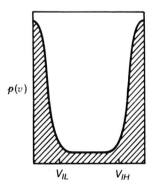

TABLE 6.3 Later probability distribution

Voltage range	Probability
1–10 μV	$0.1P_0$
10–100 μV	P_0
100–1000 μV	$10P_0$
1000–10,000 μV	$100P_0$

seconds later it is still just as likely to be at V_{MS} as at $V_{MS} + 1$ mV, even though the likelihood of the voltage being at either point is orders of magnitude smaller than it was earlier.

This is significant because, if we assume the voltage can be anywhere about V_{MS} at the sampling time, then, if it is in the linear region, it can still be anywhere for any $T > 0$. We are now in a position to understand the effects of noise. Noise of small amplitude will act as an offset bias. But it is just as likely to hurt as to help since it moves the system among identical scenarios. It might, for instance, move the metastable voltage up one millivolt. On the other hand, we have just shown that this does not change anything.

As a final argument for why noise does not matter, consider that if noise did matter then we could build a better synchronizer. No one has.

When the noise is of moderate amplitude, it can actually hurt the synchronizer time by moving the system into a region where τ_{switch} is not as short. When the noise is of high enough amplitude we have Glasser's perfect arbiter—one which is perfectly arbitrary. There is one case in which an arbiter need never take an unbounded length of time to decide: when it never looks at the input data. The arbiter time can be made very short in this case.

6.6   Global clocking issues

The correct resolution of global clocking issues depends on the total systems context. We examine several common scenarios. The most straightforward case is when one is the master of the universe. You generate the "tune" with an on-chip clock generator and the rest of the system must dance to it. Another fairly straightforward case is when a good full swing clock is generated off-chip. This system clock should be the same type of clock used on all VLSI chips in the system (for instance, two-phase nonoverlapping). Since no on-chip clock buffers are needed, skew between chips is minimized. Skew between the clock phases is still a problem since these skews accumulate and, in a high-performance system, can get to be a significant fraction of the period. A skew of 5 ns on a board is quite good and 50 ns of skew in a machine that is several racks large is quite common. Note that it is not just distributing the clocks that is the issue, but distributing the data lines as well. After all, it is the timing of the clock relative to the data signals that determines the interpretation of the voltages on the data wire. Another problem involves bonding wire inductance.[7] The worst case is when a low-voltage

---

[7] Inductance is a problem with on-chip clock generation as well. Instead of inductance on the clock wire being a problem, the inductance on the V_{DD} wire is a problem. $L\frac{di}{dt}$ on the ground wire is always a problem.

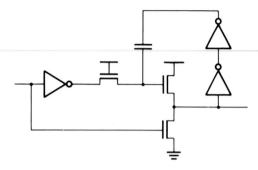

FIGURE 6.56   A late booting buffer.

(read "TTL") clock is brought in from the outside world. Then the clock must be buffered and the system timing margins are reduced. Often a late booting buffer is used to get the clock swing started as early as possible. A late booting buffer is illustrated in Fig. 6.56.

Often the output signal skew for a chip is specified in relation to the input clock. In this case it is profitable to partition the chip into two timing regimes as shown in Fig. 6.57. Skew between $\phi_{external}$ and $\phi_{internal}$ will typically be large, and this skew comes directly out of the speed budget for the chip. On the other hand, one is at least able to closely time the output signal by using a signal that has very little skew with respect to $\phi_{external}$. The differences in loads that the clock buffers must drive cause the differences in skew. The skew between $\phi_{internal}$ and the external clock tends to be large because all the capacitance it must drive necessitates large buffer delays. These delays are process sensitive.

FIGURE 6.57   Deskewing a chip interface. There is internal time and external time.

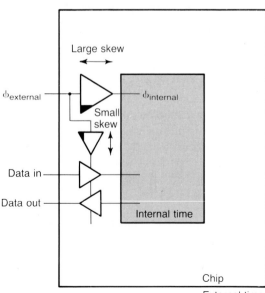

■■■■■ EXAMPLE 6.6

Microprocessor system timing

The block diagram of a simple microprocessor system is illustrated in
Fig. 6.58. The microprocessor reads the register file on ϕ_1, does an ALU
operation on ϕ_2, and writes the register file on ϕ_3. We could choose to
do I/O on either ϕ_1 or ϕ_3. If we choose ϕ_1 and the ALU is used, we
would waste two clock phases. On the other hand, if we choose ϕ_3, and
the register file is used, but not the ALU, then one clock phase is wasted.
We have arbitrarily chosen to do I/O on ϕ_3 in this example.

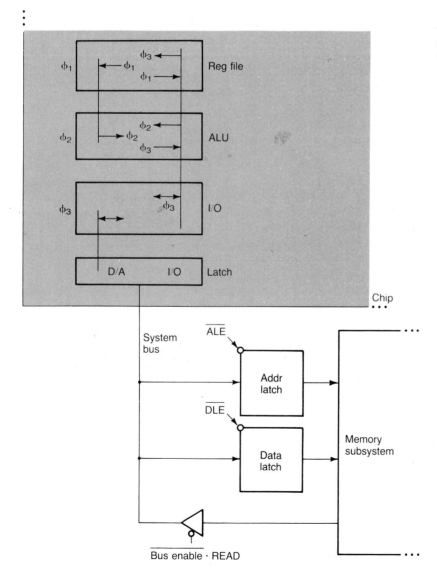

FIGURE 6.58   A system
timing example of a
microprocessor and external
memory.

Consider the case of generating an address that is an offset from the program counter (PC), where the PC and offset are both stored in the register file. The required operations are

- Simultaneously Read PC and Offset Out of Reg File;
- Add together in ALU to find the physical addr;
- Pump addr out I/O port.

These operations are finished on $\phi_1, \phi_2,$ and ϕ_3, respectively.

Assume multiplexed data and address pins. The timing diagram is illustrated in Fig. 6.59. Communication with the outside world occurs as follows.

- Master (the microprocessor) raises $\overline{\text{BUS ENABLE}}$ by end of ϕ_1, telling the world that it wants the bus. Gives the slaves one phase to get off.

FIGURE 6.59   Timing of microprocessor and external memory system.

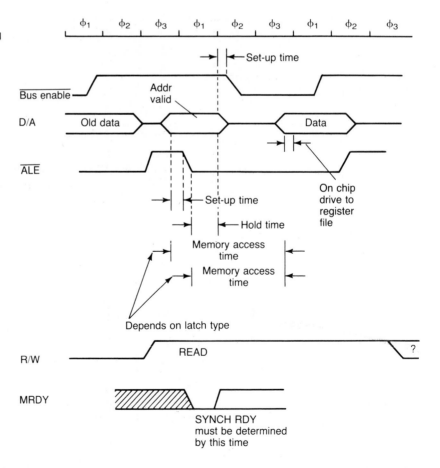

- Sometime later when the Master knows what it is doing (this timing is not very critical), it drives the R/W line to indicate either a read or write to memory. During the phase that the slaves have to get off the bus, the ALU is computing the physical address.

- During ϕ_3 the address is driven onto the system bus. If the external address latch is transparent, then the memory-access time starts when the address is valid on the system bus.

- After the address is valid on the system bus, the Master waits one setup time and drops $\overline{\text{ALE}}$, the address latch enable signal. The setup time is the time from when the address is valid until the $\overline{\text{ALE}}$ signal leaves a valid one level (V_{OH}).

- At the beginning of ϕ_2, the Master tristates the system bus drivers. At this point the address is no longer valid. The time from when $\overline{\text{ALE}}$ reached a valid low level until the address first becomes invalid is the hold time. Notice that the time during which $\overline{\text{ALE}}$ was dropping is neither hold nor setup time. It is wasted time. If the address latch is not transparent, then the memory access time starts from when $\overline{\text{ALE}}$ reaches a valid zero level.

- After the bus drivers get to the high impedance state, $\overline{\text{BUS ENABLE}}$ is lowered to indicate that the memory is now free to grab the bus. The setup time between when the address becomes invalid and $\overline{\text{BUS ENABLE}}$ leaves the high state is usually zero.

- The memory system does its job and sometime before the end of ϕ_3, valid data are on the system bus. The microprocessor has the rest of ϕ_3 to drive this data from its input pins into the register file, probably through a mux. (The data could alternatively be kept in the I/O latch. This would mean that if the next operation used this data in the ALU, the data would need to be driven into the ALU on ϕ_1 as if it came from a register.) The memory access time is complete when the data is valid on the system bus.

- If the master would like another bus cycle, it raises $\overline{\text{BUS ENABLE}}$ and $\overline{\text{ALE}}$ before the end of the next ϕ_1.

If the microprocessor would like to write rather than read memory, another signal, data latch enable (DLE), would be needed.

The situation is somewhat more complicated if asynchronous signals are included. For instance, one might have different classes of addresses— some that are in fast memory and others that are on disk. Or one might have error correcting codes (ECC), where the accesses of perfect data might be fast while that of corrupted data slow. Let us examine the case when normal memory accesses correspond to the timing already developed. But in special cases the memory access might take an arbitrarily longer time. We will need to include one more signal, RDY, and a synchronizer.

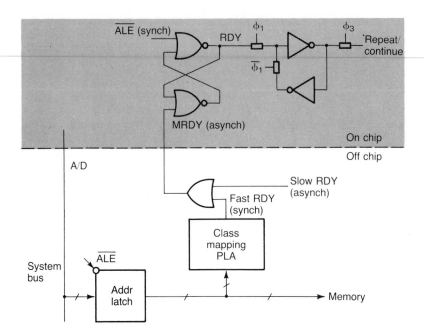

FIGURE 6.60 Handling asynchronous events. Most references are synchronous.

To minimize the probability of a synchronizer failure, we design the memory system so that the MRDY line is guaranteed to be low if the memory is not ready, but it can go high at any time. Moreover, in the case of a normal access, the memory is guaranteed to be able to determine before ϕ_2 that it will be able to meet the normal timing constraints. For instance, the determination of the class of memory access is determined in the class mapping PLA. If the address is in fast memory, then MRDY is issued during ϕ_1. The RDY signal is sampled at the end of ϕ_1, but not used until ϕ_3. This allows one third of a microcycle, or more, for the synchronization latch to settle. The RDY signal is used during ϕ_3 to determine whether or not the microcycle should continue or be repeated (to wait for the memory to be ready). This is illustrated in Fig. 6.60. ■

6.7 Perspective

The interpretation or meaning of a signal depends on its relative position in time and space. In this chapter we focused our attention on the time axis. A digital signal cycles through a sequence of interleaved times when it is correct, then incorrect, then correct again, to interpret the signal voltage as representing valid logic values. We must discover the times when a signal is valid, but we must not confuse or mix up these many different times.

For a machine to compute a meaningful function, it is important that the entire system have a unique sense of time. At a minimum, this may be nothing more than a partial ordering of events (A must occur before B and C), but usually we exploit the structure provided by a clock. When a single time sense does not exist, then problems such as clock skew and synchronizer failure occur.

Delay uncertainties must be removed from the system. This is done by delaying signals at registers to make all delays equally bad. Clocks are typically used to control these registers. Clocks are also used in various circuit techniques, and they ease the constraints on circuit delay. If one does not want constraints on the minimum delay in a clocked system, one must use a clock with either two or more nonoverlapping phases, or three or more overlapping phases. If additional clocks are needed, they may be generated in a self-timed manner by logically detecting the completion of some event. To obtain good component efficiency in the generation of these extra clocks, we exploit the tracking of component values on the integrated circuit.

Problems

6.1 Draw the constraint space of a two-phase clocking methodology in which ϕ_1 falls before ϕ_2 rises, but ϕ_1 rises again before ϕ_2 falls. Does this methodology generate one- or two-sided constraints on the delay through a logic module?

6.2 Draw the output waveform from the circuit illustrated in Fig. 6.61.

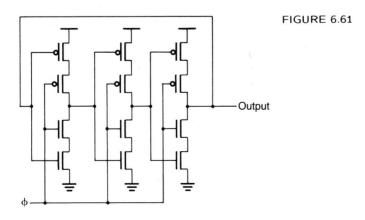

FIGURE 6.61

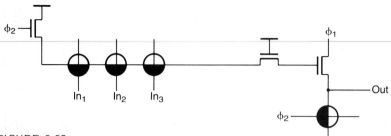

FIGURE 6.62

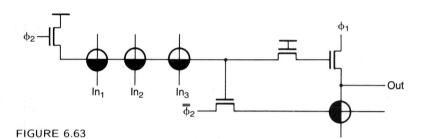

FIGURE 6.63

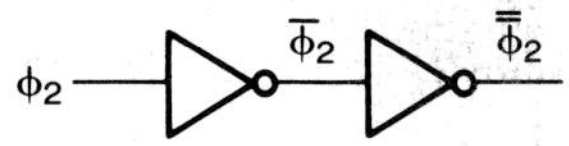

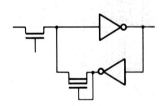

FIGURE 6.64

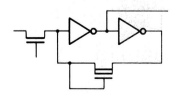

FIGURE 6.65

FIGURE 6.66

6.3 The circuit in Fig. 6.62 evaluates in the dead time between when ϕ_2 falls and ϕ_1 rises. It has been modified as shown in Fig. 6.63, where $\overline{\overline{\phi_2}}$ is a delayed version of ϕ_2. (See Fig. 6.64.) What useful property does the circuit in Fig. 6.63 have that the circuit in Fig. 6.62 lacks? Why do we use $\overline{\overline{\phi_2}}$ rather than ϕ_2?

6.4 Show that the latch circuit in Fig. 6.65 is superior to that in Fig. 6.66.

6.5 Show how the large NOR gate in the binary counter of Fig. 1.58 can be precharged. Check for charge sharing, sneak paths, and correct timing.

6.6 Figure 6.67 illustrates a bootstrapped clock buffer. What is the purpose of M_X? Discover and describe the critical timing and bootstrapping issues.

6.7 While an arbiter in an asynchronous system is deciding which of two inputs arrived first, it is important that the output of the arbiter not be uncertain. Assume that the arbiter has two output wires, one that indicates input A arrived first and one that indicates that input B arrived first. It is critical that we never have the condition that both A and B are both high simultaneously. Design a circuit, called a mutual exclusion circuit, that has two inputs and two outputs. If only A_{in} is true, then only A_{out} will be true, and the same for B. If A_{in} and B_{in} are both false, then both outputs must be low, but *never* should both outputs be high.

6.8 The shift register cell in Fig. 6.68 is generally preferred over those in Figs. 1.48 and 1.52. Why?

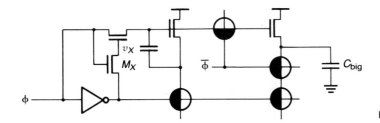

FIGURE 6.67

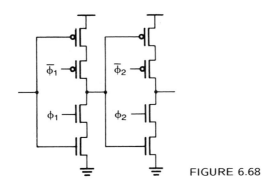

FIGURE 6.68

References

[Anceau 82] F. Anceau, "A Synchronous Approach for Clocking VLSI Systems," *IEEE J. Solid-State Circuits* **SC-17**: 51–56, 1982. (On-chip phase-lock loop to eliminate skew due to clock buffers)

[Bazes 83] M. Bazes, J. Nadir, D. Perlmutter, B. Mantel, and O. Zak, "A Programmable NMOS DRAM Controller for Microcomputer Systems with Dual-Port Memory and Error Checking and Correction," *IEEE J. Solid-State Circuits* **SC-18**: 164–172, 1983. (Single-phase clocking and distributed TTL input buffers)

[Boysel 70] L. L. Boysel and J. P. Murphy, "Four phase LSI logic offers new approach to computer design," *Computer Design*: pp. 141–146, April, 1970.

[Couranz 75] G. R. Couranz and D. F. Wann, "Theoretical and Experimental Behavior of Synchronizers in the Metastable Region," *IEEE Trans. Computers* **C-24**: 604–616, 1975.

[Ellul 75] J. P. Ellul, M. A. Copeland, and C. H. Chan, "MOS Capacitor Pull-up Circuits for High-Speed Dynamic Logic," *IEEE J. Solid-State Circuits* **SC-10**: 298–307, 1975.

[Gardner 79] F. M. Gardner, *Phaselock Techniques*, Wiley, New York, 1979.

[Goncalves 83] N. F. Goncalves and H. J. De Man, "NORA: A Racefree Dynamic CMOS Technique for Pipelined Logic Structures," *IEEE J. Solid-State Circuits* **SC-18**: 261–266, 1983.

[Heller 84] L. G. Heller and W. R. Griffin, "Cascade Voltage Switch Logic: A Differential CMOS Logic Family," *IEEE International Solid-State Circuits Conf.*, pp. 16–17, San Francisco, Calif., 1984.

[Kinniment 76] D. J. Kinniment and J. V. Woods, "Synchronization and Arbitration Circuit in Digital Systems," *Proc. IEE* **123**: 961–966, 1976.

[Krambeck 82] R. H. Krambeck, C. M. Lee, and H.-F. S. Law, "High-Speed Compact Circuits with CMOS," *IEEE J. Solid-State Circuits* **SC-17**: 614–619, 1982. (Domino CMOS)

[Kung 82] R. I. Kung, S. T. Flannagan, and J. N. Spitz, "An 8K × 8 Dynamic RAM with Self-Refresh," *IEEE J. Solid-State Circuits* **SC-17**: 863–871, 1982. (Uses an arbiter)

[Man 78] H. J. De Man, C. J. Vanderbulcke, and M. M. Van Cappellen, "High-Speed NMOS Circuits for ROM-Accumulator and Multiplier Type Digital Filters," *IEEE J. Solid-State Circuits* **SC-13**: 565–572, 1978.

[Mead 80] C. A. Mead and L. Conway, *Introduction to VLSI Systems*, Addison-Wesley, Reading, Mass., 1980.

[Suzuji 73] Y. Suzuji, K. Odagawa, and T. Abe, "Clocked CMOS Calculator Circuitry," *IEEE J. Solid-State Circuits* **SC-8**: 462–469, 1973.

[Veendrick 80] H. J. M. Veendrick, "The Behavior of Flip-Flops Used as Synchronizers and Prediction of Their Failure Rate," *IEEE J. Solid-State Circuits* **SC-15**: 169–176, 1980.

[Ward 83] S. Ward, private communication, 1983.

[Watkins 67] B. G. Watkins, "A Low-Power Multiphase Circuit Technique," *IEEE J. Solid-State Circuits* **SC-2**: 213–220, 1967.

7 CIRCUIT TECHNIQUES FOR ARRAY STRUCTURES

In Chapter 5, we investigated the design of specialized small integrated circuits such as drivers and sense amplifiers. In Chapter 6, we examined clocking techniques. We now put these subjects together and apply them to the design of large regular structures.

7.1 Decoders

In the class of regular combinational logic structures, the simplest is the decoder. It is composed of only one level of logic. We saw, in Chapter 1, that the decoder could be implemented in several different ways. For a simple static implementation, the NOR form has a higher performance in nMOS technology than the NAND form. This is because the pulldowns are wired in parallel rather than in series. The same argument applies to the ratioed CMOS circuit in Fig. 7.1. In this circuit, p-channel devices with grounded gates, which mimic depletion loads, are used for the pullups. For a classical CMOS implementation, the NAND form is preferred because this allows the series-connected transistors to be of the higher performance variety (n-channel). That is, one would prefer to have five series n-channel devices limiting the performance of a circuit rather than five series p-channel devices. Note that it is always possible

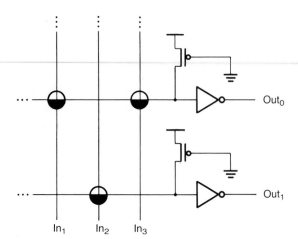

FIGURE 7.1   Ratioed
CMOS NOR form decoder.

In_1 In_2 In_3

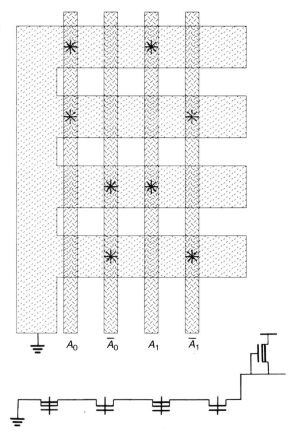

FIGURE 7.2   Implant
programmed NAND form
decoder.

A_0 $\overline{A}_0$ A_1 $\overline{A}_1$

to reduce the fan-in of a logic gate to two by partitioning the function into a cascade of several low fan-in gates. But even with a fan-in of five, it is probably still marginally better to use only one logic gate.

When performance is a lesser issue than compactness, the NAND form may even be preferred in nMOS because the personalization of the array can be accomplished by implants. An implant programmed decoder, where the depletion device is viewed as an "always on" transistor, is illustrated in Fig. 7.2. Note how compactly the circuit lays out since there are no contacts to metal internal to the array. The layout pitch (repeat distance) is determined by the wire widths and the implant surround rules. There are twice as many series transistors as in the normal NAND case. The voltage waveforms out of implant programmed decoders exhibit a pronounced rippling. This is due to series depletion devices with their gates at ground. We saw this cascade connection when studying single-ended sense amplifiers (see Fig. 5.29). It was also used in the clock buffer of Fig. 5.16 (M_{10}) in order to sharpen the down-going transition out of the delay line. In the implant programmed decoder, however, this circuit only gets in the way of the performance because the source voltage on the series depletion device must fall all the way to about 2.5 V before the drain even starts to move. Some processes have a deep depletion transistor type, with a threshold near $-V_{DD}$, to increase the performance of this sort of circuit.

Because of the special logic regularity of the decode function, other circuit implementation structures are possible, such as the tree decoder in Fig. 1.65. One extremely useful special structure exploits the virtual ground technique illustrated in Fig. 7.3. This decoder is much like the NOR decoder, except that ground is selected to be either the even or odd rows, as a function of A_0. Instead of one series pulldown, as in the case of the NOR decoder, there may be two. As an example of a worst-case scenario, M_1 can be required not only to pull down C_1 but also to pull down C_2 through the transistor M_4. The array's performance can also be degraded by diffusion resistance since the horizontal wires are often run in n+. Thus there can be a large series resistance to the source terminal of M_4. For these, possibly small, sacrifices in performance, one typically gains over 30% in layout density. Connected to the output lines of the precharged virtual ground circuit shown in Fig. 7.3 is a simple nMOS bootstrapped buffer.

In all arrays, coupling noise is a major concern because of the coherent noise effects of many lines all changing at once in the same direction. Thus many small noise sources, which in random logic would not pose a problem, are a serious concern in tightly packed arrays. On the other hand, because coupling in regular arrays does behave in a predictable manner, one can analyze the noise and either control its amplitude with circuit techniques (such as limiting voltage swings and

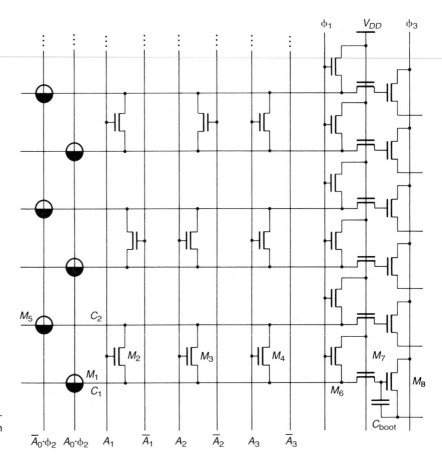

FIGURE 7.3 Two-for-one dense decoder with bootstrapped outputs.

precharge currents or generating differential signals) or, at the very least, predict the required noise margins. Since there is a trade-off between noise margin and speed, being able to estimate the required noise margin goes a long way toward improving the speed. Of primary importance are the coupling between adjacent lines, crossing groups of lines, and the resistance of diffused and poly wires. Current spikes due to precharging are also of great importance.

Many noise issues are manifested in the circuit of Fig. 7.3. Most center around the lack of noise robustness in the bootstrapped driver stage. This circuit is poor from a noise standpoint because precharged lines (C_1, C_2, etc.) are followed by circuits with low noise robustness. M_6 is responsible for precharging C_1. Let us first examine the case when M_8 should bootstrap. For M_8 to bootstrap, the voltage on C_1 must stay above V_{DD} minus the threshold of M_7. Note, however, that if M_6 and M_7 are the same size, then they have the same threshold voltage. Worse yet, because enhancement-mode n-channel devices make such poor pullups, it is unlikely that M_6 will really do that good a job of pulling C_1 all the way up to where M_7 turns off.

There are several possible solutions to this problem. One is to draw M_7 longer than M_6. This will cause the threshold of M_7 to be higher than the threshold of M_6. Alternatively, we could make M_6 a zero threshold device. Yet another technique is to notch the V_{DD} voltage down just before ϕ_3 goes high. Remember that one source of noise is V_{DD} bounce or "slewing." If V_{DD} bounces high just before ϕ_3 goes high, then the bootstrapping action could be lost because M_7 would be on. Note that even the falling of the precharge line ϕ_1 is in the direction to hurt rather than help the bootstrapping action. Of greater concern, however, is the movement of the numerous address lines after ϕ_1 falls. Falling address lines will pull C_1 and C_2 low. This is yet another way to kill the bootstrapping. Thus it is important to have the address line stable before ϕ_1 drops. (In this circuit, where each address line is paired with its complement, this effect is not as strong as in other contexts, assuming symmetry in the layout capacitances. Of course, this symmetry must exist even under misalignment in order for one to count on it.)

All of the discussion so far has dealt with ensuring that bootstrapping action occurs when it should. A second important case is preventing the bootstrapping action when it should not happen because we want the output to remain low. Address decoders such as the one in Fig. 7.3 often drive precharged circuits. Thus it is important for the output not to bounce. This requires that M_7 be wide enough so that the gate/drain overlap capacitance of M_8 does not yank up the gate of M_8 to the point where it turns on. This is particularly important in this decoder because there is no transistor holding the source of M_8 down. Thus if the source voltage moves up at all, it will not recover.

In a CMOS implementation, an inverter could be used as the buffer, or the precharged circuit in Fig. 7.4 could be used. (Watch out for the logic inversion.) In this circuit, $NR_H = |V_{TP}|$, so all the sorts of noise issues seen in the circuit of Fig. 7.3 must be considered in this circuit

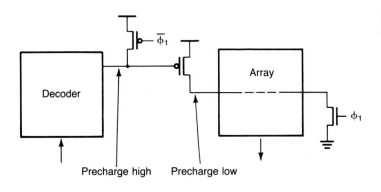

FIGURE 7.4 Precharged CMOS decoder.

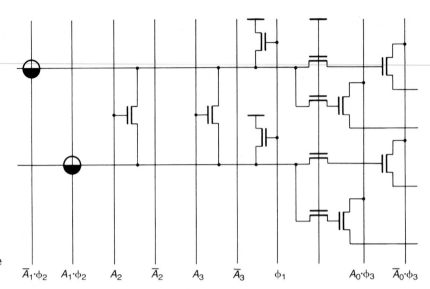

FIGURE 7.5   Four-for-one dense decoder.

$\overline{A}_1 \cdot \phi_2$ $A_1 \cdot \phi_2$ A_2 $\overline{A}_2$ A_3 $\overline{A}_3$ ϕ_1 $A_0 \cdot \phi_3$ $\overline{A}_0 \cdot \phi_3$

also. Figure 7.5 illustrates another decoder, designed in the same idiom as the one in Fig. 7.3. Additional decoding is done in the output buffer.

7.2   ROMs and PLAs

ROMs and PLAs are the two most common array structures for the implementation of combinational logic. In the study of these arrays we will see the interaction of many different domains of abstraction.

Issues in ROM design

Near the end of Chapter 2 we did a detailed analysis and simulation of an HROM. The circuit issues we confronted in Chapter 2 were poly propagation times in the word line and noise on the ground wires. In Chapter 4, we revisited the problem and discovered coupling between the bit lines to be a major issue. In Chapter 5, we examined sense amplifiers appropriate for placing on the output of the ROM. We have not, however, exhausted the set of issues. In this section, we consider three more: the use of virtual grounds, the XROM layout technique, and logic optimization.

The virtual ground technique we saw in the decoder designs can be applied to the ROM bit lines, as shown in Fig. 7.6. Transistors in the ROM are addressed by raising one of the word lines and using the Y_i lines to select the correct bits. For instance, the transistors in column B_4

are addressed by raising Y_1. This output is directed to output Z_1. At the same time, the programming in the B_3 column is directed by the selector to the output Z_0. All columns are precharged high on ϕ. In the virtual ground ROM circuit, each bit requires an average of just one horizontal and one vertical wire. In the nonvirtual ground circuit, we require 1 1/2 by 1 wires per bit because of the necessity for the shared ground wire.

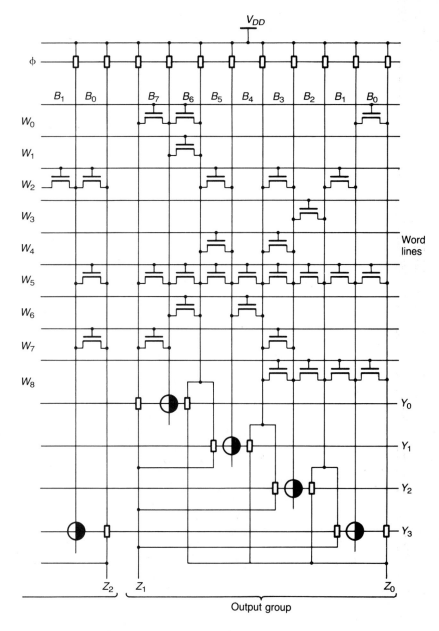

FIGURE 7.6   Part of virtual ground ROM.

A major issue in the virtual ground ROM is sneak paths. For instance, raising Y_2 and W_5 causes all of the bit lines in the group to discharge. This is the cause of the complexity of the output selection circuitry. The space between B_0 and B_7 is there to halt the sneak path. If the space was not there, a falling bit line in one group could cause the outputs of neighboring groups to fall due to sneak paths. Note that for this to happen, however, the neighboring output would need to be pulled low through many series transistors. So rather than thwart this effect by inserting a space in the array, one can use highly resistive pullups in parallel with the precharge devices, or use a time out circuit that stops looking at the output after a certain time. Since the sneak path pulldown is so much slower than the wanted path, the wanted paths can be selected by only looking for the faster transients.

Using enhancement-mode precharge devices will, of course, work, but we should note that the use of depletion-mode pullups on the lines going to the sense amplifiers is also reasonable. They pull all the way to V_{DD} and the only drawback is that as the bit lines fall, the depletion devices turn back on. In most reasonable implementations, however, the bit lines will have been sensed before the pullups turn back on.

Another possibility is to precharge low instead of high. A high word line would then be expected to turn on a transistor whose role is to pull the bit line high. We know that the end of the pullup transition will be much slower than if the circuit was designed to pull down, but the output will probably be sensed in the first half-volt and the initial slew rates are comparable.

When ground is run parallel to the bit lines, or when the virtual ground technique is used, a layout trick is available that can substantially reduce the ROM cell area. This technique, illustrated in Fig. 7.7, results

FIGURE 7.7   XROM layout.

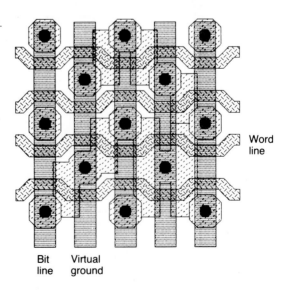

Word line

Bit line   Virtual ground

in active area patterns in the shape of an "X" rather than an "H." It is called, naturally, an XROM. Because of the tightly packed layout and strangely shaped transistors, the analysis of an XROM is quite complex. The transistors are typically very narrow and can have thresholds significantly above those used in most of the rest of the circuit. This is mainly due to the lateral diffusion of the field implant under the channel. Thus these transistors require additional characterization. XROMs are usually built using the virtual ground technique and typically exhibit a circuit speed roughly half that of the HROM implementation.

Still more density can be obtained by using implant encoding, as in the decoder of Fig. 7.2, or by adopting a design that encodes more than one bit per crosspoint. This can be done by encoding the additional information by threshold implants or transistor size encoding. The sense circuitry for ROMs that encode multiple bits per crosspoint must be quite sophisticated.

Optimization can be obtained by looking in a different dimension. One of the advantages of the ROM is that we are free to trade off the number of rows for the number of columns; that is, we can adjust the aspect ratio. If one shrinks the number of rows to 16 or 32, then one finds that a significant fraction of the columns are either completely populated or contain no programming at all. In either case, but especially in the second, it is straightforward to eliminate that column and hence shrink the size of the ROM [Guttag 82].

Issues in PLA design

Programmable logic arrays are closely related to ROMs, and many merged or bastardized structures are possible. In a pure PLA the virtual ground technique does not work. The sense amplifier design and noise calculations are also more difficult because several rows can be selected an once. On the other hand, there are additional freedoms that are not present in the ROM. Most of these occur in the logic and topology domains.

Most large arrays are precharged, and PLAs are no exception. Two precharge techniques are illustrated in Figs. 7.8 and 7.9. Both of these PLAs use two-phase nonoverlapping clocks. A single-phase PLA was illustrated in Fig. 1.70. In the circuit in Fig. 7.8, a virtual ground technique is used to keep the planes from evaluating until the inputs are stable. During ϕ_1, the AND plane is precharged high. Though the inputs may be moving during this time, the output of the AND plane is not discharged since there is no path to ground. The AND plane must be stable by the time ϕ_2 starts to rise. When ϕ_2 rises, all of the virtual ground lines in the AND plane fall. This can cause a lot of noise both inside and outside the array. The external noise is due to the current surge in the ground line. The internal noise is caused by the falling virtual ground lines, which capacitively pull down the inputs (possibly

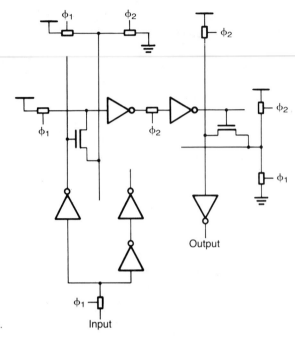

FIGURE 7.8   Virtual ground PLA.

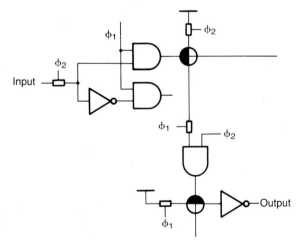

FIGURE 7.9   Precharged
two-phase PLA.

slowing down the circuit) and pull down the precharged AND plane outputs. A bootstrapped buffer in the style of Fig. 7.3 might thus be a very dangerous circuit to put between the AND and OR planes. During ϕ_2 the AND plane evaluates and the OR plane is precharged. The OR plane is evaluated during ϕ_1 while the AND plane is being precharged. The outputs of the OR plane can be brought into the AND plane at this time to implement a finite-state machine.

In Fig. 7.9 the skeleton of a second two-phase PLA is illustrated. The AND plane is precharged on ϕ_2 while the OR plane evaluates, and the OR plane is precharged on ϕ_1 while the AND plane evaluates. Instead of using the virtual ground technique, the inputs to the pulldown transistors are held low while the array is precharged.

In an nMOS implementation, clocked bootstrap buffers would be used to implement the AND gates. In the most compact implementations of these buffers (see Fig. 5.12(e), for instance), the output is precharged low but not held. Thus it is very important to examine the coupling to this line. It is possible that there will be two high-going OR plane word lines on either side of a word line that should stay low. This will tend to pull the low word line high. On the other hand, falling bit lines in the OR plane (and we should expect to see many of them) will push down on the word lines, tending to hold them low. Usually, this second effect predominates. In the AND plane this second effect certainly predominates since all but a few product terms will be low in any noncontrived example. Note that the capacitive load that the clock lines see is a function of the PLA programming.

To relieve the need to spend a whole phase time precharging, a glitch precharge clock is often employed. A glitch precharge clock usually occurs near the start of a phase and lasts only a small fraction of the total phase time. It is used to precharge those components that will be evaluated later that phase.

PLAs can be optimized in many dimensions. We will now examine a few examples. The most obvious technique is straight logic minimization. For even modest size input and output vectors, minimization of the number of product terms[1] (rows) is computationally intractable. A number of heuristic techniques have, however, been developed to somewhat (10 to 30%) reduce the number of product terms [Hong 74]. For instance, the two sum of product expressions

$$X = \overline{A}\,\overline{B}\,\overline{C} + \overline{A}B\overline{C} + \overline{A}BC + ABC \qquad (7.1)$$

and

$$Y = \overline{A}\,\overline{B}\,\overline{C} + \overline{A}\,\overline{B}C + A\overline{B}C, \qquad (7.2)$$

representing six out of a possible eight product terms, can be reduced to

$$X = \overline{A}\,\overline{B}\,\overline{C} + \overline{A}B + BC \qquad (7.3)$$

and

$$Y = \overline{A}\,\overline{B}\,\overline{C} + \overline{B}C. \qquad (7.4)$$

Equations (7.3) and (7.4) require only four product terms to implement.

---

[1] Using, say, the Quine-McCluskey technique.

Note that rewriting X as

$$X = \overline{A}\,\overline{B}\,\overline{C} + \overline{A}B + ABC \qquad (7.5)$$

would have required one more transistor, but would not change the size of the array because it would not change the number of product terms.

The usual way of constructing AND planes is somewhat transistor-inefficient. Take the example of two input strings A and B. If these strings are interleaved, then the inputs to the AND array of the PLA are

$$(a_0, \overline{a}_0, b_0, \overline{b}_0, a_1, \overline{a}_1, b_1, \overline{b}_1, a_2 \ldots). \qquad (7.6)$$

There are four array inputs per bit position and they take the form

$$(a_i, \overline{a}_i, b_i, \overline{b}_i). \qquad (7.7)$$

For each two variables, a_i and b_i, there are four places for transistors in the AND plane per product term. This means that there are 16 possible programming combinations. It also turns out that there are 16 possible logical functions of two variables. This is a wonderful coincidence, but unfortunately one does not take advantage of it when the inputs are brought into the array in the form given in Eq. (7.7). This is because there are many programming combinations that are equivalent. For instance, $a_i + \overline{a}_i$ is the same (uninteresting) function as $b_i + \overline{b}_i$. We would ideally like to bring into the array four functions of a_i and b_i that allow all 16 functions to be programmed in each product term. The correct functions to bring into the array are the elements of the Karnough map; they are

$$(a_i b_i, a_i \overline{b}_i, \overline{a}_i b_i, \overline{a}_i \overline{b}_i). \qquad (7.8)$$

This is done by placing logic gates, rather than simple inverters, at the inputs to the array. All of the functions one could previously implement in a single product term can still be implemented in a single product term. In addition, there is a whole class of useful functions that were previously very difficult, but that can now be implemented in a single product term. For instance, PLAs are notoriously bad at checking the equality of two input strings A and B, yet with the encoded input, this test can be performed in a single product term. This is because functions such as $a_i \overline{b}_i + \overline{a}_i b_i$ are much easier to implement. Note that a_i is implemented as $a_i b_i + a_i \overline{b}_i$, and $a_i + \overline{b}_i$ is implemented as $a_i b_i + a_i \overline{b}_i + \overline{a}_i \overline{b}_i$.

In yet another dimension, we can look at the partitioning and folding of PLAs. In PLA folding techniques, one takes a standard PLA, such as that in Fig. 7.10, and reduces its size by cutting out unused portions,

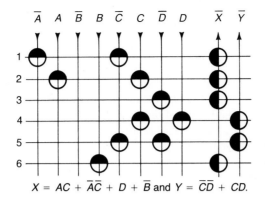

FIGURE 7.10 PLA that implements $X = AC + \overline{A}\,\overline{C} + \overline{B} + D$ and $Y = \overline{C}\,\overline{D} + CD$. [Hatchell 82] (©1982 IEEE)

$X = AC + \overline{A}\,\overline{C} + D + \overline{B}$ and $Y = \overline{C}\,\overline{D} + CD.$

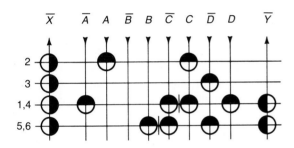

FIGURE 7.11 Row-folded version of the PLA in Fig. 7.10. [Hatchell 82] (©1982 IEEE)

then filling the spaces in by folding the array in various creative ways. Two folded arrays, which perform the same function as that illustrated in Fig. 7.10, are illustrated in Figs. 7.11 and 7.12 [Hachtel 82]. The main drawback of this technique is that the positions of the inputs and outputs are now scattered, perhaps to the detriment of the floor plan. Changes are also difficult to implement. Note that folding can be done in conjunction with input encoding.

Storage logic arrays (SLAs) are designed with the floor plan very much in mind [Patil 79]. PLAs can be folded because they are typically sparsely populated. This is particularly true if the PLAs are algorithmically generated from a higher level representation by a computer program. In most large systems there is a locality of communication. By this we mean that very few wires are global. Most go to only a few places. (What would a chip with 100,000 gates look like if every gate was connected to every other gate?) This is, in part, why PLAs are sparsely populated. Thus one would hope that by rearranging the parts of a PLA, as we did in folding, but only more so, that a regular structure without long wires could be obtained. In particular, OR plane outputs may be interleaved with the AND plane inputs. This

FIGURE 7.12 Column-folded version of the PLA in Fig. 7.10. [Hatchell 82] (©1982 IEEE)

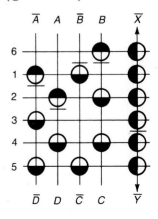

is the philosophy behind SLAs. Registers are typically interleaved with the fragmented logic arrays in order to provide greater design freedom.

Even in the circuit domain, the sparseness of a PLA personalization can be exploited. One of the limitations to the speed of the PLA (or ROM) is the size of the transistors that can be placed at the crosspoints. If the PLA is sparsely populated and one has a clever rearrangement technique, then one can place transistors that are larger than one array site at the crosspoints by making sure the neighboring crosspoints are unprogrammed. In the few cases when rearrangement is unsuccessful, the array must be enlarged.

We have seen several dimensions of PLA optimization. We have certainly not exhausted the space of excellent work on logic arrays [for example, Weinberger 67, Lopez 80]. Our main purpose is to give a flavor of the modes of thinking that have proved profitable. Note the many domains—layout, topology, circuit, logic, and so on—that come into play. These domains are not orthogonal. Optimizations in one domain influence the potential for optimization in other domains. It is, as of this writing, a completely open question as to how these domains interact. Excellent optimization programs have been written in individual domains, but coordinated optimization techniques spanning multiple domains of abstraction remain unrealized, waiting to challenge future researchers and computers.

7.3 Random access memory

There are many types of random access memory (RAM) that may be implemented on a VLSI chip. In general, the fewer transistors there are in the memory cell, the more involved and extensive the peripheral circuitry. We will start our discussion with the memory cells with the least complicated peripheral circuitry.

The eight-transistor RAM

The simplest RAM cell design uses a single bus and cross-coupled inverters. Two cell designs are shown in Fig. 7.13. The cell in Fig. 7.13(a) is static and requires a precharged bus. The time to write the cell includes the time to propagate a signal around the inverter loop. This time can be long since one typically uses very weak devices in the cell so that it can be overwritten by the bit line; and one also wants to save power and area. The cell in Fig. 7.13(b) is semistatic (the clock can only be

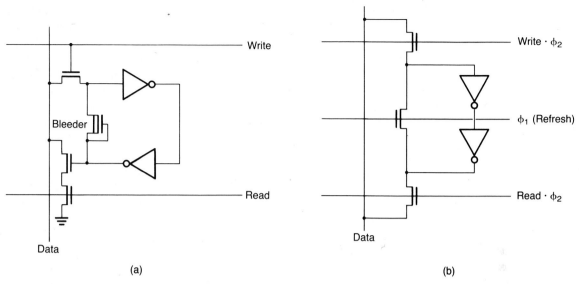

FIGURE 7.13   The "eight" transistor RAM: (a) static; (b) semistatic.

stopped with ϕ_1 high). It uses one fewer transistor than the first design, but needs one more control wire.

These cells can be made multiported simply by attaching additional data and control wires, as illustrated in Fig. 7.14. Of course, the cell must now drive two buses, putting a further burden on the weak cell transistors.

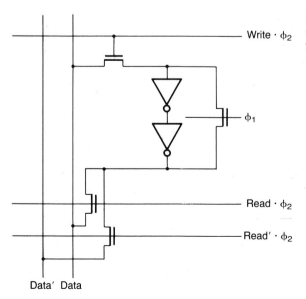

FIGURE 7.14   Two-port version of the "eight" transistor RAM.

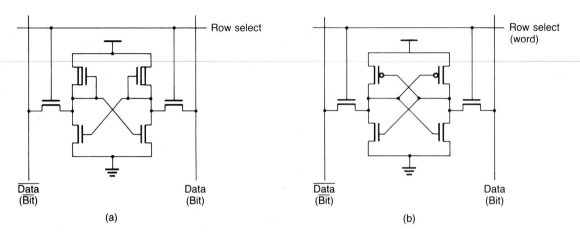

FIGURE 7.15   The six-transistor RAM: (a) nMOS; (b) CMOS.

The six-transistor RAM

Fig. 7.15 illustrates two six-transistor RAM cells. These cells can be quite fast in large arrays because they generate a differential output signal. The existence of a differential signal eliminates a whole class of common mode noise sources. This allows us to use a sense amplifier with a lower noise margin. Lower noise margins allow higher speeds. In addition, the speed at which the cell can be written is limited by the speed of the transistors in the write amplifier rather than by the speed of the transistors in the cell, as was a limitation of the cell in Fig. 7.13.

Writing is done by raising row select and lowering one bit line while raising the other bit line. This will force one of the sides of the selected cell low, storing the data. A read is performed by raising both bit lines high and selecting the desired word. The cell will pull one of the bit lines low. When it has moved sufficiently, a differential sense amplifier is activated. The overall organization of a typical static RAM is illustrated in Fig. 7.16. There are, of course, many variations on this theme. One issue concerns the placement of the column selectors[2] with respect to the sense amplifiers. Another issue concerns the number of columns among which the sense amplifiers are shared. When sense amplifiers are shared, the differential input signals must be carefully routed in such a way that they see the same noise and load capacitance. (The capacitances can never be exactly matched since they are somewhat dependent on the data in the array.)

In the RAM cell itself, the pulldown transistors are typically two to three times as wide as the transfer devices. The pullups are a problem

---

[2] The column selectors are often called Y-decoders; the row decoders are called X-decoders.

in nMOS because of the power budget. The pullups need only replace the current lost through leakage and α-particle hits. We would prefer megohms rather than kilohms. The cheapest step in that direction is the use of a weak depletion implant. (This is the sort of device we might get in the 2 μm example process if both an enhancement- and depletion-mode implant were used on the same transistor.) Commercial static RAM chips use poly resistors, which require extra processing steps, with resistances in the megohm/□ range. Of course, p-type pullups, illustrated in Fig. 7.15(b), are excellent in terms of power dissipation, but area is a problem. As discussed in Section 5.4, to prevent the information in the cell from being lost, the loop gain in the cell must not drop below unity. During a transient in a static RAM, two word lines can be on at the same time and this can be a worst-case in terms of pulling the bit lines low and degrading the loop gain. Assuming that the pullups are weak, when the word line is high, the cell (nMOS or CMOS) looks like cross-coupled enhancement-load nMOS inverters.

A fundamental difference between clocked and unclocked static RAM designs is the types of sense amplifiers, row drivers, and bit line

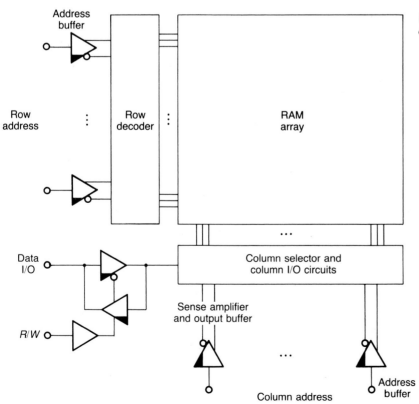

FIGURE 7.16   Organization of a RAM array.

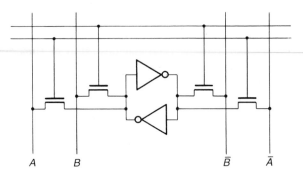

FIGURE 7.17 Dual-port version of the six transistor RAM.

A　B　$\bar{B}$　$\bar{A}$

pullups that are employed. Enhancement-mode bit line pullup devices are used in the fully static (unclocked) case. This is because we want to limit the bit line voltage swing in order to speed the recovery time between two successive reads. Clocked static RAM designs have been found to be of higher performance than unclocked designs. It may be important to put bleeder pulldowns on the bit lines to keep subthreshold effects from making the precharge voltage an uncontrolled function of the time between accesses.

Because we are using a differential scheme for reading the cells, we are most concerned with differential-mode noise. One large source of differential noise is neighboring bit lines. We are also sensitive to common mode noise to the extent that the sense amplifier speed is impacted by common mode signals and by mechanisms, such as differential load capacitances, that convert common mode noise to differential-mode noise. Other sources of differential-mode noise include the patterning effects of reading a 1 and then a 0, and mismatches between the sense amplifier input transistors.

V_{DD} and ground noise can be quite large in RAM arrays since the power supply lines are often run in diffusion. For some CMOS designs V_{DD} has actually been run in poly. Current spikes caused by reading a word in the array will appear as voltage noise. Note that the noise implications are quite different depending on whether the ground lines are run parallel to the word or bit lines. In one case accessed cells are disturbed, and in the other unaccessed cells are disturbed. As we saw in the ROM example of Chapter 2, surge currents can cause the ground line voltage to increase significantly and this can cause an increase in the access time.

Dual-porting the six-transistor RAM is done in the obvious way, as illustrated in Fig. 7.17. Dual-porting halves the effective pullup resistance of a cell in which both pairs of access transistors are on, thus lowering the loop gain of the cross-coupled inverters and increasing its sensitivity to disturbances.

The four- and three-transistor RAMs

Commercial memories are fabricated with special processes. Static RAM processes, unlike dynamic RAM processes, are very similar to the processes for implementing logic. Nevertheless, static RAM processes often have, for instance, two poly layers and special poly resistors. When designing a RAM as part of, say, a microprocessor, high-resistance poly loads are usually not a process option. Depletion or p-type loads consume enormous quantities of area and, perhaps, power. One alternative is simply to leave the pullups out of the circuit. This makes the memory dynamic. As pointed out by Wooley [Wooley 80], in ROM designs we go to dynamic techniques to save power, but in RAM designs we use dynamic techniques to save area as well as power.

Figure 7.18 illustrates a four-transistor RAM cell. Additional issues brought up by this design include the need to refresh the cell (on the order of once per millisecond), the lower noise margins in the cell (the high side pulls only to $V_{DD} - V_T$), and the need to restore the bit line high voltage when writing back to the cell. That is, the high level in the cell must be restored somehow, and since there are no pullups in the cell, the refresh must be done external to the cell.

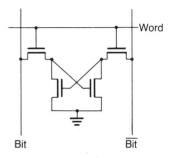

FIGURE 7.18 The four-transistor RAM.

■■■■■■■■■■ EXAMPLE 7.1

The three-transistor RAM cell applied to a shift register

Lyon has used the three-transistor dynamic RAM cell to build a very long shift register for signal processing applications [Lyon 80]. The system we will examine, which uses the same general organization as Lyon's, consists of four parts (not including some fancy control).

Central to the design is a three-transistor RAM cell. Two three-transistor RAM cells are illustrated in Figs. 7.19 and 7.20. The bit lines

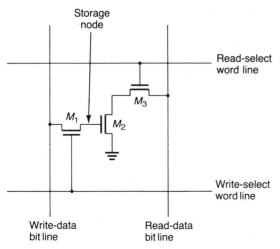

FIGURE 7.19 A three-transistor RAM with separate READ/WRITE bit lines.

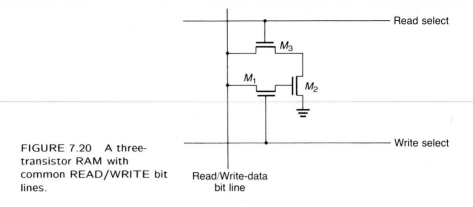

FIGURE 7.20 A three-transistor RAM with common READ/WRITE bit lines.

FIGURE 7.21 Organization of a shift register based on the three transistor RAM cell.

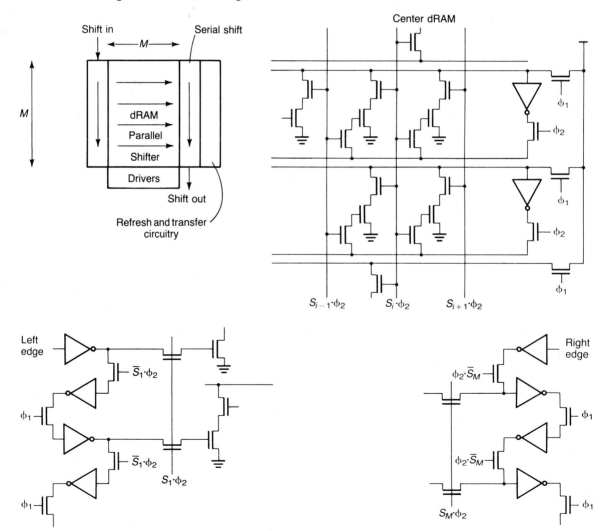

must be precharged to do a read cycle. Information is stored dynamically on the gate node of the transistor with the grounded source. Information is gated onto the storage node by controlling M_1. A read is performed by turning on M_3 and determining whether or not the precharged bit line was discharged. These cells can be fairly nasty to design because of the extreme importance of parasitics. The principal concern is storing a low voltage on the gate of M_2. C_{DG} coupling on M_2 during a read can yank this voltage high. This is especially critical because the input capacitance of M_2 is extremely low when M_2 is below threshold; it may not take much coupling to turn M_2 on. (Note that SPICE2G.5 overestimates C_{GB} below threshold and hence overestimates the robustness of the cell.)

To build a long shift register, we start with two M bit serial shift registers of the form shown in Fig. 7.21. These form the input and output circuitry. Between these serial shift registers is an M by M bit array of three transistor dynamic RAM (dRAM) cells in which M bit parallel shifts occur at $\frac{1}{M}$ times the rate that the bits come into the serial shift registers. Data come into the serial shift register at a rate of one bit per clock period. After M bits have entered the array they are shifted in parallel to the dRAM array. To make the timing work out, there is one parallel shift of a column of bits every period. The columns are shifted to the right starting at the right and proceeding left until all columns have been moved. The action then starts again at the right. Note that each column is refreshed as it is shifted. The refresh circuitry is the fourth part of the system. This design uses much less area and power than a conventional nMOS shift register design.   ■

The one-transistor RAM

The one-transistor memory cell is at once the simplest and most complex of memories [Dennard 68]. It represents what is perhaps the most sophisticated corner of VLSI circuit design. Although the details of one-transistor dRAMs are beyond the scope of this text, we will give the reader a literacy in the subject.

A schematic for a 1T dynamic RAM is illustrated in Fig. 7.22. In commercial RAMs, the cell array often occupies less than 50% of the area. The rest is consumed by clocks, sense amplifiers, drivers, and so on. Four issues are closely coupled: cell layout, fabrication technology, circuit design, and array organization.

The cell itself consists of a transistor and a capacitor. Information, in the form of charge, is stored on the capacitor. Approximately one million electrons differentiate between a stored high and a stored low. A write is performed by placing the data on the bit line and gating this data into the capacitor with the word line. A read is performed by precharging the bit line and raising the word line. Charge sharing between the cell capacitance and the bit line change the voltage on the bit line by between 5 and 30%, while the information in the cell is destroyed. To consistently

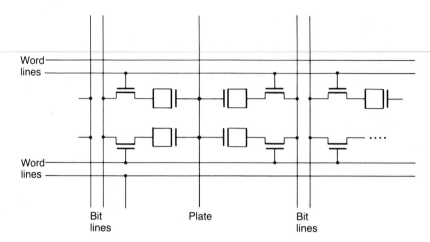

FIGURE 7.22   A one-transistor RAM.

sense the change in the bit line charge, a sense amplifier compares two bit lines. The charge in a dummy cell is placed on one bit line, while the other bit line is connected to the accessed cell. The dummy cell should store a charge corresponding to the average charge of a cell with a stored high and stored low.

We will concentrate on the two central issues of cell layout and circuit design. Of key importance in the layout are the cell area needed to store the required charge and the pitch of the bit lines, which dictate the difficulty of the sense amplifier layout. From a circuits perspective, we are most concerned with the efficiency of the charge transfer from the cell to the sense amplifier and RC time constants in the array access wires. The longer the bit lines, and the more cells they connect, the worse their RC time constant and transfer efficiency. This encourages one to split the array into several small segments, which, however, increases the area taken up by the overhead circuitry.

Let us confront some of the issues by examining two different novice cell layouts done in our 2 μm technology. The first cell, illustrated in Fig. 7.23, uses a metal word line. Since the word lines connect to the gates of the transistors, metal-to-poly contacts are needed in the cells. To form the storage capacitor, we build a MOS capacitor. Most of the charge is stored on C_{OX}, though some charge is stored in the depletion region. (The very thick depletion region of the example nMOS process would probably be very good at collecting the charge due to α-particle hits!) The plate that forms one end of the storage capacitor must be tied to V_{DD} unless we use a depletion implant. Care must be taken to ensure that the MOS capacitor does not venture below threshold. Diffusion is used for the bit lines. For this reason, both the capacitance and the resistance of the bit line are issues. The bit lines could be run in poly instead of diffusion by adding a buried contact. This technique has been used to advantage on some chips.

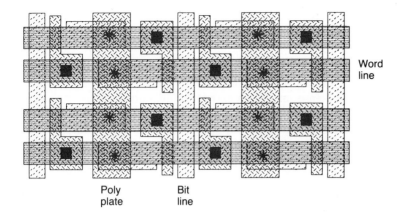

FIGURE 7.23   Layout of a one-transistor RAM with metal word lines.

Poly plate

Bit line

Word line

A second cell, with roughly the same capacitance, is shown in Fig. 7.24. Instead of metal word lines, it uses metal bit lines. The area is roughly half that of the previous design, but there are other issues. It is clear, for instance, that poly propagation on the word lines will be quite deadly. The width of the poly is modulated to lower the RC effects. The RC time constant will have a large influence on the number of segments, and hence overall size, of the array. Two-level metal can help somewhat by allowing one to strap the poly every few cells to short out RC effects. Nevertheless, because the pitch on second-layer metal is much worse than poly or metal one, some area is lost with this approach. One can see that there is not a lot of extra area left over in the cell to place additional contacts. Low resistance silicides would help a great

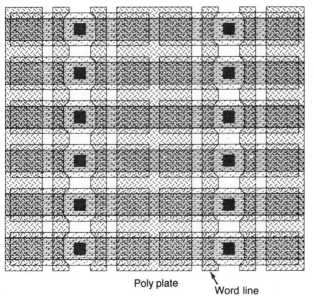

FIGURE 7.24   Layout of a one-transistor RAM with metal bit lines.

Bit line

Poly plate

Word line

deal in this circuit. Note also that, with our design rules, we do not have room to place a depletion implant in the MOS capacitor so the plate will need to be attached to V_{DD}. This limits the range of voltages that can be stored on the capacitor. Another issue concerns the pitch of the bit lines that has been reduced from 30 μm to 6 μm. This makes the layout of the sense amplifiers extremely difficult.

In commercial memories there are many additional issues. For instance, because the array cell is replicated so many times, it pays to push its design rules. This, however, can cause yield problems. Details such as the number and nature of the metal steps in the cell are routinely considered in the cell design.

Figure 7.25 illustrates a sense amplifier connected to two bit lines, with one dummy cell each. In this illustration, the bit lines go to opposite sides of the array. If the bit lines are folded so they run side by side to the same half of the array, then some of the common mode noise will be reduced. We can go one step further by using two cells to store each bit in a differential manner. This leads to very high noise margins, but takes more area. Dummy cells, at least, can be dispensed with.

In Fig. 7.25, the cell is read by first precharging the bit lines and turning on the word line of the cell to be sensed and the corresponding dummy. One then activates the sense amplifier, amplifying the difference to a full swing, and finally writes back the restored voltage into the accessed cell.

Another issue concerns the placement of the output selectors in relation to the sense amplifiers. Are the sense amplifiers to be placed before or after the column selector? The more sense amplifiers, the larger the chip power dissipation; the fewer the sense amplifiers, the lower the signal levels. Low signal levels can be sensed, but because the low voltages must be amplified to full logic swings, the gain-bandwidth product limits the overall speed. As in the ROM, row accesses take much longer than column accesses.

FIGURE 7.25 Bit line organization of a one-transistor RAM.

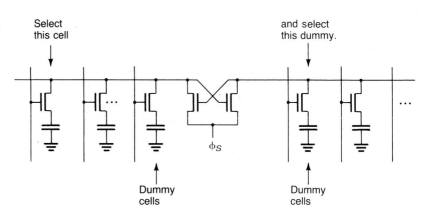

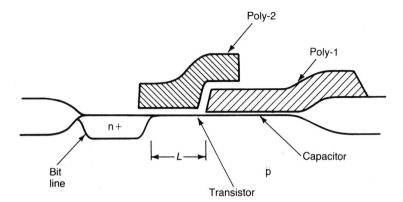

FIGURE 7.26   Cross section of a one-transistor RAM cell with double poly.

Critical to the speed and reliability of the RAM are the signal levels stored on the capacitor. To increase the stored voltages, the word lines are sometimes bootstrapped above V_{DD}. Other techniques are more in the technology domain. Double poly allows one to tighten the layout and use an extremely thin oxide for the storage capacitor. This is illustrated in Fig. 7.26. The High-C cell uses a heavy p+ implant under the storage transistor to decrease $X_{P\mathrm{max}}$ and increase Q_B. To compensate for the upward threshold shift, a shallow n+ implant is also required. Another technique for increasing the stored charge is to increase the dielectric constant of the capacitor oxide. $\mathrm{Si_3NO_4}$ ($\kappa \doteq 7.5$, where κ is the relative dielectric constant) and $\mathrm{Ta_2O_5}$ ($\kappa = 22$) have both been tried.

Commercial dRAMs must draw zero static power. For this reason, and to a lesser extent for reasons of speed, dozens of clocks are generated on-chip (by timing chains) to activate and deactivate the sense and drive circuitry.

The circuit techniques used for logic parts often follow a few years behind those of commodity RAMs. Two disparities in style that are evident, at this writing, are the more extensive use of self-timed clocking in commercial RAMs and the use of sensing techniques that are more robust to movements in V_{DD}. This means that either the reference voltage must be sampled or the unknown signal must be turned into a differential signal. In a static RAM, a differential signal is available, but in cases where only a single-ended signal is available, the signal can be compared to its value at a previous time. For instance, one can precharge to some voltage (and here there is little advantage in precharging to V_{DD}), sample that voltage, and then compare that sampled value to the voltage after the signal has developed. One can imagine using techniques of this sort to, for instance, eliminate the dummy cells in a dRAM. One can precharge both bit lines attached to a common sense amplifier to approximately $V_{DD}/2$, and then turn on one word line that will allow the accessed cell to pull its bit line either higher or lower than its initial precharged value. Since the bit line attached to the other half of the sense amplifier has no accessed cells in this case, it can be used as the reference voltage since its final value is the same as the initial value of the activated bit line. The precharge voltage is re-established by shorting the bit lines, which have been driven to zero and V_{DD}, together.

7.4 Systolic arrays

Systolic arrays are a methodology for combining logic and memory to create powerful regular structures [Kung 79, 79a, 82, Leiserson 81]. A systolic array is a network of small processors wherein the communication and processor activity is organized harmoniously so that processor and communication resources can be fully utilized. Communication is usually local. The word "systolic" comes from the word "systole," meaning contraction. In a biological context, this term refers to the contraction of the heart, which pumps the blood through the system. Similarly, a systolic VLSI system rhythmically pumps information through the array. In a pure systolic system[3] all interprocessor communication goes through a register. In a semisystolic system this constraint is relaxed. Well over a hundred applications have been found to have systolic solutions, including digital filtering, matrix multiplication, matrix LU-factorization, regular expression recognition, sorting queues, and priority queues. Generalizing on a class of these, H. T. Kung has reported on the design of a programmable systolic chip [Fisher 83].

Pattern matching array

Foster and Kung [Foster 80] have designed a pattern matching chip that illustrates many of the issues involved in the design of a systolic subsystem. We will look at their design as a way to get a feel for systolic design.

The pattern matcher takes two inputs and produces one output. The inputs are the pattern P and the string S. The object is to find words in the input string that match the entire pattern P. Thus the string $asbksskkkc$ produces the output 0001001100 when matched against the pattern $s\Omega k$, where Ω can stand for any letter.

One algorithm for performing a pattern match is to slide the string past the pattern and, by comparing every letter of the pattern to the segment of the string next to it, discover when the entire pattern matches

---

[3] Pipelined logic arrays and cellular or tesselation automata are special cases of systolic arrays.

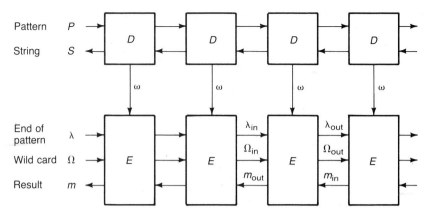

Pattern P

String S

ω

End of pattern λ

λ_{in} λ_{out}

Ω_{in} Ω_{out}

Wild card Ω

Result m

m_{out} m_{in}

FIGURE 7.27 General organization of a systolic pattern matcher. [Foster 80] (©1980 IEEE)

a string segment. The problem with this technique is that one must decide, in a global manner, whether or not there is a complete match.

A better technique is to collide the pattern and string, as shown in Fig. 7.27. The information about matches can be accumulated locally, without global communication. Since both the pattern and the string move, we must slow them each down so that a new input comes with every second clock tick; otherwise, they will miss each other. This means that, on any given clock tick, half the comparator cells D are idle.

To expedite the generation of the output string m, a symbol λ accompanies the last letter in the pattern. Looking at the bottom cells E in Fig. 7.27, we find that on each clock tick, we have

$$\lambda_{in} \rightarrow \lambda_{out}, \tag{7.9}$$

$$\Omega_{in} \rightarrow \Omega_{out}, \tag{7.10}$$

$$\text{TEMP} \cdot \lambda_{in} + m_{in} \cdot \overline{\lambda_{in}} \rightarrow m_{out}, \tag{7.11}$$

$$\lambda_{in} + \text{TEMP} \cdot (\Omega_{in} + \omega_{in}) \rightarrow \text{TEMP}. \tag{7.12}$$

TEMP is a state variable that stores the information about the letters matched so far. TEMP = 1 indicates that, so far, there is a match. When $\lambda_{in} = 1$, indicating the end of a pattern, TEMP is moved to the output string m and reset.

Three conditions can lead to a true value of TEMP on the next clock tick:

- TEMP is reset by the end of the pattern marker λ;
- TEMP is true, indicating a partial string match, and the character Ω is true; and
- TEMP is true, indicating a partial string match, and ω_{in} is true, indicating that the next letter also matches.

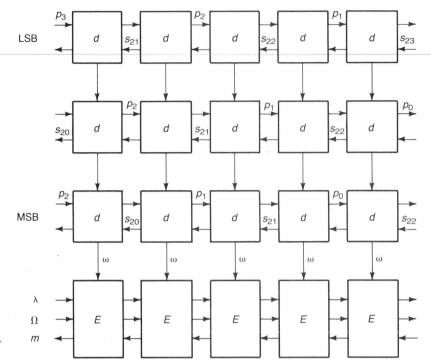

FIGURE 7.28 Details of a systolic pattern matcher. [Foster 80] (©1980 IEEE)

We could implement the system in such a way that each cell D compares two letters. Since letters are made of several bits, this comparison implies mildly complex computation in each cell and this may slow the system down. One alternative is to pipeline the comparison of the letters. This is illustrated in Fig. 7.28. The active d subcells form a checkerboard. On one tick all of the "red squares" are used, and on the next tick all of the "black squares" are used. The least significant bits (LSBs) are input first in the top row, then the next LSB is input in the second row, and so on.[4] The equations for the comparison cells are

$$p_{\text{in}} \rightarrow p_{\text{out}}, \tag{7.13}$$

$$s_{\text{in}} \rightarrow s_{\text{out}}, \tag{7.14}$$

$$\omega_{\text{in}} \cdot (p_{\text{in}} = s_{\text{in}}) \rightarrow \omega_{\text{out}}. \tag{7.15}$$

Both the d and E blocks can be thought of as simple finite-state machines. They are sufficiently simple that they could conveniently be implemented in "random" logic rather than in a PLA.

---

[4] In the absence of good computer animation, the best way to understand systolic systems is to draw the system and data components on different vugraphs, in unique colors, and move them against each other to observe the flow of data operations.

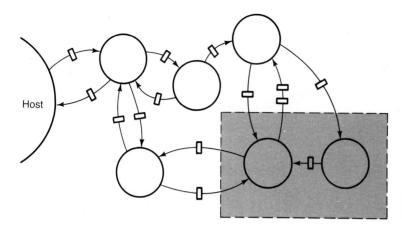

FIGURE 7.29   A systolic system.

Array optimization

Given a logic structure with systolic potential, there are a number of transformations that can be used to manipulate and optimize it [Leiserson 83]. We will now examine a few of these operations. Be aware that efficient algorithms have been developed to optimally solve many of the problems we will pose.

One of the techniques is retiming. It is analogous to the skew transformations we saw in Chapter 6. A system is shown in Fig. 7.29 in which circles represent logic and rectangles represent registers. We can draw a surface around one part of the system, as illustrated. In Chapter 6, we observed that we could add delay to all the inputs of a subsystem and subtract that delay from all the outputs, thus obtaining an exactly equivalent external system. We can generalize this concept so that if we add P registers to each input, and remove P registers from each output, then we will maintain an equivalent system. This is retiming. Note that the functionality and timing of the retimed system are exactly the same as far as the external system is concerned. Let us look at the practical example of a finite impulse response (FIR) digital filter; its schematic is illustrated in Fig. 7.30. The filter implements the equation

$$y_t = \sum_{k=0}^{N-1} a_k x_{t-k}, \qquad (7.16)$$

where y_t is the output at time t, x_t is the input at time t, and the a_k are the filter coefficients. We implement Eq. (7.16) with an array of "inner-product" processors. Each processor implements the functions

$$C + AB \to C' \qquad (7.17)$$

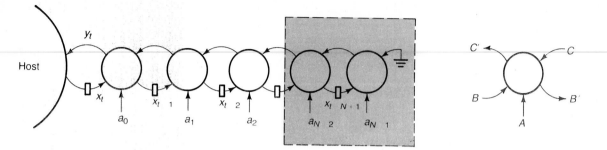

FIGURE 7.30 A semisystolic implementation of a FIR filter.

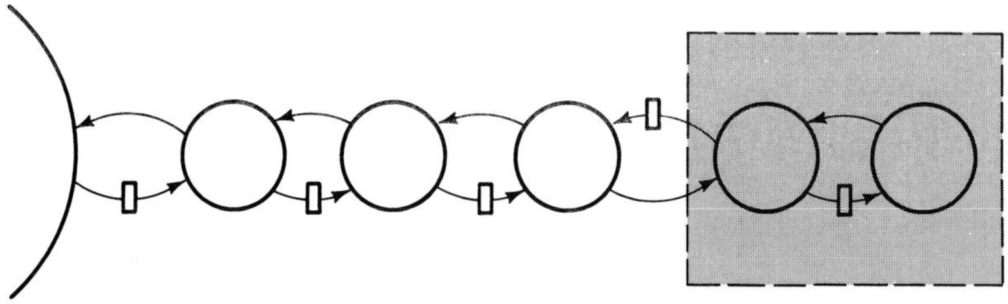

FIGURE 7.31 A retimed version of the FIR filter.

and

$$B \rightarrow B',\qquad(7.18)$$

where B' and C' are outputs and A, B, and C are inputs.

The filter implementation shown in Fig. 7.30 is quite straightforward, but the performance is poor because y_t must ripple through N combinational logic blocks. For the sake of clarity, let us pick $N = 5$ in this example. In Fig. 7.30, we select the last two inner-step processors for retiming. We remove one register from the input and add one register to the output, as illustrated in Fig. 7.31. The functionality is identical except that now the critical path is through three rather than five inner-step processors. Thus if it takes 50 ns for a signal to propagate through each processor, the maximum clock rate has been increased from 4 MHz to 6.7 MHz by retiming. So for essentially no cost, we have gained almost a factor of two in the maximum clocking rate. Figure 7.32 illustrates a better retiming, which reduces the maximum critical path delay to two modules—a performance gain of 2.5 and a clock frequency of 10 MHz. Constant inputs are suppressed in this figure. This is the optimal retiming, but not necessarily the best we can do.

In addition to retiming, there are two other operations we can perform: "slow-down" and "hold-up." Both of the operations maintain the functionality of a circuit, but alter the clocking. We will examine slow-down first.

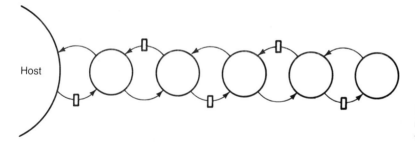

FIGURE 7.32   An optimally retimed version of the FIR filter.

Slow-down, like retiming, is a trivial operation with potentially profound consequences. To slow down a network we multiply the number of registers. Let us take the original FIR filter implementation and multiply the number of registers by two. The result, shown in Fig. 7.33, is a naive sort of pipelining. Each result takes twice as many clock ticks to reach the output as before. The critical path remains the same. Note, however, that the signals entering on the odd clock ticks never interact with the signals entering on the even clock ticks. It seems like two separate filters instead of one.

The throughput of a network is the rate at which information appears at the output; the latency is the time it takes for information from the input to propagate to the output. There is, unfortunately, some ambiguity in these definitions. For the filters we are considering, we take the throughput as the single channel I/O rate. This is not necessarily the clocking rate or the rate at which data appear at the output. This is because slow-down does not change the system clocking rate or the rate at which information appears at the output, but it lowers the rate at which any single data stream (channel) may be processed.

If we slow down a system by multiplying the number of registers by S, then the number of channels is multiplied by S and the (single channel) throughput is divided by S. We measure latency in terms of an integral number of clock ticks. For the filter, we define the latency as the time from when we input x_n to the time we receive y_n. This means

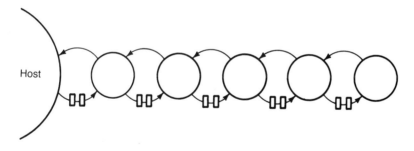

FIGURE 7.33   A slowed down version of the original FIR filter.

there is some ugliness having to do with the startup of the system. Unless we assume that the initial conditions are properly set up, the first few clock ticks may produce invalid data. We do not include this time in our calculation of the latency because in most signal processing and control applications it is the steady state (once the pipeline is full) response that is important. Latency is the time required to get the first result (assuming initial conditions were properly set up). Throughput is the rate at which results arrive thereafter (in response to a single data stream input). In Fig. 7.30, the throughput and clock rate are 4 MHz. In Fig. 7.33, the clock rate is 4 MHz, while the throughput is 2 MHz for each of two identical channels. In Fig. 7.30, with $N = 5$, the latency is 250 ns. In Fig. 7.33, the latency is 500 ns. In Fig. 7.32, the throughput and clock rate are 10 MHz, while the latency is 100 ns. (It would be 50 ns, but we require the latency to be measured in units of clock periods. We must clock the circuit once with a 10-MHz clock, hence the 100-ns result.)

After applying slow-down in Fig. 7.33, we can again retime the circuit. Fig. 7.34 illustrates the optimal retiming. The critical path has been reduced to the propagation time through one processor. This doubles the best previous result and is five times better than the original solution. Here, the clock rate is 20 MHz, the throughput is 10 MHz for each of two channels, and the latency is again 100 ns.

Another implementation can be obtained by studying the internals of the inner-product processor. The generation of B' from B requires only a wire, while the generation of C' requires one addition and one

FIGURE 7.34 An optimally slowed down and retimed version of the FIR filter.

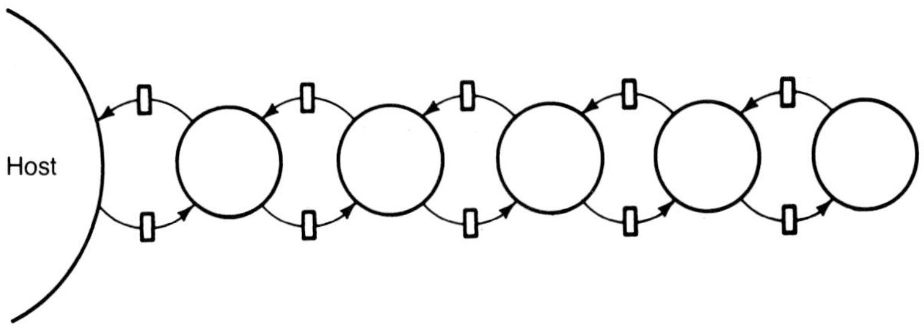

FIGURE 7.35 Another retimed version of the FIR filter.

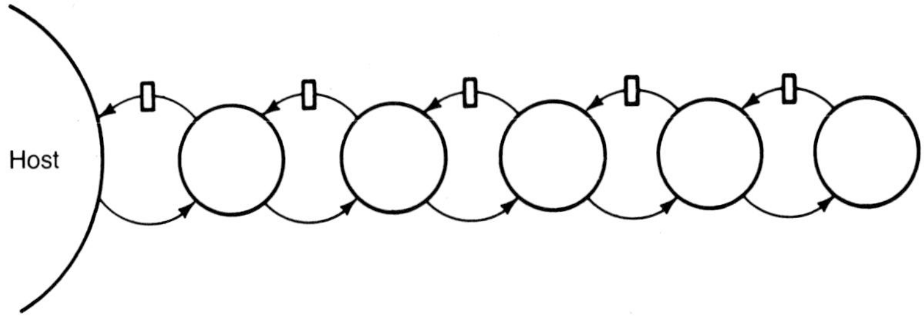

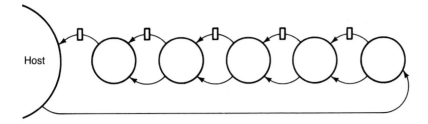

FIGURE 7.36   Changing the topology of the filter.

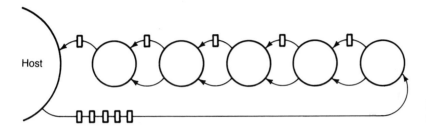

FIGURE 7.37   Holding up the filter.

multiplication. Therefore, if in the original implementation a signal must be global, it should be x and not y. Retiming will help us here also. A retimed circuit with one global wire is shown in Fig. 7.35.

A wire is only a wire. It does not have a unique direction in which signals must flow. Figure 7.36 illustrates another equivalent system, where x flows from right to left rather than left to right. (Trees or other topologies could have been used to distribute x.) To pipeline this system we require a third operation: hold-up. In hold-up we place registers at the input or output of a system to increase the latency. Figure 7.37 illustrates this operation. Retiming results in a pipelined system, as shown in Fig. 7.38. In this circuit both the clock rate and the throughput have achieved their optimal values of 20 MHz. To achieve this increased throughput, the latency grew to 500 ns. In transforming from Fig. 7.30 to Fig. 7.32 we gained without cost because the original system was suboptimal. In the optimal circuits of Figs. 7.32, 7.34, and 7.38, we

FIGURE 7.38   Hold-up plus retiming for maximum throughput.

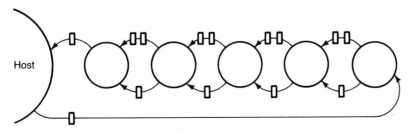

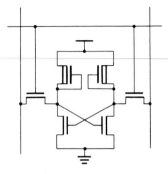

TABLE 7.1 FIR filter characteristics

Figure number	Clock rate	Throughput	Latency	Number of channels
7.30	4 MHz	4 MHz	250 ns	1
7.32	10 MHz	10 MHz	100 ns	1
7.33	4 MHz	2 MHz	500 ns	2
7.34	20 MHz	10 MHz	100 ns	2
7.38	20 MHz	20 MHz	500 ns	1

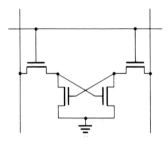

see the trade-offs among the number of channels, the throughput per channel, and latency. Table 7.1 summarizes some of the key results.

Other transformations, such as logic duplication, resource sharing, and so forth, augment the three transformations we have discussed. One way to exploit these techniques is by an appealing methodology propounded by Charles Leiserson [Leiserson 83]. He suggests designing complex machines of tailored performance by first designing an intuitively appealing and straightforward machine that is sure to work, and then manipulating that design, using proven transformations, into the optimal machine for the task.

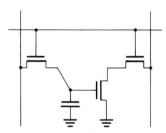

7.5 Perspective

The design of large arrays spans many levels of abstraction. Techniques we have investigated include

- The X-cell layout,
- Virtual grounds,
- Implant encoding,
- Logic minimization in PLAs,
- PLA and ROM folding,
- Input encoding, and
- Retiming, slow-down, and hold-up of systolic systems.

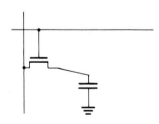

FIGURE 7.39 A fanciful evolution of the RAM cell. What will be next? (K. U. Stein, A. Sihling, and E. Doering, "Storage Array & Sense/Refresh Circuit for Single-Transistor Memory Cells," IEEE J. Solid State Circuits, Vol. SC-7, p. 336, 1972.) (©1972 IEEE)

The optimization of large arrays is extremely important because they often end up being critical in a VLSI system in terms of both area and speed. In the design of memory arrays, optimization of the cell size causes the complexity and area of the peripheral circuitry to grow. An intriguing, though *a posteriori*, view of the evolution of the RAM cell from the six-transistor static to the one-transistor dynamic is illustrated in Fig. 7.39. One wonders what will be next.

Problems

7.1 Design a circuit that will raise its output when there is a transition on its input. The circuit must work independently of the speed of the input waveform. Include a reset input.

7.2 Design a sense amplifier for the ROM of Example 2.9, using the schematic illustrated in Fig. 7.40. $\triangle$

7.3 A general problem with precharged nMOS circuits is that the exact value of the high voltage depends on how long the precharge has been on. Subthreshold leakage is a real culprit. This problem also shows up when simulations are done. Discuss some places that this phenomenon can get one into trouble. $\triangle$

7.4 One technique for empirically determining the sensitivity to noise of a memory part is to quickly slew the power supply from, say, 4 V to 6 V. This generally discovers parts of the design that are not very robust. What sorts of problems might this technique uncover? $\triangle\triangle$

7.5 Implement a binary-to-BCD decoder with a PLA. Use folding and logic minimization techniques to reduce the size of the array. $\triangle$

7.6 A content addressable memory (CAM) is an array that, conceptually, works in just the opposite fashion from a normal RAM. Instead of outputting the cell contents when given an address, the CAM outputs the address that matches the given contents. The key part of the CAM is the match cell. Figure 7.41 illustrates a dynamic CAM cell. Explain how this CAM cell can be read and written.

7.7 Draw a transistor level CMOS circuit for each of the two pattern matcher cells discussed in Section 7.4.

7.8 Reimplement the logic of Fig. 7.10 in a PLA with input encoding. Use a minimum number of product terms. Examine the opportunities for row and column folding.

7.9 Retiming can be used to produce a nontrivial, high-performance design from a conceptually simple, low performance design.[5] Consider the simplest type of $N \times N$ parallel multiplier, which performs $N - 1$ repeated shifts and adds of the multiplicand, each time propagating the carry the entire length of the word. An implementation of such a multiplier, for the case of $N = 3$, is shown in Fig. 7.42, using the cell in Fig. 7.43. Seven registers have been placed on each of the inputs to allow pipelining. Retime the multiplier in Fig. 7.42 to obtain a pipelined parallel multiplier that accepts a 3 b multiplier and a 3 b multiplicand, and produces a 5 b product on each clock tick. Maximize the throughput.

7.10 A five-pole infinite impulse response (IIR) filter with the transfer function

$$H(z) = \frac{1}{1 - \sum_{i=1}^{5} a_i z^{-i}}$$

is implemented in the time domain by solving the recursion

$$y_n = a_1 y_{n-1} + a_2 y_{n-2} + \ldots + a_5 y_{n-5} + x_n,$$

---

[5] Problems 7.9 and 7.10 due to C. Hauck.

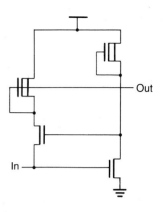

FIGURE 7.40

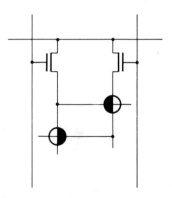

FIGURE 7.41

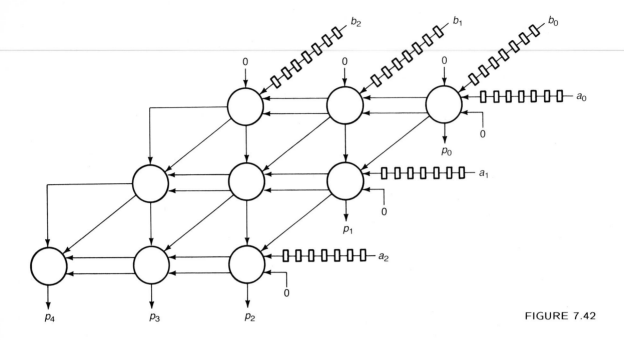

FIGURE 7.42

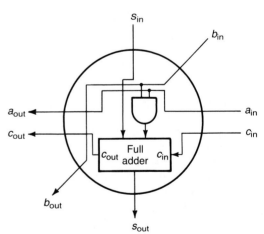

FIGURE 7.43

where x_n and y_n are the filter input and output, respectively. The recursion can be implemented with the circuit in Fig. 7.44, which uses a five-point FIR filter in a feedback loop. The input and output to the feedback elements are y_n and w_n, where

$$w_n = \sum_{i=1}^{5} a_i y_{n-i}.$$

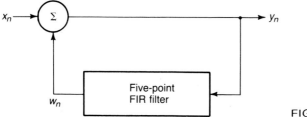

FIGURE 7.44

Figures 7.32 and 7.38 give two possible circuits for realizing the feedback element. Choose the circuit that maximizes the throughput of the IIR filter; calculate this maximum throughput; and briefly explain the limitations imposed on the IIR filter throughput by the latency and throughput of the FIR feedback element. Ignore register delays and assume all cells, including the adder in Fig. 7.44, have 50 ns propagation delays. △

References

[Anderson 82] J. M. Anderson, B. L. Troutman, and R. A. Allen, "A CMOS LSI 16 × 16 Multiplier/Multiplier-Accumulator," *IEEE International Solid-State Circuits Conf.*: 124–125, San Francisco, Calif., 1982.

[Ayres 83] R. F. Ayres, *VLSI Silicon Compilation and the Art of Automatic Microchip Design*, Prentice-Hall, Englewood Cliffs, N.J., 1983.

[Cook 79] P. W. Cook, S. E. Schuster, J. T. Parrish, V. DiLonardo, and D. R. Freedman, "1 μ MOSFET VLSI Technology: Part III," *IEEE Trans. Electron Devices* **ED-26**: 333–345, April 1979.

[Cook 79b] P. W. Cook, C. W. Ho, and S. E. Schuster, "A Study in the Use of PLA-Based Macros," *IEEE J. Solid-State Circuits* **SC-14**: 833–840, October 1979.

[Dennard 68] R. H. Dennard, "Field-Effect Transistor Memory," U.S. Patent 3.387.286, June 4, 1968.

[Fisher 83] A. L. Fisher, H. T. Kung, L. M. Monier, H. Walker, and Y. Dohi, "Design of the PSC: A Programmable Systolic Chip," *Proc. Third Caltech Conf. on VLSI*: 287–302, Pasadena, Calif., 1983.

[Fleisher 75] H. Fleisher and L. J. Maissel, "An Introduction to Array Logic," *IBM J. Res. Develop.* **19**: 98–109, 1975.

[Foster 80] M. J. Foster and H. T. Kung, "The Design of Special-Purpose VLSI Chips," *IEEE Computer* **13**: 26–40, 1980.

[Guttag 82] K. M. Guttag, J. F. Sexton, K. S. Chang, "A 16b Microprocessor with a 152b Wide Microcontrol Word," *IEEE International Solid-State Circuits Conf.*: 120–121, San Francisco, Calif., 1982. (TMS 99000)

[Hachtel 82] G. D. Hachtel, A. R. Newton, and A. L. Sangiovanni-Vincentelli, "An Algorithm for Optimal PLA Folding," *IEEE Trans. Computer-Aided Design* **CAD-1**: 63–77, 1982.

[Hardee 81] K. C. Hardee and R. Sud, "A Fault-Tolerant 30 ns/375 mW 16K×1 NMOS Static RAM," *IEEE J. Solid-State Circuits* **SC-16**: 435–443, 1981. (A number of amazing circuits)

[Hardee 84] K. Hardee, M. Griffus, and R. Galvas, "A 30 ns 64 K CMOS RAM," *IEEE International Solid-State Circuits Conf.*: 216–217, San Francisco, Calif., 1984.

[Hong 74] S. J. Hong, R. G. Cain, and D. L. Ostapko, "MINI: A Heuristic Approach for Logic Minimization," *IBM J. Res. Develop.* **18**: 443–458, 1974.

[Hudson 82] E. L. Hudson and S. L. Smith, "An ECL Compatible 4K CMOS RAM," *IEEE International Solid-State Circuits Conf.*: 248–249, San Francisco, Calif., 1982. (This paper contains some "typos" concerning the types of bipolars used, so be careful.)

[Isobe 81] M. Isobe, Y. Uchida, K. Maeguchi, T. Mochizuki, M. Kimura, H. Hatano, Y. Mizutani, and H. Tango, "An 18 ns CMOS/SOS 4 K Static RAM," *IEEE J. Solid-State Circuits* **SC-16**: 460–465, 1981.

[Itoh 84] K. Itoh, R. Hori, J. Etoh, S. Asai, N. Hashimoto, K. Yagi, and H. Sunami, "An Experimental 1Mb DRAM with On-Chip Voltage Limiter," *IEEE International Solid-State Circuits Conf.*: 282–283, San Francisco, Calif., 1984.

[Jaing 81] C.-L. Jaing and R. Plachno, "A 32K Static RAM Utilizing a Three-Transistor Cell," *IEEE International Solid-State Circuits Conf.*: 86–87, New York, N.Y., 1981.

[Kang 81] S. Kang, "Synthesis and Optimization of Programmable Logic Arrays," Technical report no. 216, Computer Systems Laboratory, Stanford University, 1981.

[Kawagoe 76] H. Kawagoe and N. Tsuji, "Minimum Size ROM Structure Compatible with Silicon-Gate E/D MOS LSI," *IEEE J. Solid-State Circuits* **SC-11**: 360–364, 1976.

[Kertis 84] R. A. Kertis, K. J. Fitzpatrick, and Y.-P. Han, "A 59ns 256K DRAM using LD^3 Technology and Double Level Metal," *IEEE International Solid-State Circuits Conf.*: 96–97, San Francisco, Calif., 1984.

[Kung 79] H. T. Kung and C. E. Leiserson, "Systolic arrays (for VLSI)," In I. S. Du and G. W. Stewart (eds.), *Sparse Matrix Proceedings 1978*, Society for Industrial and Applied Mathematics, pp. 256–282, 1979. (An earlier version appears in Chapter 8 of Mead and Conway under the title, "Algorithms for VLSI Processor Arrays.")

[Kung 79a] H. T. Kung, "Let's Design Algorithms for VLSI Systems," *Proceedings of the Caltech Conference on Very Large Scale Integration*, Charles L. Seitz (ed.), Pasadena, Calif., 55–90, 1979.

[Kung 82] H. T. Kung, "Why Systolic Architectures," *IEEE Computer Magazine*: 37–46, 1982.

[Leiserson 81] C. E. Leiserson, *Area-Efficient VLSI Computation*, Ph.D. dissertation, Department of Computer Science, Carnegie-Mellon University, October 1981. Published in book form as part of the ACM Doctoral Dissertation Award Series by the MIT Press, Cambridge, Mass., 1983.

[Leiserson 83] C. E. Leiserson and J. B. Saxe, "Optimizing Synchronous Systems," *J. VLSI and Computer Systems* **1**: 41–67, 1983.

[Lewyn 84] L. L. Lewyn and J. D. Meindl, "Physical Limits of VLSI DRAMs," *IEEE International Solid-State Circuits Conf.*: 160–161, San Francisco, Calif., 1984.

[Lopez 80] A. D. Lopez and H. F. Law, "A Dense Gate Matrix Layout Style for MOS LSI," *IEEE International Solid-State Circuits Conf.*: 212–213, San Francisco, Calif., 1980.

[Lund 70] D. Lund, C. Allen, S. Anderson, and G. Tu, "Design of a Megabit Semiconductor Memory System," *AFIPS Fall Joint Computer Conf.*: 53–62, 1970. (4T RAM)

[Lyon 80] R. F. Lyon, private communication, 1980.

[Mashiko 84] K. Mashiko, T. Kobayashi, W. Wakamiya, M. Hatanaka, and M. Yamada, "A 70ns 256K DRAM with Bitline Shielding Structure," *IEEE International Solid-State Circuits Conf.*: 98–99, San Francisco, Calif., 1984.

[Masuoka 84] F. Masuoka, S. Arizumi, T. Iwase, M. Ono, and N. Endo, "An 80ns 1Mb ROM," *IEEE International Solid-State Circuits Conf.*: 146–147, San Francisco, Calif., 1984.

[May 76] P. May and F. C. Schierick, "High-Speed Static Programmable Logic Array in LOCMOS," *IEEE J. Solid-State Circuits* **SC-11**: 365–369, 1976.

[Minato 81] O. Minato, T. Masuhara, T. Sasaki, Y. Sakai, and K. Yoshizaki, "A High-Speed Hi-CMOSII 4K Static RAM," *IEEE J. Solid-State Circuits* **SC-16**: 449–453, 1981.

[Minato 84] O. Minato, T. Masuhara, T. Sasaki, Y. Sakai, and T. Hayashida, "A 20 ns 64 K CMOS SRAM," *IEEE International Solid-State Circuits Conf.*: 222–223, San Francisco, Calif., 1984.

[O'Connell 77] T. R. O'Connell, J. M. Hartman, E. D. Errett, G. S. Leach, and W. C. Dunn, "A 4 K Static Clocked and Nonclocked RAM Design," *IEEE International Solid-State Circuits Conf.*: 14–15, Philadelphia, Pa., 1977.

[Ochii 77] K. Ochii, Y. Suzuki, M. Ueno, K. Sato, and K. Asahi, "C^{2}MOS 4K Static RAM," *IEEE International Solid-State Circuits Conf.*: 18–19, Philadelphia, Pa., 1977.

[Patil 79] S. S. Patil and T. A. Welch, "A Programmable Logic Approach for VLSI," *IEEE Transactions on Computers* **C-28**: 584–601, 1979. (SLAs)

[Remshardt 76] R. Remshardt and U. G. Baitinger, "A High Performance Low Power 2048-Bit Memory Chip in MOSFET Technology and Its Application," *IEEE J. Solid-State Circuits* **SC-11**: 352–359, 1976.

[Regitz 70] W. M. Regitz and J. Karp, "Three Transistor Cell, 1024 bit, 500ns MOS RAM," *IEEE J. Solid-State Circuits*, pp. 42–43, 1970.

[Schmidt 65] J. S. Schmidt, "Integrated MOS Random-Access Memory," *Solid-State Design*, pp. 21–25, 1965.

[Schnookler 80] M. S. Schnookler, "Design of Large ALUs Using Multiple PLA Macros," *IBM J. Res. Develop.* **24**: 2–14, January 1980.

[Sud 81] R. Sud and K. C. Hardee, "Designing Static RAMs for Yield as Well as Speed," *Electronics* **54**: 121–126, July 28, 1981.

[Taguchi 84] M. Taguchi, S. Audo, S. Hijiya, T. Nakamura, S. Economo, and T. Yabu, "A Capacitance-Coupled Bit-Line Cell for Mb Level DRAMs," *IEEE International Solid-State Circuits Conf.*: 100–101, San Francisco, Calif., 1984.

[Terman 71] L. M. Terman, "MOSFET Memory Circuits," *Proc. IEEE* **59**: 1044–1058, 1971.

[Wada 78] T. Wada, O. Kudoh, Y. Nagahashi, and S. Matsue, "A 15-ns

1024-Bit Fully Static MOS RAM," *IEEE J. Solid-State Circuits* **SC-13**: 635–639, 1978.

[Weinberger 67] A. Weinberger, "Large-Scale Integration of MOS Complex Logic: A Layout Method," *IEEE J. Solid-State Circuits* **SC-2**: 182–190, 1967.

[Wong 81] J. Wong, M. Ebel, and P. Siu, "A 45-ns Fully Static 16K MOS ROM," *IEEE J. Solid-State Circuits* **SC-16**: 592–594, 1981.

[Wooley 80] B. A. Wooley, "Course notes for EECS 243," University of California, Berkeley, unpublished, 1980.

[Yamada 84] J. Yamada, T. Mano, J. Inoue, S. Nakajima, and T. Matsuda, "A Submicron VLSI Memory with a 4b-at-a-Time Built-in ECC Circuit," *IEEE International Solid-State Circuits Conf.*: 104–105, San Francisco, Calif., 1984.

[Zippel 83] R. E. Zippel, "A Survey of Static Memory Techniques," unpublished, 1983.

THE MICRO-ARCHITECTURE OF VLSI SYSTEMS

8

Microarchitecture spans the domain between the macroarchitecture (the lowest-level hardware visible to the user) and the implementation technology (MOS VLSI). This chapter is an introduction to some of the important issues that apply to digital systems designs implemented in MOS VLSI. Systems architecture cuts across many disciplines. Although each is as worthy of development as detailed as that given MOS circuits in the previous chapters, this chapter is mostly overview and examples. The reader is encouraged to investigate the bibliography for sources covering digital systems architecture, logic design, and microprogramming.

Architectural considerations are the rightful domain of the circuit designer, just as circuit considerations are clearly appropriate to the study of computer architecture. In their book, *Introduction to VLSI Systems*, Carver Mead and Lynn Conway describe the process of designing a VLSI chip [Mead 80, p. 89].

> Designs are then done in a "top down" manner but with full understanding by the architect of the successive lower levels of the hierarchy.... For example, the activity "logic design" in integrated systems might best be conceptualized as the search for techniques and inventions that best couple the physical, topological, and geometric properties of integrated devices and circuits with the desired properties of digital VLSI systems.

This is a holistic view of VLSI system design. The success of a VLSI machine implementation depends on the harmonious interactions of

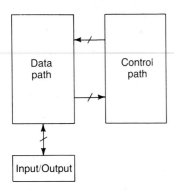

FIGURE 8.1 A simple microprocessor consisting of a data path and a control path.

wire, transistors, logic, and architecture. This can best be accomplished with an integrated approach to system design.

There is a conflict in the way we partition and think about systems. While the effective engineer must exploit the powerful intellectual mechanism of abstraction to divide a problem into pieces small enough to be conveniently digested, he or she must be ever aware that abstractions are synthetic intellectual compartments. Indeed we purposefully restrict and limit our design style so we can form clean abstractions. That is what a methodology is all about. Abstractions have an essential purpose, but they are only a means to an end. Certainly, our intellectual barriers do not necessarily reflect the total harmony of nature. Abstraction, then, is both enabling and confining. In many cases, the most profound insights and inventions cross the boundaries of our established conventions—often creating new ones in the process. In the first seven chapters of this text we investigated many levels of abstraction. But if we examine these levels carefully, we find that each interacts with the others.

VLSI grants a designer new freedoms, both economic and architectural. Unfortunately, no one yet understands how to fully use these freedoms. Even in the arena of conventional computers composed of a data path plus a control path, the merits of various architectures are still energetically debated. We will use techniques that have arisen from competitive implementations of conventional machines to establish a groundwork for further efforts in VLSI microarchitecture.

Figure 8.1 illustrates the canonical microprocessor, consisting of a data path, a control path, and an I/O section. We will begin our investigation with the data path since this is the simplest component.

8.1 Data-path design

In this, and in all subsequent analyses in this chapter, there is a tacit assumption that the overriding design goal is to achieve maximum computational power with minimum chip resources, principally chip area and power. (Design resource limitations can also be a consideration.) Performance and resources form two axes of a design space. For a fixed set of resources there is an optimum attainable performance. As a function of the chip resources, the optimum achievable performance forms a curve in the design space. Resource or performance constraints define the target point for a particular design.

The principal microarchitectural variables in data-path design are the functional capability and types of the individual logic elements, the quantity of elements of each functional type, and the communication paths among elements. Examples of functional elements include ALUs, shifters, and register files. ALU parameters include the types of functions the ALU can perform, the width (bit serial, 8b, 16b, and so on), and the operating speed. For a shifter, the width, the number of bits that can be shifted left, and the number of bits that can be shifted right are important design parameters. In a register file, we are concerned with the width, the number of registers, the number of read ports, the number of write ports, and the number of read/write ports.

Arithmetic and logic units

Virtually all microprocessors contain a central ALU for performing the logic and arithmetic functions essential to basic computing. In addition, many processors also contain less general elements for performing more specialized functions, such as incrementation, addition, or comparison.

A general-purpose ALU is typically composed of three basic elements: the logic function generator, the carry chain, and a final XOR. The data input to an ALU is usually the positionally paired contents of two registers. The purpose of the logic function generator is to provide inputs to the carry chain and XOR. These inputs represent various functions of the two input variables. There are sixteen possible logic functions of two variables. Four control wires are needed to select among the sixteen functions. A common implementation of the function generator is shown in Fig. 8.2. The two outputs, P and G, represent the propagate and generate inputs to the carry chain for arithmetic operations and can be manipulated to achieve other logic functions using the four control wires.

The functional block can be thought of as a special case of the more general technique of table look-up. Tables are required, for instance, in logarithmic arithmetic. A constant generator is a special case of table look-up, where important numbers (such as 1, 0, −1) and special addresses are stored in a PLA or ROM for easy access by the control path.

A bit decoder serves a similar purpose, while being both more compact and less general. In a 32 bit data path, a bit decoder would take in five address lines to select which of the 32 data-path lines should be set. The inverse function is performed by a priority encoder that examines the data path for a set bit and encodes its position for use by the control path. The purpose of these functional units is to lower the number of wires passing between the control and data paths.

The barrel shifter is found in many data-path implementations. A schematic of a simple barrel shifter is illustrated in Fig. 8.3. The barrel shifter rotates the data-path word by an arbitrary number of bits.

The carry chain is typically the most speed-critical element in an ALU since it must pass data horizontally across the width of the ALU. The standard implementation in MOS logic uses the Manchester configuration with pass devices. We observe that if we wish to add two

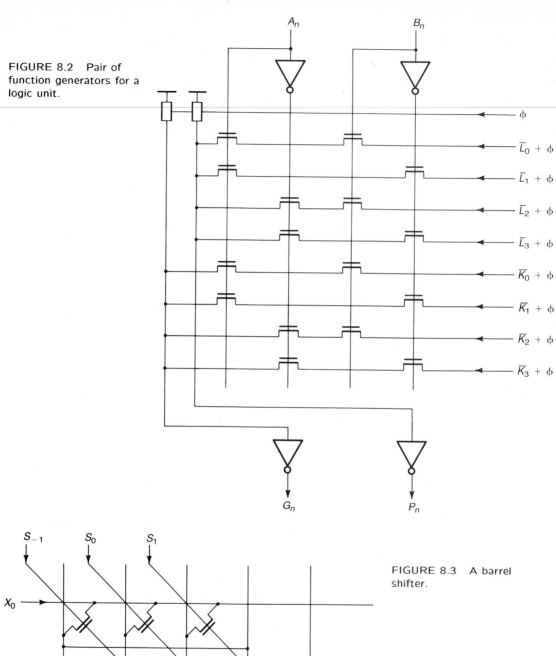

FIGURE 8.2 Pair of function generators for a logic unit.

A_n

B_n

ϕ

$\overline{L}_0 + \phi$

$\overline{L}_1 + \phi$

$\overline{L}_2 + \phi$

$\overline{L}_3 + \phi$

$\overline{K}_0 + \phi$

$\overline{K}_1 + \phi$

$\overline{K}_2 + \phi$

$\overline{K}_3 + \phi$

G_n

P_n

S_{-1} S_0 S_1

FIGURE 8.3 A barrel shifter.

X_0

X_1

X_2

Y_0 Y_1 Y_2

numbers A and B, then at each bit position we have both the possibility of a carry input and a carry output. If A_i and B_i are 1, there will be a carry out, independent of the carry input. In this case, it is possible to "generate" a carry out without waiting for the carry in. If A_i and B_i are both 0, the carry out will be 0, again independent of the carry input. In this case, it is safe to "kill" the carry output immediately. If $A \oplus B$ is 1, the carry out will be 1 if and only if the carry input is 1. That is, the carry is "propagated." Table 8.1 summarizes these conditions.

We have

$$K_i = \overline{A}_i \overline{B}_i, \tag{8.1}$$

$$G_i = A_i B_i, \tag{8.2}$$

and

$$P_i = A_i \oplus B_i. \tag{8.3}$$

The sum S_i is given by

$$S_i = P_i \oplus C_{i-1}, \tag{8.4}$$

where C_{i-1} is the carry in.

A typical implementation of the Manchester carry chain using precharging is shown in Fig. 8.4. To combat the quadratic delay behavior of a series of pass devices, buffers are inserted at regular intervals. Figure 8.5 shows a buffer configuration that avoids adding the buffer delay in series with the critical path, but accomplishes the desired level restoration function. Using this approach, it is beneficial to insert buffers quite often (every two to four stages) for maximum performance.

A technique that is useful for speeding up the carry propagation is carry look-ahead. This involves computing carries for groups of bit positions rather than just one at a time. A MOS implementation that utilizes the efficiency of the Manchester scheme to achieve group carry look-ahead is shown in Fig. 8.6. The size of the grouping can be optimized for the particular application, but experience has shown that

TABLE 8.1 Manchester carry conditions

A_i	B_i	Action
0	0	kill
0	1	propagate
1	0	propagate
1	1	generate

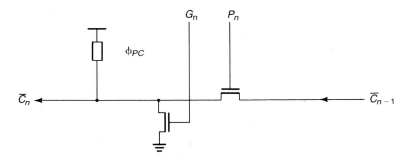

FIGURE 8.4 One stage of a precharged Manchester carry chain.

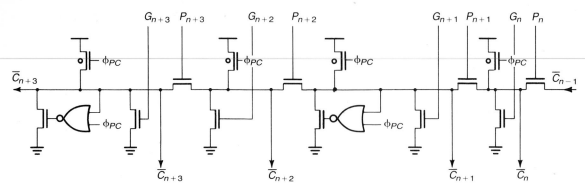

FIGURE 8.5 Signal-restoring buffers for the carry chain that are not in the critical path.

a group size of four is generally optimum, especially if carry buffers are not used in the look-ahead generation path.

One quite general technique, useful for control as well as data, is the use of condition selectors. If we need to take one action when a binary input is high, and another action when it is low, then one possibility is to do both now (in separate hardware), and when the input finally shows up, select the correct answer. It is often possible to share some of the hardware that computes the two possibilities. There are limitations to

FIGURE 8.6 Group carry look-ahead using the Manchester scheme.

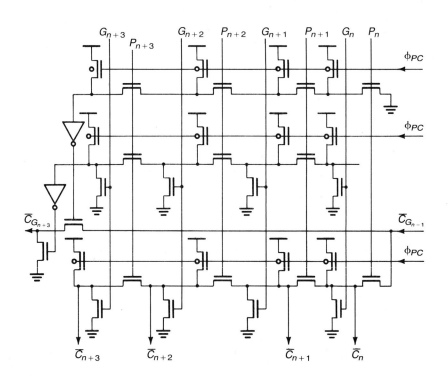

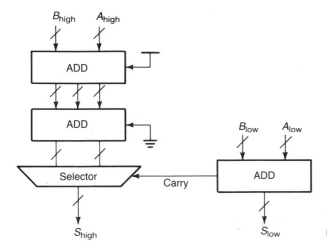

FIGURE 8.7 Carry select adder.

this, and it gets exponentially out of hand with the number of input bits, but in many cases it maps well into a VLSI implementation. Selectors are inexpensive and the duplication of hardware is straightforward. Figure 8.7 illustrates this technique applied to an adder, where the middle carry bit is assumed both ways, and the correct answer selected.

The final XOR always presents a challenge to the circuit designer because, unlike the function generator and the carry chain, precharging is not generally possible since the output can glitch between the time of the initial P and G evaluation and the time of the final local carry input. If the appropriate clock edge is available, the XOR can be strobed after the inputs are set up. If, however, the ALU must be flow-through, as is often the case, the designer must choose among static implementations. Figure 8.8 shows a static implementation using inverters and pass devices. The critical path includes only the local carry inverter because the inverter on P is normally well set up.

The apparent width of the ALU is usually proscribed by the high-level architecture of the system. However, it is often possible to emulate this "virtual" ALU with a physical ALU which is smaller but requires several passes to perform the same operation as the virtual device. For example, bit serial ALUs were used in the early 4b microprocessors and 8b ALUs have been used in some 16b processor units. The most extreme example of this technique is found in handheld calculators that invariably use bit serial arithmetic to perform their precise computations. Area and power are nearly proportional to ALU width, and this argues in favor of narrow ALUs. However, the delay through a wide ALU need only be logarithmically dependent on the width. This argues in favor of one pass through a wide ALU rather than several passes through a narrower ALU.

Subsets of the full ALU function are often useful for special functions in the data path. A register incrementer is an example of such a function. Figure 8.9 shows an implementation of an incrementing register that

FIGURE 8.8 Static implementation of the final XOR gate in an ALU.

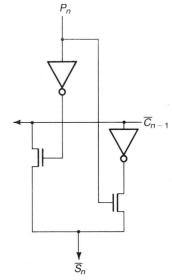

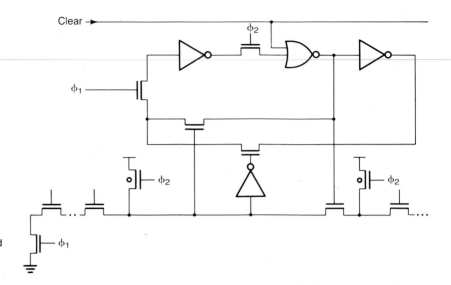

FIGURE 8.9   A precharged incrementer cell similar to the one in Fig. 1.58.

does not require a ripple except in the pass device stack. The carry propagation through this stack could be speeded up with the carry buffers shown in Fig. 8.5.

Register files

Register files in microprocessors are similar to the RAMs discussed in Chapter 7. However, the constraints are often quite different. For example, the pitch of the bit lines may be limited by the ALU such that the layout of the RAM cell is suboptimal.

A major consideration in the design of the file is the number of ports required. Each port imposes new requirements on the basic cell as well as using additional decoders and I/O circuitry. Many microprocessors use a dual-port read, single-port write register file. This is based on the fact that the ALU accepts two inputs and generates one output per cycle.

Often the size (number of words) as well as the performance requirements dictate the type of cell and sense technique. For a small file, it is often possible to make the individual cells large and powerful such that no special sense amplifier is required. As the number of cells in the array grows, the overhead of sense amplifiers decreases in proportion to the cell area, and the cost of large cells increases. Dual versus single rail porting is another issue that must be addressed. If the pitch allows, dual porting can save power (write amplifiers can do the work of writing and sense amplifiers the work of reading) and enhance speed. However, for dual- or triple-ported arrays, the pitch seldom allows all ports to be dual rail. Figure 8.10 shows a cell configuration that has been used in several microprocessors.

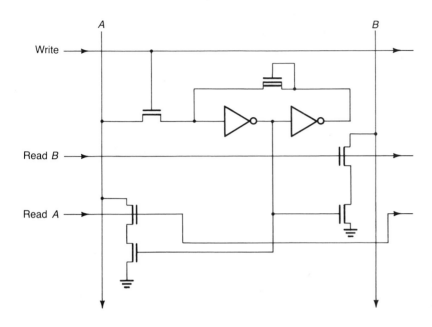

FIGURE 8.10   A dual-port read, single port write, register cell.

Intrachip communication

A complete data path consists of functional elements, registers, and the buses used to interconnect them. Common buses are an extremely efficient structure for achieving communication between the data-path elements. Unfortunately, a single bus can handle only one transfer at a time (multiple destinations are possible, of course.) If there are many functional units connected to a common bus and they could otherwise operate concurrently, the bus can be a bottleneck to performance. Thus multiple buses can offer a major performance incentive.

A technique used in all MOS chips to minimize interconnect area is to build logic functions (registers and combinational logic) under the data buses to the greatest extent possible. The most efficient layout results in having control and data lines running at right angles. In addition, to avoid wasting interconnect area, data line pitch must be matched from section to section. Generally, the elements requiring the greatest pitch dictate the pitch of the entire data structure. Most data paths have at least one functional unit that requires two inputs per bit (for instance, the ALU). Thus two buses seem to be a natural choice for a VLSI data path. The T11, LSI-11/23, MC68000, HP 9000, and many other microprocessors have a two-bus structure. The iAPX 432 has a three-bus architecture on the instruction execution chip while the NSC 16032 has only one data bus.

To avoid wasted space, the layout should have a linear organization, if possible. The important area parameter is the number of buses

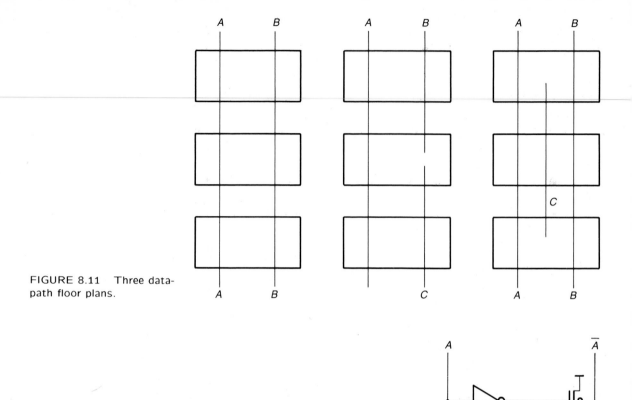

FIGURE 8.11   Three data-path floor plans.

FIGURE 8.12   Circuit for restoring the high levels on a dual rail bus.

FIGURE 8.13   Data-path organization of the MC68000. Note the transmission gate connections between buses. [Stritter 79] (©1979 IEEE)

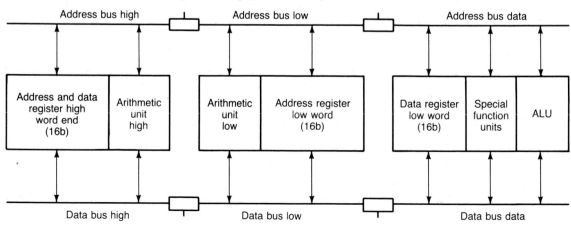

that must be supported simultaneously in the layout. For example, in Fig. 8.11, the left and center layouts are area equivalent with two and three buses, respectively. The layout on the right also has three buses but would be larger.

In general, for a fixed area, more buses provide potentially more concurrency, but fewer (or more indirect) communication paths because the buses must be shorter. Thus in the center layout in Fig. 8.11, we may simultaneously perform three local data transfers, but in the left layout two global transfers may be accomplished that are not possible in the center layout. More buses also require more from the control path.

It is useless to duplicate functions if there are data dependencies that preclude parallel operations. One pair of orthogonal functions that occur in most machines are addressing operations and data operations (which normally share the same data path in 8 and 16 bit machines). Separate arithmetic units for these operations are often employed.

Both single and dual rail buses have been used in microprocessors' data paths. Invariably, the buses operate using a precharge technique. Fig. 8.12 shows a technique that can be used to reinforce the "1" level on a precharged bus that might be subject to noise coupling or charge splitting, or that might be required to hold its state for an extended period of time. Figure 8.13 shows the busing structure used on the MC68000. The structure provides dual-port capability along with potential concurrency between address and data elements, without sacrificing communication paths.

■■■■■■ EXAMPLE 8.1

Accumulator circuit
We examine the design of a fast adder circuit to accumulate a continuous stream of numbers. The key to doing this is to avoid the carry propagation in the feedback loop. A reasonable solution is shown in Fig. 8.14, which uses a carry save adder in the feedback loop and performs the carry propagate additions in a pipelined manner, external to the loop. The maximum clock rate is determined by the delay through an adder cell. (Actually, the maximum clock rate is probably determined by the I/O bandwidth of the pads.)

This technique can be applied to recursive digital filters. Carry propagate additions (or multiplications) need only be done at the output, just before the data are used.   ■

FIGURE 8.14   A circuit for accumulating a long stream of numbers.

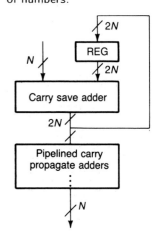

8.2   Implementation of the control structure

The control structure is that logic that sequences the machine through a set of logic states. These states may have a hierarchy of levels, that is, major state X may have minor associated states $x_1, x_2, \ldots, x_n$, down to any level. We also include in the control section the logic that interprets the state needed to produce activation signals. In most microprocessors, there is about an equal balance between control circuitry and data circuitry. In other digital applications, where the complete machine is primarily a controller, this is reflected down to the microarchitecture, which can be almost entirely control circuitry. On the other hand, a dedicated signal processor might be mostly data path.

The two major alternative implementation techniques for the control section are random logic and structured logic. Random logic implementations are characterized by complex networks of simple gates and storage elements. This technique can generate the most transistor-efficient implementation if and only if the control function is simple or great effort is expended in optimization. The major disadvantages of this technique include increased logic, circuit, and layout design time; increased effort required to correct mistakes; and a lack of flexibility for engineering changes.

Most early microprocessors, including most of the 8 bit machines, used random logic control. Most of the 16 bit and 32 bit microprocessor designs are built with structured techniques. The remainder of the chapter is devoted to structured techniques.

The canonical structured controller

All of the structured controllers that we will discuss are variations on the theme of a canonical state machine consisting of combinational logic and a state register as shown in Fig. 8.15. The variations involve the implementation of the combinational logic and the organization of the state register.

Two levels of logic, in the form of a PLA, can implement any general control function. However, in the interest of structured logic design and overall efficiency, it is important to restrict the operation to a relatively small set of primitives and to build complex control functions out of these primitive operations. This simplifies not only the design of the controller but also of the data structure being controlled. Microprogramming can achieve regularity in both the layout and digital design domains. Programming is the most powerful technique known for the construction of complex control structures.

We refer to the state register as the microinstruction register (MIR). Major variations on the state register include different address sequencing techniques, branching schemes, state hierarchies, and pipelining techniques. Variations on the combinational logic include ROM versus PLA, a horizontal versus vertical control word, and the number of levels and separate instances of ROM or PLA occurrences. We will see that many of the issues are simply a matter of how to handle context switches efficiently, with regard to both area and speed.

FIGURE 8.15 Combinational logic and register of a finite state machine.

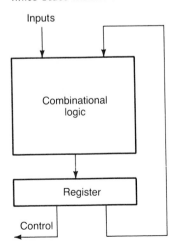

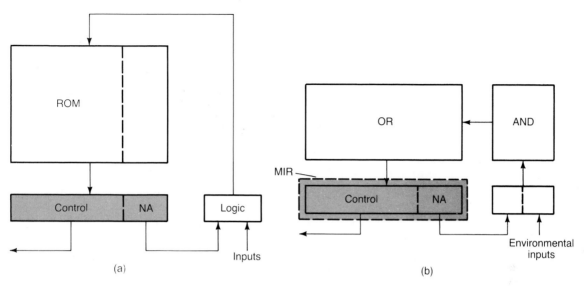

FIGURE 8.16   The use of a next address field and (a) external logic plus a ROM, or (b) a PLA, both used to build a control machine.

Address sequencing techniques

A simple and effective technique for providing the sequencing function is to incorporate a next address (NA) field in the microinstruction word. This field serves as a pointer to the next word to be fetched. Conditional branching may be accomplished by having the environmental inputs modify the next address field, as shown in Fig. 8.16(a). When the array is implemented as a PLA, the merging of the environmental inputs can be done in the AND plane, as shown in Fig. 8.16(b).

In many applications there are cases where subsequences of microinstructions can occur in separate control paths. It is desirable to maintain only one copy of the subsequence. To allow subflows to merge back into the main flow, it is necessary to provide some kind of subroutine return mechanism. It is not efficient to keep this information in the main control register because it gets updated on each cycle. Some additional special storage is needed. While this can be as fancy as a hardware stack, in many applications a much simpler technique suffices.

In the T11 chip, which uses a next address field, the macroinstruction register (IR) provides the necessary branching and return information to the microsequencer for the multiple use of microwords. No other linkage mechanism is necessary. This structure is shown in Fig. 8.17. This is especially valuable for operand addressing operations, which are common to all instructions and must precede actual instruction execution. For instance, the instructions ADD, SUB, and EQUAL? all require

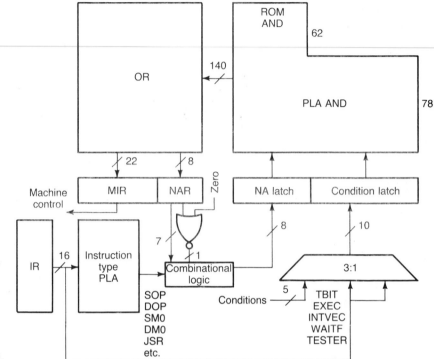

FIGURE 8.17 T11
microsequencer logic.

two operands. These operands must be fetched from memory. In the
T11, the instruction-type PLA determines which addressing flow to
follow, in this case the routine that fetches two operands. At the end
of the addressing flow, the two operands are available and the machine
must branch to the execution flow for the particular instruction. The
last microinstruction for the addressing flow has a next-address field
designed to work in conjunction with the IR to determine the next
address corresponding to that instruction.[1]

An alternative to the next-address field is to use a program counter
approach. Its principal advantage is to eliminate the next-address bits
from the OR plane. The savings in bits is $K \log_2 K$, where K is the
number of words in the control store. This must be balanced against
a more complex sequencer and some timing disadvantages. Generally,
when this approach is taken, a ROM is used rather than a PLA. In
order to accommodate the environmental inputs, a combinational logic
block is always necessary in the sequencer. This is the approach used
in the Cal Tech OM2, as described in Mead and Conway [Mead 80,
Chapt. 6]. The HP 9000 CPU chip apparently uses a program counter

---

[1] This technique is not to be confused with structured programming.

technique to find its way through the 9K, 38 bit microinstructions. A hardware stack is used for subroutine return as in the OM2.

One modification that can improve the efficiency of either structure is a multiplexer on the environmental inputs. (Abbreviated "mux," multiplexer is another word for a selector.) By controlling the mux from the previous microword, the conditions to be used in determining the next state can be preselected. This allows the PLA inputs to be confined to only those signals that might affect the next microinstruction. This drastically reduces the number of inputs required in the AND plane. The HP 9000 CPU chip selects between 55 separate condition sets generated throughout the machine. The multiplexer is distributed throughout the machine as part of a bus. In the logic domain, we are expanding the number of levels of logic, past the required two, in order to minimize area. This type of control bus consists of the data lines and control lines that determine which sources should be active for a particular cycle. The control lines may be explicitly routed along with the data bus or they may be implicitly represented in the microinstruction word itself.

Conditional branching schemes

We have already touched on the subject of functional or unconditional branching. Conditional branching is a special case because the timing can be in the critical path of the machine. Most microprogrammed machines have, at a minimum, an overlap of microinstruction fetch with data-path execution, as shown in Fig. 8.18. Thus while the data path is executing an operation controlled by microinstruction N, the control path is fetching the microinstruction needed by the data path to control operation $N + 1$.

From the figure it might appear that a conditional branch depending on the result of data-path operation N could not affect the fetch of microinstruction $N + 1$ because they occur at the same time. Thus the conditional branch would not have an effect until the execution of instruction $N + 2$. As a result, slot $N + 1$ would probably have

FIGURE 8.18   Overlapped microinstruction FETCH and EXECUTE timing.

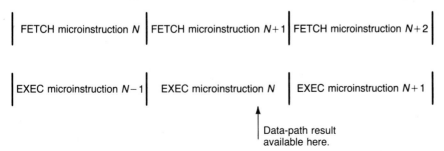

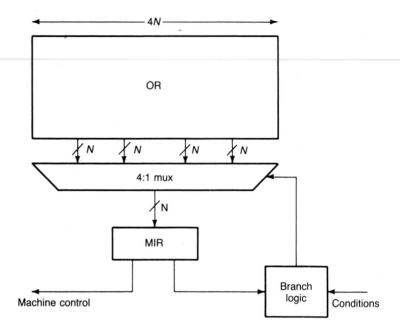

FIGURE 8.19 Page-mode ROM access where branch logic selects which of four possible microinstructions are loaded into the MIR.

to be wasted. In a conventional machine this argument is surely true, however, in an LSI implementation a simple trick is available. Assume that the microstore ROM is organized such that each access brings two or more candidate microinstruction words to the column selector. If the conditional branch can be confined to selection at the column selector level, then the actual selection can be delayed until late in the cycle. We are taking advantage of the fact that column selection is a much faster operation than selecting a new row in the ROM.[2] Both the Motorola 68000 and the HP 9000 microcomputer have reported using this scheme. The MC68000 brings four words to the mux for branch alternatives, while the HP machine brings 32 words to the mux. This general scheme is illustrated in Fig. 8.19 with a 4:1 multiplexer. Note how circuit and microarchitecture issues interact.

Another approach that can minimize the effect of conditional branches at the various levels of the control hierarchy is to assume a branch along the most probable path. Consider, for example, branching in a program loop. In this case, for the majority of branches, no penalty is incurred if one assumes the branch destination is the beginning of the loop.

[2] This is generally called page-mode access. Rows define different pages, and columns define position in the page.

Horizontal versus vertical control word

The distinction between horizontal and vertical is relevant to the bit width of the microinstruction. The lowest-level control function is the binary control point, which is a Go/NoGo command to an elementary logic function. Ultimately, all control words must accomplish these basic tasks. The completely horizontal control word has one bit position for each elementary control point. Recent microprocessors have about 100 to 300 control points, implying a 100 to 300 bit MIR width. If all combinations of bits were used, then no further encoding would be possible. Fortunately, this is not true, and encoding can be used. (A 10 MHz processor would need 4×10^{15} years to exercise all combinations possible with a 100 bit wide MIR.) It is generally the degree of encoding that is at issue in the horizontal/vertical distinction. Most computers, and virtually all microprocessors, subdivide the microword into fields corresponding to the control of individual logical units—for example, register file, ALU, bus control, and so on. Each of the units would have its own decoder for converting the control field into elementary control signals. The important design parameter is usually the number of fields that can occur simultaneously in one microinstruction. If each field controls a logical unit, then the number of fields determines the maximum number of units that can operate simultaneously. This feature, plus the buses available for data operations inside the machine, determines the maximum concurrency available at the microcode level.

The ultimate vertical encoding of the microsequencer is to collapse all the fields into one binary coded field. The microcode would then be used by a nanocode ROM that would select one of V possible control words for the machine, as illustrated in Fig. 8.20. Thus the microcode ROM would have $\log_2 V$ outputs, which would address V words in the nanocode ROM. These words would require minimal further decoding. Microcode/nanocode structures are very common in computers. They were used in microprocessors as far back as the LSI-11. Feedback can be used in the nanocode ROM to allow it to interpret, rather than just decode, the microcode ROM outputs. This technique has advantages when a very wide control word is desired.

When feedback is not used in the nano-ROM, we can compare the size of the one- and two-level implementations. Figure 8.21 illustrates a simple microcode PLA with K words of N bits. The area A_1 taken up by the one-level scheme is roughly

$$A_1 = K(N + 3\log_2 K). \tag{8.5}$$

In the two-level scheme, illustrated in Fig. 8.22, the area A_2 taken up by the circuitry is

$$A_2 = K(\log_2 V + 3\log_2 K) + V(2\log_2 V + N). \tag{8.6}$$

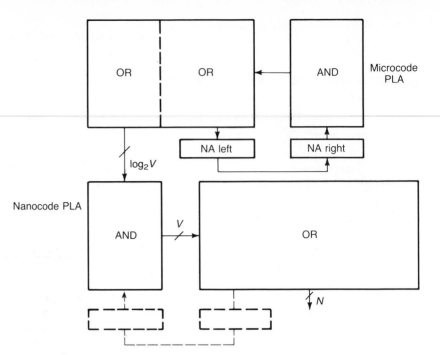

FIGURE 8.20 Two-level
microcode/nanocode control
path.

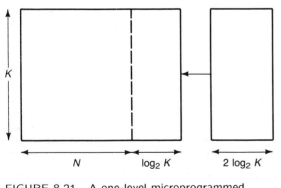

FIGURE 8.21 A one-level microprogrammed
control machine.

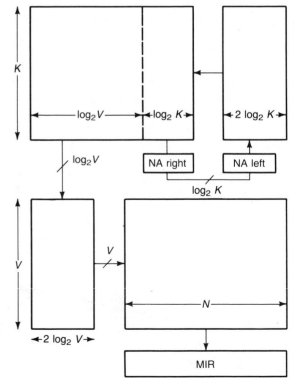

FIGURE 8.22 A two-level
nanocoded machine with the
same number of instructions
and same MIR width as in
Fig. 8.21.

The sharing ratio ρ is defined as the average number of times the words in the nanostore are shared by the words in the microstore. We have

$$\rho \equiv \frac{K}{V}. \tag{8.7}$$

$K - V$ output words in the microcode PLA are duplicated. Rewriting A_2, we have

$$A_2 = K\left(4\log_2 K - \log_2 \rho\right) + \frac{K}{\rho}\left(N + 2\log_2 \frac{K}{\rho}\right). \tag{8.8}$$

The value of N for which $A_1 = A_2$ is given by

$$N = \frac{\rho + 2}{\rho - 1}\log_2 \frac{K}{\rho}. \tag{8.9}$$

When ρ is low, N must be large for this technique to be area efficient. For instance, at $K = 512$ words, the break-even values for N are 70 and 32 for ρ values of 1.4 and 2, respectively. The power of microcode/nanocode structures arises from the storage of very wide control words only once in the nanostore despite their use in multiple control flows.

A variant of this technique is used on the MC68000. The structure in the MC68000 is more complicated because the arrays are merged, allowing inputs into the first PLA AND plane to address words directly in the nanostore. A structure similar to that in the MC68000 is illustrated in Fig. 8.23. Note how selectors are used to match the widths of the nanoarrays and microarrays.

A final consideration in the nanostore is the distinction between central and distributed decode. We have seen that the nanostore can be justified only if it outputs relatively wide (highly horizontal) control words. This implies that there are a large number of wires to route from the nanostore and that little additional decoding will be done at the destinations. Thus the nanostore offers the advantage of an efficient array approach to the decoder but the disadvantage of more global wires to be routed throughout the chip. Individual circumstances will dictate which offers a net advantage.

Hierarchy of states

We often want to maintain some control state in addition to the microinstruction register. In a microprocessor this includes, at a minimum, the macroinstruction register. Previously, we mentioned the use of the IR to provide a context to allow shared subflows of the microcode to dispatch to the proper next address. The IR provided major state information, which controlled the minor state sequencing. The major state information can also control the interpretation of the activation

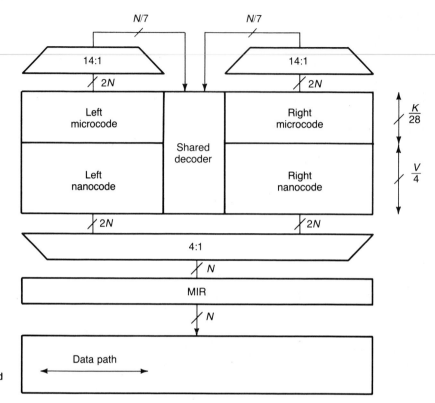

FIGURE 8.23 Control machine of the type found in the MC68000.

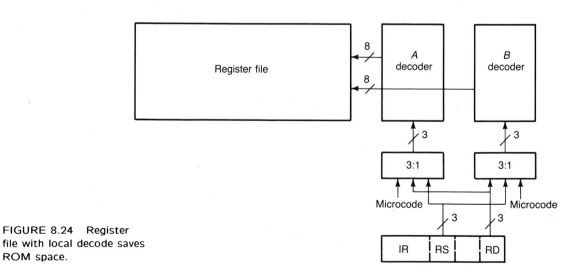

FIGURE 8.24 Register file with local decode saves ROM space.

signals generated by the minor state sequencing. Again, using the IR as the major state register, we can see how the major state can directly influence the control of the machine. In computers with fixed instruction addressing formats like the PDP-11 or the Motorola 68000, the address fields of the macroinstruction can be used to directly address local registers in the machine. The microinstruction merely provides a connection path from the proper field in the macroinstruction to the register file decoder as illustrated in Fig. 8.24. This technique was discussed in Section 5.9. The issues here are succinct communication and short critical paths.

In the case when nanocode operations interpret microcode instructions, the microcode/nanocode structure presented earlier in this section is another example of hierarchical states.

A set of major operations common to all macroinstructions are

1. Instruction fetch,
2. Operation decode,
3. Operand fetch (possible multiple), and
4. Execute.

These major operations can be equated with the major states of the machine. Thus, in addition to the macroinstruction register, it is usually helpful to have a register that keeps track of the current major state. To allow sharing of operand fetch microcode for multiple operand instructions, a counter can be employed to distinguish between operand 1, operand 2, and so on. For executions such as multiply and divide, which require microinstruction loops, a loop counter is expedient. The lesson here is that the natural hierarchy and partitioning of the problem often suggest good hardware implementations.

Pipelining in the control path

In general, MOS logic lends itself to pipelining techniques because of the small incremental cost of data latches. Pipelining is extremely effective in increasing the amount of work that can be done concurrently without adding parallel hardware. The major disadvantage is that when there are data dependencies among levels of the pipe, timing problems develop. If separate hardware modules are used instead of pipelining, these timing problems show up as wiring problems, and cases where the pipeline needs to be flushed show up as cases where most of the hardware modules are idle. We saw a simple example of pipelining with the overlap of microinstruction fetch with instruction execution, resulting in a problem with conditional branches.

Most high-speed computers use an instruction prefetch pipeline; recent high performance microprocessors are no exception. To minimize the size of the control store and allow overlap, a separate mapping PLA is used to find the starting microcode address for each macroinstruction. Thus the prefetch pipeline is roughly as follows.

1. Prefetch macroinstruction,
2. Use mapping PLA to find microcode starting address,
3. Fetch microinstruction, and
4. Execute.

Not all these operations will take the same amount of time (especially the prefetch itself), so more than four microcycles may be required. Note that a macroinstruction branch requires the complete pipeline to be flushed and restarted. In some machines, a macroinstruction branch occurs on an average of one out of every three or four instructions. This is because as more parallelism is built into the hardware, branches become more common. A probabilistic prefetch algorithm for branches is especially helpful in keeping the prefetch performance high.

It is often useful to view the instruction prefetch unit as a semiautonomous machine that addresses memory to fetch (probably) pending instructions as often as possible. It can then stuff these unexecuted instructions in a queue for use by the execution unit. Unanticipated branches require flushing the queue.

To extract maximum performance from instruction prefetch, a very high memory bandwidth is required. One possibility is to pipeline the memory operation itself so that prefetch and data operations on memory may operate simultaneously. The HP 9000 chip set has integrated a pipeline memory architecture in its CPU and custom memory chips. A high-speed tightly coupled cache memory is another possibility. This avoids some of the complications of the pipeline approach. Both techniques have been necessary in mainframes. At very high levels of integration, an on-chip memory cache becomes extremely attractive since it can potentially remove off-chip delays from the critical path of the majority of operations. The Z80000 reports the use of on-chip cache for both programs and data. The MC68020 reports the use of on-chip program cache.

ROM versus PLA for microsequencer combinational logic

In a custom MOS implementation, one is free to choose either a ROM or PLA for the microsequencer. Each has pros and cons and each has been used in microprocessor implementations. As we will see, the PLA generally has the advantage in small control applications, while the ROM is more suitable for large implementations. First we consider the actual circuit and layout implications of each.

Logically, a ROM can be considered a subset PLA with the AND plane completely programmed (no "don't care" conditions). Because the ROM AND plane is maximally encoded, it is more transistor efficient than the PLA AND plane. This can be a factor in overall efficiency. However, a ROM usually requires external circuitry to encode its inputs to be binary compatible. This extra level of encoding can cost time as well as area. Often, for small arrays, the PLA is overall more transistor efficient. In large arrays, the ROM AND plane savings can more than compensate for the area of the input encoder; however, the delay penalty remains.

A technique that minimizes the PLA inefficiency while retaining the two-level logic structure has been used in the LSI-11/23 and the T11. These chips use a next-address sequencing scheme in conjunction with modifiers (called "translators") to provide branching. Only the instructions that are branch destinations are placed in the PLA; the remainder are in pure ROM using the NA field as a binary-coded address. The resulting physical structure can be merged, as is done in the T11 chip. This technique is illustrated in Fig. 8.17.

Structurally, the ROM does provide the designer with some unique options. ROM decode may be partitioned arbitrarily with respect to rows and columns. This may be important in chip layout design and has advantages in minimizing the impact of conditionals if the branch choice can be deferred to the column decoders, as we described earlier in this section.

■■■■■■■■■■ EXAMPLE 8.2

The Connection Machine
There have been many attempts to explode the Von Neumann memory bandwidth bottleneck through the use of unconventional architectures. One general strategy is to distribute the processor throughout the memory, which then becomes "smart memory." The MIT Connection Machine is one example of this [Hillis 81]. It consists of a large collection of small processors, each with slightly fewer than 1 K bits of memory and a bit serial ALU. One thing that is noteworthy from our perspective is an interconnection scheme due to Carl Feynman. While the connection machine has a general communication mechanism, it was thought that one would, for certain well-structured problems, like to form the machine as a giant mesh. Data flow would be in the direction of any of the four winds. However, the number of pins per chip would be very large because there are 64 processors per chip. An alternate scheme was developed which uses 25% fewer pins. The conventional wiring strategy is shown in

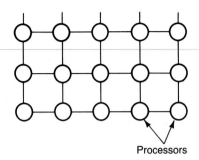

FIGURE 8.25 Conventional wiring of a processor array.

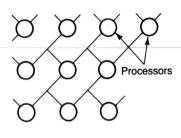

FIGURE 8.26 Connection-machine processor organization.

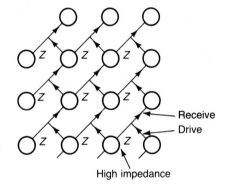

FIGURE 8.27 Scheme for shifting data from south to north in the connection machine.

Fig. 8.25. Only half of the wires are used at any one time if we constrain data to flow in only one direction at a time. The Connection Machine technique is illustrated in Fig. 8.26. All of the drivers are tristate I/O types. Figure 8.27 shows how data are moved from south to north. This is an example of how topology can influence the architecture of a machine at a very high level. ∎

8.3 Test and debug

The design of a VLSI circuit is not complete until it has been debugged, verified, and successfully transferred to manufacturing. The extensive use of computer-aided design tools during the design phase makes this task less painful. Essential tools include circuit simulators, switch or logic level simulators, register transfer level simulators, design rule checkers, and continuity checkers. A host of additional programs have been written to increase the designer's productivity and probability of success. These programs address the tasks of process and device modeling, symbolic layout compaction, automatic placement and interconnect, schematic entry, circuit topology and parameter extraction, logic verification, methodology verification, timing analysis, circuit optimization, logic optimization, floor planning, microcode assemblers and compilers, statistical simulation, systolic array optimization, and so forth. Indeed, one of the more challenging design automation tasks is unifying these programs and the underlying database so they can be united in a single problem. The success of these efforts in computer-aided design is attested by the increasing number of chips that work the first time, despite ever-increasing chip complexity.

Even when chips are designed correctly, a certain fraction of them will contain manufacturing imperfections, resulting in imperfect yield. The good chips must be sorted from the bad. Two distinct problems must be addressed in the test phase. The first is debug, wherein we examine a chip for design errors. The second is what the manufacturing community has traditionally called testing, in which we search for errors in the manufacturing process. These problems are quite different, though both benefit from foresight. A chip that has been designed without regard to the challenges of debug and test will most likely be a nightmare in both regards. It is possible to design a chip with testing issues in mind. This is called designing for testability.

Debug

When the first chips emerge from the fabrication line, their problems are unknown. Problems can include design bugs, fabrication skews, and manufacturing defects. The first task is to eliminate the possibility of severe fabrication errors. This is done by examining parametric test structures that have been placed on the wafer. Process test structures [NBS 83] should include, as a minimum:

- Transistors of each type and of various geometries,
- Breakdown and capacitance structures for diffusion,
- Capacitors of all oxide types and electrode types to check dielectric thickness and integrity,
- Contact chains to determine contact integrity and resistance,
- Resistance structures for all signal conductors, including metal, and
- Ring oscillators or other performance structures that are designed to integrate the process characteristics into a performance figure of merit.

Having ascertained that the process is probably all right, the next step is to find a chip free of manufacturing defects. The best scenario involves discovering two chips from different parts of the wafer (to eliminate some of the types of mask defects) that operate identically. If these chips operate correctly, the debugging has ended. Otherwise, one must enter the debugging cycle. The key aim in debugging is to find the maximum possible number of errors per mask revision. Errors are typically nested. For instance, a short between V_{DD} and ground will hide an error in which two signal wires are crossed. It is extremely important to cut apart the V_{DD} and ground short so this second error can be found. The crossed signal lines might in turn be hiding a microcode bug. Every time one goes back to the stage of generating a new mask, a great span of time is lost. The following techniques are useful in the debugging cycle.

- Cutting wires with an ultrasonic cutter or laser zapper. This allows one to isolate sections of the chip and to remove shorts.

- Probing points internal to the chip. Probes with as little capacitance as 0.1 pF to 0.01 pF are available. One is able to probe any visible metal line on the chip. Probing poly is possible but extremely difficult. It is important to bring all nodes one might want to probe up to the top layer of metal. (Scanning the chip in voltage contrast mode with an electron beam machine is an alternate technology.)

- Injecting signals. This allows one to bypass nonfunctional parts of the circuit and find nested bugs.

- Holding wafers prior to each of the major processing steps so that one can recover quickly from an error (either processing or design) found in one of the later mask levels.

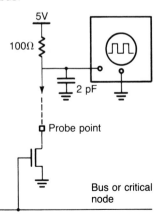

FIGURE 8.28 Technique for probing a high-speed bus.

- Placing open drain transistors below critical nodes, as shown in Fig. 8.28. The drain should be brought to the highest level of metal, and the source should be grounded. By placing a low-impedance external pullup on this node, one can sense very fast voltages with even a 2 pF scope probe (on, say, the 5 mV/div scale). The RC time constant of a 2 pF capacitor and 100 Ω load, for instance, is 0.2 ns.

- Using static or semistatic latches so the machine can be single stepped.

- Using special debug logic features. One example involves the ability to force addresses directly into a control store ROM.

- Using process skewing to purposely achieve worst-case corners of process parameters for quick diagnosis of problems caused by marginal design.

Manufacturing test

The goal of a manufacturing test is to discover the simplest set of input stimuli that demonstrates that every possible fault has been successfully avoided. A fault is a logic or switch level concept. Imperfections in the manufacturing process can, for instance, cause the failure of some physical component. This may show up as a fault in the logic. A fault model is an idealized fault, used so we can say something formally about the problem. The most popular fault model is the "stuck-at" model. In the single stuck-at fault model, one assumes that if a machine is nonfunctional, the cause of its failure is a single node in the machine that is either stuck at 1 or stuck at 0. This model is obviously limited. For instance, it does not cover the case of two signal wires shorted together. Experience has shown, however, that if an LSI system has passed a test designed to discover a high percentage of all single stuck-at faults, then it is probably good in all respects. Only time will tell whether this will continue to be true at ever higher levels of integration.

Tests are most economically applied to a VLSI chip while the wafer is still whole. Probes are brought down onto each chip and a vector of stimuli is applied. To minimize test time, testing is ordered such that the most easily discovered faults are discovered first. Generally, gross parameter tests, such as I_{DD} current and pin leakage, are done first. When the chip fails a test, it is marked with a drop of ink, so that later, when the wafer is cut up, it can be identified for disposal. The time required to apply a set of tests to a VLSI chip can represent a significant fraction of the cost of the final product. It costs several cents per second to test an LSI chip. Nevertheless, it is generally much cheaper to test than not to test. The testing folklore is that detecting a fault in a packaged chip costs ten times as much as detecting a fault at the wafer level, another ten times more at the board level, and ten times that at the systems level. And if a faulty system is shipped, it costs yet another order of magnitude more to find and correct the fault. Economy clearly mandates early fault detection. The time required to apply a test scales somewhere between linearly and quadratically with the number of transistors.

The first step in testing is to choose either a set of random test inputs or a set of functional tests. Randomly generated test patterns can reveal 55 to 90% of the faults in a piece of combinational logic; however, for a circuit like a microprocessor designed without scan techniques, randomly generated patterns will uncover only 20 to 30% of the faults. For this sort of circuit, test patterns based on the functional specification of the module are used. These test patterns start out with perhaps 75 to 90% fault coverage. The percentages increase as the test program is augmented to include the results of field failures resulting from poor screening [Goel 84].

When designing for testability, the key concepts are observability and controllability. One has the possibility of a stuck-at fault for each internal node in the circuit. To test for stuck-at 1 and stuck-at 0 faults on a node, one must be able to control the node; that is, set it to a high and a low. Nodes that are difficult to control are very difficult (expensive) to test. For instance, the high-order bits of a counter might require a very long input sequence in order to set. They would therefore be considered difficult to control. To control an internal node, one sets up a sensitized path from the inputs to the node. This requires setting the inputs of various logic gates on the path to the appropriate value. (The "other" inputs to a NAND gate go high, to a NOR gate they go low.) The other half of the problem is observability. Not only do we need to control each node but we need to observe that we have done so. A sensitized path to the output (which is not in conflict with the inputs required to sensitize the input path) must also be discovered. Algorithms to discover the correct inputs for combinational logic blocks include the D-algorithm [Roth 80] and PODEM [Goel 82]. Nodes that are too hard to test are left untested.

Generating a good test program for a chip involves both generating a test pattern and verifying the test. Verification is done by repeated fault simulation. Fault simulation involves inserting a fault in the model of the circuit and then demonstrating that the test vector does indeed discover the fault. This must be done for each possible fault. The amount of computer time spent on these tasks goes roughly as N^K, where $2 \leq K \leq 3$ and N is the number of transistors. The maximum practical size circuit is roughly 2000 gates if sequential and 5000 gates if combinational. For this reason, one must be able to partition the circuit into small independent blocks. It is best if these blocks are purely combinational. The test verification of a MC68000 class machine can take weeks of minicomputer time.

One must design a circuit that is testable. This often requires some compromise of the design methodology. There are both formal and ad hoc techniques for doing this. The key is to partition the circuit into small combinational logic blocks, where the inputs and outputs are both easily controllable and observable. This might involve the insertion of additional test points or test modes. At a higher level, for instance, it is clearly essential to easily control and observe the contents of the MIR in a microprocessor. It must be easy to read the contents of all ROMs and PLAs. Being able to control and observe all latches in a VLSI chip is an important part of design for testability. Two general techniques for making a chip testable are signature analysis and scan design.

In scan design, all registers (except those that are easily controlled and observed, such as those in the data path) of a VLSI circuit are strung together in a long shift register that can be activated in a test mode. This essentially reduces the rest of the circuit to combinational logic blocks.

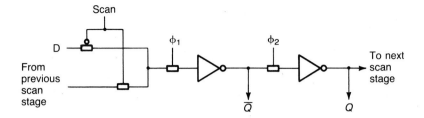

FIGURE 8.29 Scan register implemented with dynamic logic.

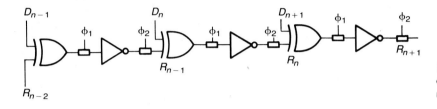

FIGURE 8.30 Circuit for collecting the signature of a number of test points D_n.

The input signals can be shifted in (for control), the system clocked, and then the output values shifted out (for observation). Test patterns for systems of this sort are reasonable to generate. The area overhead required for this technique has been quoted as anywhere from 4 to 20%. The extensive use of dynamic storage, typical of MOS design styles, does not favor scan design. Figure 8.29 shows a possible implementation of a scan-style register.

In signature analysis, one exploits the pseudorandom sequence generator of Example 1.4. The sequence generator can perform two functions. It can provide random inputs to internal nodes that are difficult to access from external pins. Also, a modified version, as shown in Fig. 8.30, can be used to collect a signature that results from another generator or from external inputs. Typically, the output is shifted out in bit serial manner continuously as the signature is collected.

Other techniques are variants or combinations of these. The interested reader is referred to the literature.

8.4 Perspective

A well-designed VLSI machine reflects the capabilities of the implementation medium. Most VLSI systems can be partitioned into a data path and a control path. In the control path, the key design decisions revolve around the kind of array structures to use, the degree to which the control is centralized, and the degree of pipelining. In the data path, key decisions include the numbers and types of functional elements and the buses connecting them. When partitioning a VLSI system, it is not always obvious which functional modules belong where. The key to resolving these issues is to look at the communication implications of the various alternatives. High performance implies a tight coupling of control and data paths—this is one of the reasons systolic systems are so powerful.

While we have spent a great deal of effort on the study of optimization, remember that the real power of MOS VLSI is economic. When compared to other technologies, MOS may not be as fast in terms of raw speed, but in the ratio of speed per unit cost, MOS excels. The power of a MOS VLSI machine is dependent on the cleverness and creativity with which the engineer solves a twofold problem: the physical problem of planning the geometric placement of millions of transistors; and the strategic problem of harnessing their strength.

Problems

8.1 Show that the pulldown string required in a complex gate implementation of a carry look-ahead adder is roughly equivalent to the chain of pass transistors needed in a precharged Manchester carry chain implementation.

8.2 In CMOS combinational logic circuits, certain types of faults can turn a combinational network into a sequential network. Give an example of such a fault.

8.3 We discussed, in Section 8.2, how using a ROM in page mode can help the system adapt to conditional branches. Explain how page mode ROM access can speed the execution of straight line code.

8.4 For a given instruction set, it appears that the performance of a microprocessor versus the number of transistors looks something like the curve shown in Fig. 8.31 (assuming one uses those transistors intelligently). Note that the microcode store (if one is used) and any local registers and caches are included in the calculation. Speculate on the origin of $N_{\min}$ and $P_{\max}$. $\triangle\triangle$

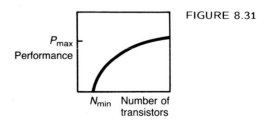

FIGURE 8.31

8.5 In the two-level microcode/nanocode calculations, we assumed a PLA was used for the nanocode store. Redo these calculations, assuming that the nanocode is stored in a ROM array.

8.6 Redesign the accumulator circuit of Example 8.1 with pipelined carry propagate adders in the loop in such a way that it is just as fast as the carry save implementation illustrated in the example.

8.7 Figure 8.32 illustrates a bit slice of a simple ALU. Two function blocks are used. Explain how this ALU would be used to perform the logical operation A AND B. How would one perform the arithmetic two's complement operation $A - B$?

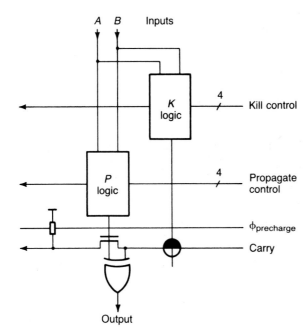

FIGURE 8.32

8.8 If we use a precharged Manchester carry chain circuit, show that the probability that a significant carry signal is propagated between the 15th and 16th bits of a 32 bit adder is 0.25, assuming random inputs.

References

[Alpert 83] D. Alpert, D. Carberry, M. Yamamura, and P. Mak, "32-bit Processor Chip Integrates Major System Functions," *Electronics* **56**: 113–119, July 14, 1983. (The Z80000)

[Anceau 80] F. Anceau, "Architecture and Design of Von Neumann Micro-processors," *NATO Advanced Study Institutes Series*, series E., no. 47: 301–327, July, 1980.

[Bayliss 81] J. A. Bayliss, J. A. Deetz, C-K Ng, S. A. Ogilvie, C. B. Peterson, and D. K. Wilde, "The Interface Processor for the Intel VLSI 432 32-Bit Computer," *IEEE J. Solid-State Circuits* **SC-16**: 522–530, 1981. (The i43203)

[Bayliss 81b] J. A. Bayliss, S. R. Colley, R. H. Kravitz, G. A. McCormick, W. S. Richardson, D. K. Wilde, and L. L. Wittmer, "The Instruction Decoding Unit for the VLSI 432 General Data Processor," *IEEE J. Solid-State Circuits* **SC-16**: 531–537, 1981. (The i43201)

[Beck 84] J. Beck, D. Dobberpuhl, M. J. Doherty, E. Dorenkamp, B. Grondalski, D. Grondalski, K. Henry, M. Miller, B. Supnik, S. Thierauf, and R. Witek, "A 32b Microprocessor with On-Chip Virtual Memory Management," *IEEE International Solid-State Circuits Conf.*: 178–179, San Francisco, Calif., 1984. (A single chip VAX)

[Bell 71] C. G. Bell and A. Newell, *Computer Structures: Readings and Examples*, McGraw-Hill, New York, 1971.

[Bell 78] C. G. Bell, J. C. Mudge, and J. E. McNamara, *Computer Engineering*, Digital Press, New York, 1978.

[Bennetts 82] R. G. Bennetts, *Introduction to Digital Board Testing*, Crane Russak, New York, 1982.

[Bennetts 84] R. G. Bennetts, *Design of Testable Logic Circuits*, Addison-Wesley, Reading, Mass., 1984.

[Beyers 81] J. W. Beyers, L. J. Dohse, J. P. Fucetola, R. L. Kochis, C. G. Lob, G. L. Taylor, and E. R. Zeller, "A 32-Bit VLSI CPU Chip," *IEEE J. Solid-State Circuits* **SC-16**: 537–542, 1981. (The HP 9000 CPU)

[Bhavsar 81] D. K. Bhavsar and R. W. Heckelman, "Self-Testing by Polynomial Division," *IEEE Test Conference*: 208–216, Philadelphia, Pa., 1981.

[Budde 81] D. L. Budde, S. R. Colley, S. I. Domenik, A. L. Goodman, J. D. Howard, and M. T. Imel, "The Execution Unit for the VLSI 432 General Data Processor," *IEEE J. Solid-State Circuits* **SC-16**: 514–521, 1981. (The i43202)

[Byrne 80] W. J. Byrne, "Three Decapsulation Methods for Epoxy Novalac Type Packages," *Proc. IEEE International Reliability Physics Symposium*: 107–109, 1980.

[Curtis 81] S. Curtis, T. Lake, G. Simpson, "A Magical Mystery Tour of the Motorola MC68000," unpublished, 1981.

[Dang 77] L.-G. Dang, P. B. Ashkin, R. Yee, and M. O'Brein, "A CMOS/SOS 16-Bit Parallel μ CPU," *IEEE International Solid-State Circuits Conf.*: 134–135, 238, 1977.

[Donath 79] W. E. Donath, "Placement and Average Interconnection Lengths of Computer Logic," *IEEE Trans. Circuits and Systems*: 272–277, 1979.

[Flores 63] Ivan Flores, *The Logic of Computer Arithmetic*, Prentice-Hall, Englewood Cliffs, N.J., 1963.

[Goel 82] P. Goel and B. C. Rosales, "PODEM-X: An Automatic Test Generation System for VLSI Logic Structures," *18th Design Automation Conference*: 260–268, Nashville, Tenn., 1982.

[Goel 84] P. Goel, private communication, 1984.

[Haviland 80] G. L. Haviland and A. A. Tuszynski, "A CORDIC Arithmetic Processor Chip," *IEEE J. Solid-State Circuits* **SC-15**, 1980.

[Heller 78] W. R. Heller, W. F. Mikhail, W. E. Donath, "Prediction of Wiring Space Requirements for LSI," *Journal of Design Automation and Fault Tolerant Computing*: 117–144, 1978.

[Hennessy 81] J. L. Hennessy, N. Jouppi, F. Baskett, and J. Gill, "MIPS: A VLSI Processor Architecture," *Proc. CMU Conf. VLSI Systems and Computations*: 337–346, Pittsburgh, Pa., 1981.

[Hiatt 80] J. Hiatt, "Microprobing," *Proc. IEEE International Reliability Physics Symposium*: 116–120, 1980.

[Hill 74] F. J. Hill and G. R. Peterson, *Introduction to Switching Theory and Logical Design*, Wiley, New York, 1974.

[Hillis 81] W. D. Hillis, "The Connection Machine," AI memo no. 646, Massachusetts Institute of Technology, 1981.

[Hwang 79] K. Hwang, *Computer Arithmetic, Principles, Architecture and Design*, Wiley, New York, 1979.

[Johnson 84] W. N. Johnson, "A VLSI Superminicomputer CPU," *IEEE International Solid-State Circuits Conf.*: 174–175, San Francisco, Calif., 1984. (VAX V11)

[Kaminker 81] A. Kaminker, L. Kohn, Y. Lavi, A. Menachem, and Z. Soha, "A 32-Bit Microprocessor with Virtual Memory Support," *IEEE J. Solid-State Circuits* **SC-16**: 548–557, 1981. (The NS16032)

[Keyes 79] R. W. Keyes, "The Evolution of Digital Electronics Towards VLSI," *IEEE J. Solid-State Circuits* **SC-14**: 193–201, 1979.

[Lutz 84] C. Lutz, S. Rabin, C. Seitz, and D. Speck, "Design of the Mosaic Element," *Proc. Advanced Research in VLSI*: 1–10, Cambridge, Mass., 1984.

[Lyon 82] R. F. Lyon and M. P. Haeberli, "Designing and Testing the Optical Mouse," *VLSI Design* **3**, no. 1: 20–30, 1982.

[Magar 82] S. S. Magar, E. R. Caudel, and A. W. Leigh, "A Microcomputer with Digital Signal Processing Capability," *IEEE International Solid-State Circuits Conf.*: 32–33, 284–285, San Francisco, Calif., February 1982. (TMS 320)

[Morse 78] S. P. Morse, W. B. Pohlman, and B. W. Ravenel, "The Intel 8086 Microprocessor: A 16-bit Evolution of the 8080," *IEEE Computer* **11**: 18–27, June, 1978.

[Murphy 81] B. T. Murphy, R. Edwards, L. C. Thomas, and J. J. Molinelli, "A CMOS 32b Single Chip Microprocessor," *IEEE International Solid-State Circuits Conf.*: 230–231, New York, N.Y., 1981. (The Bellmac 32)

[NBS 83] Standard test structures and programs for certain technologies are available from the National Bureau of Standards and the Jet Propulsion Laboratory. Address inquiries to:

VLSI Test Chip Laboratory
Jet Propulsion Laboratory
4800 Oak Grove Drive
Pasadena, CA 91109

and

National Bureau of Standards
Room B-310
Technology Building
Washington, DC 20234

[Obrebska 82] M. Obrebska, "Comparative Survey of Different Design Methodologies for Control Part of Microprocessors," Doctoral Thesis, 1982.

[Obrebska 83] M. Obrebska, "Algorithm Transformations Improving Control Part Implementation," *IEEE International Conf. on Computer Design: VLSI in Computers*: 307–310, Port Chester, N.Y., 1983.

[Patterson 82] D. A. Patterson and C. H. Séquin, "A VLSI RISC," *IEEE Computer* **15**: 8–22, September 1982.

[Richardson 81] W. S. Richardson, J. A. Bayliss, S. R. Colley, R. H. Kravitz, and G. A. McCormick, "The 32b Computer Instruction Decoding Unit," *IEEE International Solid-State Circuits Conf.*: 114–115, New York, N.Y., 1981. (The i432)

[Roth 80] J. P. Roth, *Computer Logic, Testing, and Verification*, Computer Science Press, Potomac, Md., 1980. (The D-Algorithm)

[Rowen 84] C. Rowen, S. A. Przbylski, N. P. Jouppi, T. Gross, J. Shott, and J. L. Hennessy, "A Pipelined 32b NMOS Microprocessor," *IEEE International Solid-State Circuits Conf.*: 180–181, San Francisco, Calif., 1984. (Stanford MIPS)

[Schumann 84] R. Schumann and W. Parker, "A 32b Bus Interface Chip," *IEEE International Solid-State Circuits Conf.*: 176–177, San Francisco, Calif., 1984.

[Schuster 84] M. D. Schuster and R. E. Bryant, "Concurrent Fault Simulation of MOS Digital Circuits," *Proc. Advanced Research in VLSI*: 129–138, Cambridge, Mass., 1984.

[Sherburne 82] R. W. Sherburne, Jr., M. G. H. Katevenis, D. A. Patterson, and C. H. Séquin, "Datapath Design for RISC," *Proc. Advanced Research in VLSI*: 53–62, Cambridge, Mass., 1982.

[Sherburne 84] R. W. Sherburne, Jr., M. G. H. Katevenis, D. A. Patterson, and C. H. Séquin, "A 32b NMOS Microprocessor with a Large Register File," *IEEE International Solid-State Circuits Conf.*: 168–169, San Francisco, Calif., 1984. (Berkeley RISC)

[Slager 83] J. Slager, L. Gindraux, T. Ho, G. Louie, and D. Vannier, "A 16b Microprocessor with On-Chip Memory Protection," *IEEE International Solid-State Circuits Conf.*: 24–25, New York, 1983. (iAPX 286)

[Shima 79] M. Shima, "Demystifying Microprocessor Design," *IEEE Spectrum* **16**: 22–30, July 1979.

[Stritter 78] E. Stritter and H. L. Tredonick, "Microprogrammed Implementation of a Single Chip Microprocessor," *Proceedings of the 11th Annual Microprogramming Workshop*: 8–16, 1978. (MC68000)

[Stritter 79] E. Stritter and T. Gunter, "A Microprocessor Architecture for a Changing World: The Motorola 68000," *IEEE Computer* **12**: 43–52, Feb. 1979.

[Sutton 83] J. A. Sutton, "The 80C86 CMOS Design from the nMOS Architecture," *VLSI Design* **4**, no. 8: 56–63, 1983.

APPENDIXES

APPENDIXES

A.   SPICE techniques

- The input file ends with a .END command, yet SPICE insists it isn't there. The final carriage return has been omitted.

- To fix the dc voltage on a floating node, one places a 10 MΩ resistor between that node and ground: R 100 0 10M. The problem is that M stands for milli, not mega (MEG).

- One uses Tabs in the input deck; this is a bad idea. Also, lines that are too long will be truncated without an error message. Model names, file names, and component names all have maximums on the number of characters.

- One forgets to tie the body terminal of the p-type device to V_{DD}.

- We would like to start a simulation with two nodes at the same, but unknown, voltage. Connect them together with a 1 H inductor.

- One forgets the "U" after the width or length of a transistor. In a more subtle version, write a preprocessor for SPICE that happens to output 12E-3U. This is interpreted as 12E-3.

- Set the area of the source diffusion to AS=1000U. Don't you mean AS=1000P?

- Be careful mixing level 2 and level 3 MOS models in the same circuit file. Strange effects have been reported.

- There is no convergence in dc analysis. Raise the simulation temperature, rerun the dc simulation, and use these values in

NODESET commands. Renumber the nodes so that the internal matrices in SPICE are set up differently.

- Use a current-controlled current source driving a linear capacitor in order to monitor the total charge in a network branch. This technique can be used to monitor the dynamic power dissipation of CMOS circuits. You will need some 10 MΩ resistors.

- One puts circuit information in the first line of the input file. This line is considered header information. Note that it is the first line, not the first nonblank line.

- If when plotting out a dc transfer function, one discovers a lack of convergence, lower the step size. In digital circuits, the problem usually occurs for circuits with unrealistically high gain. Lower the gain.

- Add 1 Ω series and 1 GΩ parallel resistors in various places to help "step size too small" in transient analysis. The problem is usually caused by nodes in the circuit that switch with very fast time constants. Sometimes adding capacitance to these nodes will help the simulation but not change the results.

- As a wide transistor's width is doubled, the current increases by less than a factor of two. Set RSH=0 or specify RD and RS. The default number of squares of source and drain resistance is nonzero.

B.   Nonlinear capacitance equations

$$I \equiv \frac{dQ}{dt}$$

$$Q = \int I \, dt$$

$$\Delta t = \int \frac{1}{I} \, dQ$$

$$C \equiv \frac{dQ}{dV}$$

$$I = C \frac{dV}{dt}$$

$$Q = \int C \, dV$$

$$\Delta t = \int \frac{C}{I} \, dV$$

C.   Example 2 μm nMOS process

The example 2 μm nMOS process has arsenic source/drain implants and phosphorus-doped poly. It has three transistor types, each defined by a separate implant—enhancement, depletion, and zero threshold. It is designed to operate with a -3 V V_{BB} bias.

2 μm nMOS process electrical parameters (23° C. $V_{BB} = -3$ V)

Parameter	Min	Max	Units	Comments
T_{OX}	270	330	Å	Gate oxide thickness
C_{OX}	10.45	12.77	$\times 10^{-4}$ pF/μm^2	Gate oxide capacitance
ΔL_{poly}	-0.125	0.125	μm	One side
X_J	0.27	0.33	μm	Source/drain junction depth
L_{diff}	0.19	0.23	μm	One side
ΔW	0.6	0.9	μm	One side inward
R_{poly}	15	40	$\Omega/\square$	Poly sheet resistance
R_{diff}	20	50	$\Omega/\square$	Diff (n+) sheet resistance
C_{ja}	0.2	0.3	$\times 10^{-4}$ pF/μm^2	Junction area capacitance
C_{jp}	1.0	2.0	$\times 10^{-4}$ pF/μm	Junction peripheral capacitance
C_{polyf}	0.5	0.7	$\times 10^{-4}$ pF/μm^2	Poly over field capacitance
C_{mpoly}	0.82	1.10	$\times 10^{-4}$ pF/μm^2	Metal over poly capacitance
C_{mdiff}	0.82	1.1	$\times 10^{-4}$ pF/μm^2	Metal over diff (n+) capacitance
C_{mf}	0.31	0.42	$\times 10^{-4}$ pF/μm^2	Metal over field capacitance
LRC	4	10	$\Omega/\square$	Poly/buried/n+

2 μm nMOS process device parameters[1] (23° C. $V_{BB} = -3$ V)
Minimum characterized width is 6 μm drawn.

Parameter	Enhancement Min	Max	Zero Min	Max	Depletion Min	Max	Units
V_{T0}	0.6	1.0	0.2	0.6	-2.2	-1.7	volts
μ	650	750	680	780	650	750	cm^2/(V·s)
K'	33.9	47.9	35.5	49.8	39.2	56.8	μA/V^2
N	2	5	1.5	4.0	20	50	$\times 10^{14}$/cm^3
γ	0.065	0.12	0.056	0.116	0.102	0.25	V$^{1/2}$
Theta	0.09	0.1	0.09	0.1	0.03	0.04	
Kappa	0.5	0.5	0.5	0.5	0.5	0.5	
Eta	0.22	0.22	0.22	0.22	0.22	0.22	
$\mathcal{V}_{\text{max}}$	12	17	12	17	12	17	$\times 10^4$ m/s

[1] Enhancement 20/2 Drawn, Zero 20/3 Drawn, Depletion 6/6 Drawn.

2 μm nMOS worst-case model file

```
********************************************************************
* SPICE Models for 2 um nMOS
********************************************************************
* Worst-case slow parameters
*
.MODEL NENHS NMOS LEVEL=3 RSH=0 TOX=330E-10 LD=0.19E-6 XJ=0.27E-6
+ VMAX=13E4 ETA=0.25 KAPPA=0.5 NSUB=5E14 UO=650 THETA=0.1
+ VTO=0.946 CGSO=2.43E-10 CGDO=2.43E-10 CJ=6.9E-5 CJSW=3.3E-10
+ PB=0.7 MJ=0.5 MJSW=0.3 NFS=1E10
*
.MODEL NZEROS NMOS LEVEL=3 RSH=0 TOX=330E-10 LD=0.19E-6 XJ=0.27E-6
+ VMAX=13E4 ETA=0.25 KAPPA=0.5 NSUB=40E14 UO=680 THETA=0.1
+ VTO=0.526 CGSO=2.43E-10 CGDO=2.43E-10 CJ=6.9E-5 CJSW=3.3E-10
+ PB=0.7 MJ=0.5 MJSW=0.3 NFS=1E10
*
.MODEL NDEPS NMOS LEVEL=3 RSH=0 TOX=330E-10 LD=0.19E-6 XJ=0.27E-6
+ VMAX=13E4 ETA=0.25 KAPPA=0.5 NSUB=50E14 UO=650 THETA=0.04
+ VTO=-2.078 CGSO=2.43E-10 CGDO=2.43E-10 CJ=6.9E-5 CJSW=3.3E-10
+ PB=0.7 MJ=0.5 MJSW=0.3 NFS=1E10
*
* deltaLpoly(one side)=-0.125U
* deltaW(one side)=0.9U
*
********************************************************************
* Typical
*
.MODEL NENHT NMOS LEVEL=3 RSH=0 TOX=300E-10 LD=0.21E-6 XJ=0.3E-6
+ VMAX=15E4 ETA=0.18 KAPPA=0.5 NSUB=3.5E14 UO=700 THETA=0.095
+ VTO=0.781 CGSO=2.8E-10 CGDO=2.8E-10 CJ=5.75E-5 CJSW=2.48E-10
+ PB=0.7 MJ=0.5 MJSW=0.3 NFS=1E10
*
.MODEL NZEROT NMOS LEVEL=3 RSH=0 TOX=300E-10 LD=0.21E-6 XJ=0.3E-6
+ VMAX=15E4 ETA=0.18 KAPPA=0.5 NSUB=2.75E14 UO=730 THETA=0.095
+ VTO=0.354 CGSO=2.8E-10 CGDO=2.8E-10 CJ=5.75E-5 CJSW=2.48E-10
+ PB=0.7 MJ=0.5 MJSW=0.3 NFS=1E10
*
.MODEL NDEPT NMOS LEVEL=3 RSH=0 TOX=300E-10 LD=0.21E-6 XJ=0.3E-6
+ VMAX=15E4 ETA=0.18 KAPPA=0.5 NSUB=35E14 UO=700 THETA=0.035
+ VTO=-2.231 CGSO=2.8E-10 CGDO=2.8E-10 CJ=5.75E-5 CJSW=2.48E-10
+ PB=0.7 MJ=0.5 MJSW=0.3 NFS=1E10
*
* deltaLpoly(one side)=0U
* deltaW(one side)=0.75U
```

```
*
********************************************************************
* Fast
*
.MODEL NENHF NMOS LEVEL=3 RSH=0 TOX=270E-10 LD=0.23E-6 XJ=0.33E-6
+ VMAX=17E4 ETA=0.10 KAPPA=0.5 NSUB=2E14 UO=750 THETA=0.09
+ VTO=0.612 CGSO=3.4E-10 CGDO=3.4E-10 CJ=4.6E-5 CJSW=1.65E-10
+ PB=0.7 MJ=0.5 MJSW=0.3 NFS=1E10
*
.MODEL NZEROF NMOS LEVEL=3 RSH=0 TOX=270E-10 LD=0.23E-6 XJ=0.33E-6
+ VMAX=17E4 ETA=0.10 KAPPA=0.5 NSUB=1.5E14 UO=780 THETA=0.09
+ VTO=0.179 CGSO=3.4E-10 CGDO=3.4E-10 CJ=4.6E-5 CJSW=1.65E-10
+ PB=0.7 MJ=0.5 MJSW=0.3 NFS=1E10
*
.MODEL NDEPF NMOS LEVEL=3 RSH=0 TOX=270E-10 LD=0.23E-6 XJ=0.33E-6
+ VMAX=17E4 ETA=0.10 KAPPA=0.5 NSUB=20E14 UO=750 THETA=0.03
+ VTO=-2.384 CGSO=3.4E-10 CGDO=3.4E-10 CJ=4.6E-5 CJSW=1.65E-10
+ PB=0.7 MJ=0.5 MJSW=0.3 NFS=1E10
*
* deltaLpoly(one side)=0.125U
* deltaW(one side)=0.6U
*
********************************************************************
* Noise Margin
*
* (Assumes VIN near VDD, Wpu<Wpd and Lpu>Lpd)
*
.MODEL NENHN NMOS LEVEL=3 RSH=0 TOX=270E-10 LD=0.19E-6 XJ=0.27E-6
+ VMAX=15E4 ETA=0.25 KAPPA=0.5 NSUB=2E14 UO=670 THETA=0.1
+ VTO=0.577 CGSO=2.9E-10 CGDO=2.9E-10 CJ=5.75E-5 CJSW=2.48E-10
+ PB=0.7 MJ=0.5 MJSW=0.3 NFS=1E10
*
.MODEL NZERON NMOS LEVEL=3 RSH=0 TOX=270E-10 LD=0.19E-6 XJ=0.27E-6
+ VMAX=15E4 ETA=0.18 KAPPA=0.5 NSUB=1.5E14 UO=730 THETA=0.095
+ VTO=0.365 CGSO=2.9E-10 CGDO=2.9E-10 CJ=5.75E-5 CJSW=2.48E-10
+ PB=0.7 MJ=0.5 MJSW=0.3 NFS=1E10
*
.MODEL NDEPN NMOS LEVEL=3 RSH=0 TOX=270E-10 LD=0.19E-6 XJ=0.27E-6
+ VMAX=17E4 ETA=0.10 KAPPA=0.5 NSUB=20E14 UO=730 THETA=0.03
+ VTO=-2.392 CGSO=2.9E-10 CGDO=2.9E-10 CJ=5.75E-5 CJSW=2.48E-10
+ PB=0.7 MJ=0.5 MJSW=0.3 NFS=1E10
*
* deltaLpoly(one side)=-0.125U
* deltaW(one side)=0.6U
```

(Continued)

2 μm nMOS worst-case model file *(Continued)*

```
*
*********************************************************************
* Slow Fast
*
.MODEL NENHSF NMOS LEVEL=3 RSH=0 TOX=300E-10 LD=0.23E-6 XJ=0.33E-6
+ VMAX=17E4 ETA=0.10 KAPPA=0.5 NSUB=2E14 UO=730 THETA=0.1
+ VTO=0.614 CGSO=2.9E-10 CGDO=2.9E-10 CJ=5.75E-5 CJSW=2.64E-10
+ PB=0.7 MJ=0.5 MJSW=0.3 NFS=1E10
*
.MODEL NZEROSF NMOS LEVEL=3 RSH=0 TOX=300E-10 LD=0.23E-6 XJ=0.33E-6
+ VMAX=15E4 ETA=0.18 KAPPA=0.5 NSUB=4E14 UO=700 THETA=0.095
+ VTO=0.353 CGSO=2.9E-10 CGDO=2.9E-10 CJ=5.75E-5 CJSW=2.64E-10
+ PB=0.7 MJ=0.5 MJSW=0.3 NFS=1E10
*
.MODEL NDEPSF NMOS LEVEL=3 RSH=0 TOX=300E-10 LD=0.23E-6 XJ=0.33E-6
+ VMAX=12E4 ETA=0.25 KAPPA=0.5 NSUB=50E14 UO=670 THETA=0.03
+ VTO=-2 CGSO=2.9E-10 CGDO=2.9E-10 CJ=5.75E-5 CJSW=2.64E-10
+ PB=0.7 MJ=0.5 MJSW=0.3 NFS=1E10
*
* deltaLpoly(one side)=0.125U
* deltaW(one side)=0.9U
*
*********************************************************************
```

2 μm nMOS topological design rules

Layer	Rule	Description	Dimension
1.0		Active area mask	
	1.1	Minimum width	3 μm
	1.2	Minimum spacing	2 μm
2.0		Enhancement implant mask	
	2.1	Minimum extension around gate area	2 μm
	2.2	Minimum spacing to unrelated gate area	2 μm
3.0		Depletion implant mask	
	3.1	Minimum extension around gate area	2 μm
	3.2	Minimum spacing to unrelated gate area	2 μm
4.0		Buried contact mask	
	4.1	Minimum extension around contact area	1 μm
	4.2	Minimum spacing of contact area to unrelated poly or diffusion	2 μm
5.0		Polysilicon mask	
	5.1	Minimum width	2 μm
	5.2	Minimum spacing	2 μm
	5.3	Minimum space to unrelated diffusion	1 μm
	5.4	Minimum extension beyond gate region	2 μm
	5.5	Minimum poly coincidence with active area mask to form buried contact	2 μm × 3 μm
	5.6	Minimum extension of active area mask beyond poly in field direction to form buried contact	1 μm
	5.7	Minimum extension of active area beyond gate	3 μm
6.0		Metal contact mask[2]	
	6.1	Size	2 μm × 2 μm
	6.2	Minimum extension of active area mask for n+ contact	2 μm
	6.3	Minimum spacing to poly for n+ contact	2 μm
	6.4	Minimum extension of poly mask for poly contact	2 μm
	6.5	Minimum spacing of poly contact to active gate region	2 μm
7.0		Metal mask[3]	
	7.1	Minimum width	4 μm
	7.2	Minimum spacing	2 μm
	7.3	Minimum metal extension beyond contact	1 μm
8.0		Passivation mask	
	8.1	Minimum pad window size	100 μm × 100 μm
	8.2	Minimum pad–pad spacing	75 μm
	8.3	Minimum extension of metal beyond pad	4 μm

[2]Maximum current density in contact is 0.2 mA/μm of periphery.
[3]Maximum current density in metal lines is 0.7 mA/μm of width.

D.   Example 2 μm CMOS process

The example 2 μm CMOS process has enhancement-mode p- and n-channel devices and uses n-well technology. A low resistance epitaxial layer offers good latchup protection. Plugs are generally needed only in the wells in internal circuitry. The n-channel devices have arsenic source/drain implants and the p-channel devices use boron. Field implants are used for both n- and p-channel devices.

2 μm CMOS process electrical parameters (n-well, 23° C)

Parameter	Min	Max	Units	Comments
T_{OX}	225	275	Å	
C_{OX}	12.55	15.34	$\times 10^{-4}\,\mathrm{pF}/\mu\mathrm{m}^2$	
ΔL_{poly}	−0.125	0.125	μm	one side inward
ΔW	0.6	0.9	μm	one side inward
C_{mpoly}	0.75	1.03	$\times 10^{-4}\,\mathrm{pF}/\mu\mathrm{m}^2$	plus fringing
C_{polyf}	0.47	0.53	$\times 10^{-4}\,\mathrm{pF}/\mu\mathrm{m}^2$	plus fringing
C_{mdiff}	0.75	0.93	$\times 10^{-4}\,\mathrm{pF}/\mu\mathrm{m}^2$	plus fringing
C_{mf}	0.29	0.35	$\times 10^{-4}\,\mathrm{pF}/\mu\mathrm{m}^2$	plus fringing
$R_{\mathrm{m/p+}}$		5	Ω	2×2 contact
$R_{\mathrm{m/n+}}$		10	Ω	2×2 contact
$R_{\mathrm{m/poly}}$		2	Ω	2×2 contact

2 μm CMOS process device parameters[1] (n-well, 23° C)
Minimum characterized width is 6 μm drawn.

Parameter	n-channel Min	n-channel Max	p-channel Min	p-channel Max	Units
V_{T0}	0.6	1.0	−1.0	−0.6	volts
μ	550	650	180	220	cm^2/(V·s)
K'	35	50	11	17	μmho/V
N	3	4	5	7	$\times 10^{15}$ cm^{-3}
γ	0.16	0.3	0.3	0.38	$\sqrt{\mathrm{V}}$
L_{diff}	0.1	0.15	0.3	0.4	μm
X_J	0.14	0.21	0.42	0.6	μm
R_{poly}	20	50	20	60	$\Omega/\square$
C_{ja}	1.0	1.6	6.0	7.7	$\times 10^{-4}$ pF/μm^2
C_{jp}	1.25	1.8	3.75	5.4	$\times 10^{-4}$ pF/μm
R_{diff}	20	50	80	120	$\Omega/\square$

[4] Transistors 15/2 drawn.

2 μm CMOS worst-case model file

```
********************************************************************
* 2 um n-Well Iso CMOS
********************************************************************
* Slow Slow
*
.MODEL NSS NMOS LEVEL=3 RSH=0 TOX=275E-10 LD=0.1E-6 XJ=0.14E-6
+ CJ=1.6E-4 CJSW=1.8E-10 UO=550 VTO=1.022 CGSO=1.3E-10
+ CGDO=1.3E-10 NSUB=4E15 NFS=1E10
+ VMAX=12E4 PB=0.7 MJ=0.5 MJSW=0.3 THETA=0.06 KAPPA=0.4 ETA=0.14
.MODEL PSS PMOS LEVEL=3 RSH=0 TOX=275E-10 LD=0.3E-6 XJ=0.42E-6
+ CJ=7.7E-4 CJSW=5.4E-10 UO=180 VTO=-1.046 CGSO=4E-10
+ CGDO=4E-10 TPG=-1 NSUB=7E15 NFS=1E10
+ VMAX=12E4 PB=0.7 MJ=0.5 MJSW=0.3 ETA=0.06 THETA=0.03 KAPPA=0.4
*
* deltaLpoly=-0.125um deltaW=0.9um (one sided inward)
*
********************************************************************
* Fast p-type Slow n-type
*
.MODEL NFS NMOS LEVEL=3 RSH=0 TOX=250E-10 LD=0.1E-6 XJ=0.14E-6
+ CJ=1.6E-4 CJSW=1.5E-10 UO=550 VTO=1.03 CGSO=1.33E-10
+ CGDO=1.33E-10 NSUB=4E15 THETA=0.06 KAPPA=0.4 ETA=0.14
+ VMAX=12E4 PB=0.7 MJ=0.5 MJSW=0.3 NFS=1E10
.MODEL PFS PMOS LEVEL=3 RSH=0 TOX=250E-10 LD=0.4E-6 XJ=0.6E-6
+ CJ=7E-4 CJSW=4.5E-10 UO=220 VTO=-.66 CGSO=5.5E-10
+ CGDO=5.5E-10 TPG=-1 NSUB=5E15 ETA=0.06 THETA=0.03 KAPPA=0.4
+ VMAX=17E4 PB=0.7 MJ=0.5 MJSW=0.3 NFS=1E10
*
* deltaLpoly=0um deltaWp=0.7um deltaWn=0.8um (one sided inward)
*
********************************************************************
* Fast p-type Fast n-type
*
.MODEL NFF NMOS LEVEL=3 RSH=0 TOX=225E-10 LD=0.15E-6 XJ=0.21E-6
+ CJ=1.0E-4 CJSW=1.25E-10 UO=650 VTO=0.628 CGSO=2.3E-10
+ CGDO=2.3E-10 NSUB=3E15 THETA=0.06 KAPPA=0.4 ETA=0.14
+ VMAX=17E4 PB=0.7 MJ=0.5 MJSW=0.3 NFS=1E10
.MODEL PFF PMOS LEVEL=3 RSH=0 TOX=225E-10 LD=0.4E-6 XJ=0.6E-6
+ CJ=6E-4 CJSW=3.75E-10 UO=220 VTO=-.668 CGSO=6.2E-10
+ CGDO=6.2E-10 TPG=-1 NSUB=5E15 ETA=0.06 THETA=0.03 KAPPA=0.4
+ VMAX=17E4 PB=0.7 MJ=0.5 MJSW=0.3 NFS=1E10
*
* deltaLpoly=0.125um deltaW=0.6um (one sided inward)
```

(Continued)

2 μm CMOS worst-case model file *(Continued)*

```
*
********************************************************************
* Slow p-type Fast n-type
*
.MODEL NSF NMOS LEVEL=3 RSH=0 TOX=250E-10 LD=0.15E-6 XJ=0.21E-6
+ CJ=1.0E-4 CJSW=1.5E-10 UO=650 VTO=0.626 CGSO=2E-10
+ CGDO=2E-10 NSUB=3E15 THETA=0.06 KAPPA=0.4 ETA=0.14
+ VMAX=17E4 PB=0.7 MJ=0.5 MJSW=0.3 NFS=1E10
.MODEL PSF PMOS LEVEL=3 RSH=0 TOX=250E-10 LD=0.3E-6 XJ=0.42E-6
+ CJ=7E-4 CJSW=4.5E-10 UO=180 VTO=-1.049 CGSO=4.2E-10
+ CGDO=4.2E-10 TPG=-1 NSUB=7E15 ETA=0.06 THETA=0.03 KAPPA=0.4
+ VMAX=12E4 PB=0.7 MJ=0.5 MJSW=0.3 NFS=1E10
*
* deltaLpoly=0um deltaWp=0.8um deltaWn=0.7um (one sided inward)
*
********************************************************************
```

2 μm CMOS topological design rules

Layer	Rule	Description	Dimension
1.0		n-Well mask	
	1.1	Minimum width	2 μm
	1.2	Minimum spacing, same potential	2 μm
	1.3	Minimum spacing, different potentials	2 μm
2.0		Active area mask	
	2.1	Minimum width	3 μm
	2.2	Minimum spacing	2 μm
	2.3	Minimum spacing to well edge from inside	2 μm
	2.4	Minimum spacing to well edge from outside	6 μm
3.0		Polysilicon mask	
	3.1	Minimum width	2 μm
	3.2	Minimum spacing	2 μm
	3.3	Minimum spacing to unrelated active area . .	1 μm
	3.4	Minimum extension beyond active area	2 μm
	3.5	Minimum extension of active area beyond gate .	3 μm

2 µm CMOS topological design rules *(Continued)*

Layer	Rule	Description	Dimension
4.0		p+ mask (n+ mask is complement of p+)	
	4.1	Minimum width	2 µm
	4.2	Minimum spacing	2 µm
	4.3	Minimum extension beyond p+ active area . .	4 µm
	4.4	Minimum spacing to n+ active area	4 µm
	4.5	Minimum extension beyond pMOS transistor .	4 µm
	4.6	Minimum extension to nMOS transistor . . .	4 µm
	4.7	p+/n+ shorting contact	special
5.0		Contact mask	
	5.1	Minimum contact	2 µm × 2 µm
	5.2	Maximum contact except p+/n+ shorting . .	2 µm × 2 µm
	5.3	p+/n+ shorting contact size	2 µm × 6 µm
	5.4	Minimum active area surround	2 µm
	5.5	Minimum spacing to gate area	2 µm
	5.6	Minimum poly surround	2 µm
6.0		Metal	
	6.1	Minimum width	4 µm
	6.2	Minimum spacing	2 µm
	6.3	Minimum extension beyond contact	1 µm
7.0		Passivation mask	
	7.1	Minimum pad window size	100 µm × 100 µm
	7.2	Minimum pad-pad spacing	75 µm
	7.3	Minimum extension of metal beyond pad . . .	4 µm
8.0		p+/n+ special shorting contact	
		This contact consists of a 6 × 2 µm contact cut centered in a 10 × 6 µm active area region. The p+ mask covers half the region (5 × 6 µm) and splits the contact down the middle, producing a 3 × 2 n+ and a 3 × 2 p+ contact.	
9.0		Well contact mask	
	9.1	Minimum distance between contacts	100 µm
	9.2	Maximum number of squares between contacts .	5
10.0		Guard rings, etc.	
	10.1	All internal bootstrapped nodes must be protected in the same way as I/O structure.	
	10.2	See Section 5.7.	

E. Advanced process options for the nMOS and CMOS processes

The following two process options are available (at an added cost) either singly or together.

LRP Low Resistance Poly. R_{poly} becomes 2 to 4 $\Omega/\square$ by the use of silicides.

M2 Second Layer Metal. The design rules are:

M2 contact cut (oxide thickness 1 μm)

Minimum spacing to M1 contact	2 μm
Minimum M1 contact surround	2 μm
Contact cut size	2 μm $\times$ 2 μm

M2 (metal thickness 1 μm)

Minimum spacing	3 μm
Minimum width	6 μm
Minimum contact surround	2 μm

F. Useful constants and conversions

General constants

q	1.602×10^{-19} C
ϵ_0	8.854×10^{-12} F/m
kT/q	25.86 mV
kT	0.02586 eV
μ_0	$4\pi \times 10^{-7}$ H/m
c	2.998×10^8 m/s
m_e	9.11×10^{-31} Kg
h	6.625×10^{-34} J·s
k	1.38×10^{-23} J/K

F.   Useful constants and conversions *(Continued)*

Silicon constants
(all data at 300 K = 27° C = 80.6° F)

$\kappa_{\text{thermal,SiO}_2}$	0.014 W/(cm° C)
$\kappa_{\text{thermal,Si}}$	1.412 W/(cm° C)
κ_{SiO_2}	3.9
κ_{Si}	11.7
$\kappa_{\text{Si}_3\text{N}_4}$	7.5
ϵ_{SiO_2}	3.45×10^{-11} F/m
ϵ_{Si}	1.04×10^{-10} F/m
$\epsilon_{\text{Si}_3\text{N}_4}$	6.64×10^{-11} F/m
E_G	1.124 eV, (1.170 eV at 0 K)
μ_n	1350 cm^2/(V·s)
μ_p	480 cm^2/(V·s)
$\mathcal{E}_{\text{breakdown,SiO}_2}$	3×10^7 V/m
$\mathcal{E}_{\text{breakdown,SiO}_2}$	6×10^8 V/m
χ_{Si}	4.15eV
n_i	1.45×10^{10} cm^{-3}
N_C	2.8×10^{19} cm^{-3}
N_V	1.04×10^{19} cm^{-3}

Conversions

1 mil = 25.4 μm = 10^{-3} in.
1 cm = 10^4 μm = 10^8 Å
1 eV = 1.602×10^{-19} J
0° C = 273 K

INDEX

INDEX